Unreason

Best of *Skeptical Inquirer*

VOLUME 7

EDITED BY
KENDRICK FRAZIER
BENJAMIN RADFORD

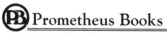 Prometheus Books

Essex, Connecticut

 Prometheus Books

An imprint of Globe Pequot, the trade division of
The Rowman & Littlefield Publishing Group, Inc.
4501 Forbes Blvd., Ste. 200
Lanham, MD 20706
www.rowman.com

Distributed by NATIONAL BOOK NETWORK

British Library Cataloguing in Publication Information Available

Library of Congress Cataloging-in-Publication Data Available

ISBN 978-1-63388-974-3 (paper : alk. paper) | ISBN 978-1-63388-975-0 (ebook)

∞™ The paper used in this publication meets the minimum requirements of American
National Standard for Information Sciences—Permanence of Paper
for Printed Library Materials, ANSI/NISO Z39.48-1992.

To Ken and Ruth Frazier,
and their daughter Michele Baldwin.

For forty-five years Ken expertly helmed the *Skeptical Inquirer,* finally succumbing to acute myeloid leukemia in November 2022, just as this collection was being completed. Ken was far more than a magazine editor; he was a writer, photographer, amateur astronomer, traveler, friend, and much more. His boundless enthusiasm for science and nature was evident; he loved ideas and thoughtful discussion, and this book reflects those passions.

Contents

Part I

SCIENCE AND BELIEF

CHAPTER 1

Science

THE GOLD STANDARD OF TRUTH

Richard Dawkins

Vol. 45, No. 2
March/April 2021

What is truth? You can speak of moral truths and aesthetic truths, but I'm not concerned with those here, important as they may be. By *truth* I shall mean the kind of truth that a commission of inquiry or a jury trial is designed to establish. I hold the view that scientific truth is of this commonsense kind, although the methods of science may depart from common sense, and its truths may even offend it.

Commissions of inquiry may fail, but we assume a truth is lurking there even if we don't have enough evidence. Juries sometimes get it wrong, and falsehoods are often sincerely believed. Scientists, too, can make mistakes and publish erroneous conclusions. That's all regrettable but not deeply sinister. What is profoundly troubling, however, is any wanton attack on truth itself: the value of truth, the very existence of truth. This is what concerns me here.

George Orwell's O'Brien held that two plus two equals five if the Party decrees it so. The "Ministry of Truth" existed for the purpose of disseminating lies. In the past four years, the U.S. government has moved in that direction. World-weary cynics sigh that all politicians lie; it goes with the territory. But normal politicians lie as a last resort and try to cover it up. Donald Trump is in a class of his own. For him, lying is not a last resort. It never occurs to him to do anything else. And far from covering up a lie, his well-named "base" will love him the more for it and will believe the lie, however far-fetched and shamelessly self-serving. Fortunately, Trump is too incompetent to fulfil Orwell's nightmare, and anyway he is on the way out—albeit kicking and screaming and trying to pull the house down with him as he goes.

A more insidious threat to truth comes from certain schools of academic philosophy. There is no objective truth, they say, no natural reality, only social constructs. Extreme exponents attack logic and reason themselves, as tools of manipulation or "patriarchal" weapons of domination. The philosopher and historian of science Noretta Koertge wrote in *Skeptical Inquirer* in 1995 (and things haven't gotten any better since):

> Instead of exhorting young women to prepare for a variety of technical subjects by studying science, logic, and mathematics, Women's Studies students are now being taught that logic is a tool of domination . . . the standard norms and methods of scientific inquiry are sexist because they are incompatible with "women's ways of knowing." The authors of the prize-winning book with this title report that the majority of the women they interviewed fell into the category of "subjective knowers," characterized by a "passionate rejection of science and scientists." These "subjectivist" women see the methods of logic, analysis, and abstraction as "alien territory belonging to men" and "value intuition as a safer and more fruitful approach to truth."

That way madness lies. As reported by Barbara Ehrenreich and Janet McIntosh in *The Nation* in 1997, at an interdisciplinary seminar the social psychologist Phoebe Ellsworth praised the virtues of the experimental method. Audience members protested that the experimental method was "the brainchild of white Victorian males." Ellsworth acknowledged this but pointed out that the experimental method had led to, for example, the discovery of DNA. This was greeted with disdain: "You believe in DNA?"

You can't *not* "believe in DNA." DNA is a fact. The DNA molecule is a double helix, a long spiral staircase with exactly four kinds of steps called nucleotides. The one-dimensional sequence of these four nucleotide "letters" is the genetic code that specifies the nature of every animal, plant, fungus, bacterium, and archaean. DNA sequences can be compared, letter for letter, between any creature and any other, much as one might compare folios of *Hamlet*. From this we can compute a numerical figure for the closeness of cousinship of any two creatures and, hence, eventually build up a complete family tree of all life. For, whether we like it or not, it is a true fact that we are cousins of kangaroos, we share an ancestor with starfish, and we and the starfish and the kangaroo share a more remote ancestor with jellyfish. The DNA code is a digital code, differing from computer codes only in being quaternary instead of binary. We know the precise details of the intermediate stages by which the code is read in our cells, and its four-letter alphabet translated, by molecular assembly-line machines called ribosomes, into a twenty-letter alphabet of amino acids, the building blocks of protein chains and hence of bodies.

If your philosophy dismisses all that as patriarchal domination, so much the worse for your philosophy. Perhaps you should stay away from doctors

with their experimentally tested medicines and go to a shaman or witch doctor instead. If you need to travel to a conference of like-minded philosophers,
you'd better not go by air. Planes fly because a lot of scientifically trained
mathematicians and engineers got their sums right. They did not use "intuitive ways of knowing." Whether they happened to be white and male, or sky-
blue-pink and hermaphrodite, is supremely, triumphantly irrelevant. Logic is
logic is logic, no matter if the individual who wields it also happens to wield
a penis. A mathematical proof reveals a definite truth, no matter whether the
mathematician "identifies as" female, male, or hippopotamus. If you decide to
fly to that conference, Newton's Laws and Bernoulli's Principle will see you
safe. And no, Newton's *Principia* is not a "rape manual" as was ludicrously said
by the noted feminist philosopher Sandra Harding. It is a supreme work of
genius by one of *Homo sapiens*'s most sapient specimens—who also happened
to be a not very nice man.

It is true that Newton's Laws are approximations that need modifying under
extreme circumstances, such as when objects travel near the speed of light. Those
philosophers of science who fixate on the case of Newton and Einstein love to
say that scientific truths are only ever provisional approximations that have so
far resisted falsification. But there are many scientific truths—we share an ancestor with baboons is one example—that are just plain true, in the same sense as
"New Zealand lies south of the equator" is not a provisional hypothesis pending
possible falsification.

The physics of the very small also goes beyond Newton. Quantum theory
is too weird for most human brains to accommodate intuitively. Yet the accuracy with which its predictions are fulfilled is shattering and beyond all doubt.
If I can't get my head around the weirdness of a theory that is validated by
such predictions, that's just too bad. There's no law that says truths about
nature have to be understandable by the human brain. We have to live with
the limitations of a brain that was built by Darwinian natural selection of
hunter-gatherer ancestors on the African savanna, where medium-sized things
such as antelopes and mates moved at medium speeds. It's actually remarkable
that human brains—even if only a minority of them—are capable of doing
modern physics at all. It is an open question whether there remain deep truths
about the universe that human brains not only don't yet understand but can
never understand. I find that open question immensely exciting, whatever the
answer to it may be.

Theologians love their "Mysteries," such as the "Mystery of the Trinity"
(how can God be both three and one at the same time?) and the "Mystery
of transubstantiation" (how can the contents of a chalice be simultaneously
wine and blood?). When challenged to defend such stuff, they may retort that

scientists too have their mysteries. Quantum theory is mysterious to the point of being downright perverse. What's the difference? I'll tell you the difference, and it's a big one. Quantum theory is validated by predictions fulfilled to so many decimal places that it's been compared to predicting the width of North America to within one hairsbreadth. Theological theories make no predictions at all, let alone testable ones.

Of course, not all the sciences can boast the formidable accuracy of physics. We biologists stand in awe of the LIGO experiments in which gravitational waves, having traveled a billion light-years, are detected by measurements accurate to less than a thousandth the width of a proton. Biological experimenters have to confront problems such as subjective bias of the experimenter—"intuitive ways of knowing." Medical scientists have perfected safeguards aimed precisely against intuitive ways of knowing, because these are highly likely to mislead. The double-blind control test has become the gold standard for demonstrating the efficacy of a medical treatment. A new drug must be compared with a placebo control, and the comparison tested statistically. Neither the patients, nor the doctors running the tests, nor the nurses administering the doses, nor the analysts evaluating the results are allowed to know which patients were given the placebo and which the drug, until all the results are in. I myself conducted a double-blind test of dowsing (water divining). It was pathetically touching to witness the sincere distress of the professional dowsers when they failed—every single one of them—to perform above chance level. The poor things had never before been tested under double-blind conditions. They had never before been deprived of whatever subliminal cues normally inform their "subjective ways of knowing." I treasure the remark of a homeopathic doctor who, when his methods failed under double-blind testing conditions, said, "You see. This is why we don't do double-blind tests any more. They never work!"

A layperson's version of the pernicious philosophy I mentioned earlier is the familiar bleat of, "Well it may not be true for you, but it is true for me." No. It's either true or it isn't. For both of us. As somebody once said (authorship multiply attributed), you are entitled to your own opinion but not to your own facts.

Some of what I have claimed here about scientific truth may come across as arrogant. So might my disparagement of certain schools of philosophy. Science really does know a lot about what is true, and we do have methods in place for finding out a lot more. We should not be reticent about that. But science is also humble. We may know what we know, but we also know what we don't know. Scientists love not knowing because they can go to work on it. The history of science's increasing knowledge, especially during the past four centuries, is a spectacular cascade of truths following one on the other. We may choose to call it a cumulative increase in the number of truths that we know. Or we can tip our hat to (a better class of) philosophers and talk of successive approximations toward yet-to-be-falsified provisional truths. Either way, science can properly claim to be the gold standard of truth.

Richard Dawkins, FRS, is a renowned evolutionary biologist and emeritus professor of the Public Understanding of Science at Oxford. He is a senior editor and columnist for *Free Inquiry* and author of many books on science and atheism.

CHAPTER 2

What Science Is and How and Why It Works

Neil deGrasse Tyson

Vol. 40, No. 5
September/October 2016

If you cherry-pick scientific truths to serve cultural, economic, religious, or political objectives, you undermine the foundations of an informed democracy.

Science distinguishes itself from all other branches of human pursuit by its power to probe and understand the behavior of nature on a level that allows us to predict with accuracy, if not control, the outcomes of events in the natural world. Science especially enhances our health, wealth, and security, which is greater today for more people on Earth than at any other time in human history.

The scientific method, which underpins these achievements, can be summarized in one sentence, which is all about objectivity: "Do whatever it takes to avoid fooling yourself into thinking something is true that is not, or that something is not true that is."

This approach to knowing did not take root until early in the seventeenth century, shortly after the inventions of both the microscope and the telescope. The astronomer Galileo and philosopher Sir Francis Bacon agreed: conduct experiments to test your hypothesis and allocate your confidence in proportion to the strength of your evidence. Since then, we would further learn not to claim knowledge of a newly discovered truth until multiple researchers, and ultimately the majority of researchers, obtain results consistent with one another. This code of conduct carries remarkable consequences. There's no law against publishing wrong or biased results. But the cost to you for doing so is high. If your research is rechecked by colleagues and nobody can duplicate your findings, the integrity of your future research will be held suspect. If you commit outright fraud, such as knowingly faking data, and subsequent researchers on the subject uncover this, the revelation will end your career.

It's that simple.

This internal, self-regulating system within science may be unique among professions, and it does not require the public or the press or politicians to make it work. But watching the machinery operate may nonetheless fascinate you. Just observe the flow of research papers that grace the pages of peer-reviewed scientific journals. This breeding ground of discovery is also, on occasion, a battlefield where scientific controversy is laid bare.

Science discovers objective truths. These are not established by any seated authority or by any single research paper. The press, in an effort to break a story, may mislead the public's awareness of how science works by headlining a just-published scientific paper as "the truth," perhaps also touting the academic pedigree of the authors. In fact, when drawn from the moving frontier, the truth has not yet been established, so research can land all over the place until experiments converge in one direction or another—or in no direction, itself usually indicating no phenomenon at all.

Once an objective truth is established by these methods, it is not later found to be false. We will not be revisiting the question of whether Earth is round; whether the Sun is hot; whether humans and chimps share more than 98 percent identical DNA; or whether the air we breathe is 78 percent nitrogen. The era of "modern physics," born with the quantum revolution of the early twentieth century and the relativity revolution of around the same time, did not discard Newton's laws of motion and gravity. What it did was describe deeper realities of nature, made visible by ever-greater methods and tools of inquiry. Modern physics enclosed classical physics as a special case of these larger truths. So the only times science cannot assure objective truths is on the pre-consensus frontier of research, and the only time it couldn't was before the seventeenth century, when our senses—inadequate and biased—were the only tools at our disposal to inform us of what was and was not true in our world.

Objective truths exist outside of your perception of reality, such as: the value of pi, $E=mc^2$, Earth's rate of rotation, and that carbon dioxide and methane are greenhouse gases. These statements can be verified by anybody, at any time, and at any place. And they are true whether or not you believe in them. Meanwhile, personal truths are what you may hold dear, but there exists no simple way of convincing others who disagree except by heated argument, coercion, or by force. These are the foundations of most people's opinions. Is Jesus your savior? Is Mohammad God's last prophet on Earth? Should the government support poor people? Is Beyoncé a cultural queen? Kirk or Picard?

Differences in opinion define the cultural diversity of a nation and should be cherished in any free society. You don't have to like gay marriage. Nobody will ever force you to gay-marry. But to create a law preventing fellow citizens from doing so is to force your personal truths on others. Political attempts to require that others share your personal truths are, in their limit, dictatorships.

Note further that in science, conformity is anathema to success. The persistent accusations that we are all trying to agree with one another is laughable to scientists attempting to advance their careers. The best way to get famous in your own lifetime is to pose an idea that is counter to prevailing research and which ultimately earns a consistency of observations and experiment. This ensures healthy disagreement at all times while working on the bleeding edge of discovery.

In 1863, a year when he clearly had more pressing matters to attend to, Abraham Lincoln—the first Republican president—signed into existence the National Academy of Sciences based on an Act of Congress. This august body would provide independent, objective advice to the nation on matters related to science and technology. Today, other government agencies with scientific missions serve similar purposes, including NASA, which explores space and aeronautics; NIST, which explores standards of scientific measurement on which all other measurements are based; DOE, which explores energy in all usable forms; and NOAA, which explores Earth's weather and climate.

These centers of research, as well as other trusted sources of published science, can empower politicians in ways that lead to enlightened and informed governance. But this won't happen until people in charge, and the people who vote for them, come to understand how and why science works.

Neil deGrasse Tyson is an astrophysicist with the American Museum of Natural History and author of numerous books. His long-running radio show *StarTalk*, with National Geographic Channel, became the first ever science-based talk show on television. He is a CSI fellow.

CHAPTER 3

Why We Need Science

Harriet Hall

Vol. 44, No. 6
November/December 2020

Most patients—and even many medical doctors and scientists—have not grasped how important it is to use rigorous science to evaluate claims for medical treatments. All too often people decide to try a treatment that is irrational, hasn't been tested, or has been tested and shown not to work. Why do they make those bad decisions?

How can you know whether a medical treatment works? If others say it worked for them, your Aunt Sally swears it cured her, there's someone in a white lab coat lecturing about it on YouTube, and you try it and your symptoms go away, you can pretty much assume it really works. Right? No, wrong! That's all the "evidence" most people need, but it's not evidence at all.

No, you can't make that assumption, because sometimes we get it wrong. For many centuries, doctors used leeches and lancets to remove blood from their patients. Everyone *knew* bloodletting worked. When you had a fever and the doctor bled you, you got better. Everyone knew of friends or relatives who had been at death's door until bloodletting cured them. Doctors could recount thousands of successful cases. George Washington accepted the common wisdom that bloodletting was effective. When he got a bad throat infection, he let his doctors remove so much of his blood that his weakened body couldn't recover, and he died. When scientists finally tested bloodletting, they found out it did much more harm than good. Patients who got well had been getting well *in spite of* bloodletting, not *because* of it. And some patients, including Washington, died unnecessarily.

People can be absolutely certain they are right when they are actually wrong. People can be fooled, including me and you. As Richard Feynman said: "The first principle is that you must not fool yourself—and you are the easiest person to fool."

Personal experience can be very impressive, but it is unreliable. Testimonials are anecdotes that can suggest which treatments might be worth testing, but they are not reliable evidence of efficacy.

"It Worked for Me"

People say, "It worked for me!" Well, maybe it didn't. They have no basis for claiming that it "worked." All they can really claim is that they observed an improvement following the treatment. That could indicate a real effect, an inaccurate observation, or a *post hoc ergo propter hoc* error—a false assumption that correlation in time meant causation. Such observations are only a starting point: we need to do science to find out what the observations mean. A commenter on the *Science-Based Medicine* blog wrote this testimonial: "I have witnessed first hand the life that begins to flow through the body upon the removal of a subluxation."

What does this even mean? Does he expect anyone to believe this just because he says it? Would he believe me if I said I had witnessed firsthand the invisible dragon in Carl Sagan's garage? Another commenter wrote that advocates of science-based medicine "seem to think that even if they see something with their own eyes that they can't believe it if there are no double blinded officially published studies to prove that what they saw actually happened."

Well, yes, that's pretty much what we do think; and we are appalled that you don't understand it yet, because it's the whole reason we have to do science. I would phrase it a bit differently: "Seeing something with my own eyes doesn't prove it's true, and it doesn't preclude the necessity for scientific testing."

Perception and Interpretation

We can't believe our own eyes. The process of vision itself is an interpretive construct by the brain. There are blind spots in our field of vision that we aren't even aware of because our brain fills in the blanks. I saw a magician cut a woman in half on stage—that was an illusion, a false perception. I saw a patient get better after a treatment—but my interpretation that the treatment caused the improvement may have been a mistake, a false attribution.

Sometimes We Get It Wrong

Before we had science to test treatments, sometimes we got it wrong. We were probably more often wrong than right, and even modern doctors sometimes get it wrong. Not long ago, doctors used to do an operation for heart disease where they opened the chest and tied off chest wall arteries to divert more blood flow to the heart. They had an impressive 90 percent success rate. Dr. Leonard Cobb wanted to make sure, so he did a sham surgery experiment where he just made the incision in the chest and closed it back up without doing anything else. He discovered that just as many patients improved after the fake surgery! Doctors stopped doing that operation.

How could so many people be so wrong? How could they believe something had helped them when it actually had done them more harm than good? There are many reasons people can come to believe an ineffective treatment works.

The disease may have run its natural course. A lot of diseases are self-limiting; the body's natural healing processes restore people to health after a time. A cold usually goes away in a week or so. To find out if a cold remedy works, you have to keep records of successes and failures for a large enough number of patients to find out if they really get well faster with the remedy than without it.

Many diseases are cyclical. The symptoms of any disease fluctuate over time. We all know people with arthritis have bad days and good days. The pain gets worse for a while, then it gets better for a while. If you use a remedy when the pain is bad, it was probably about to start getting better anyway, so the remedy gets credit it doesn't deserve.

We are all suggestible. If we're told something is going to hurt, it's more likely to hurt. If we're told something is going to make it better, it probably will. We all know this; that's why we kiss our children's scrapes and bruises. Anything that distracts us from thinking about our symptoms is likely to help. In scientific studies that compare a real treatment to a placebo sugar pill, an average of 35 percent of people say they feel better after taking the sugar pill. The real treatment has to do better than that if we're going to believe it's really effective (not to mention spend money on it).

There may have been two treatments and the wrong one got the credit. If your doctor gave you a pill and you also took a home remedy, you may give the credit to the home remedy. Or maybe something else changed in your life at the same time that helped treat the illness and was the real reason you got better.

The original diagnosis or prognosis may have been incorrect. A lot of people have supposedly been cured of cancer who never actually had cancer. Doctors who tell a patient he has only six months to live are only guessing and may be wrong. The best they can do is say the median survival for similar patients is six months—but that means half of the patients live longer than that.

Temporary mood improvement can be confused with cure. If a provider makes you feel optimistic and hopeful, you may think you feel better when the disease is really unchanged.

Psychological needs can affect our behavior and perceptions. When people want to believe badly enough, they can convince themselves they have been helped. People have been known to deny the facts—to refuse to see that the tumor is still getting bigger. If they have invested time and money, they don't want to admit it was wasted. We see what we want to see; we remember things the way we wish they had happened. When a doctor is sincerely trying to help a patient, the patient may feel a sort of social obligation to please the doctor by reporting improvement.

We confuse correlation with causation. Just because an effect follows an action, that doesn't necessarily mean the action caused the effect. When the rooster crows and then the sun comes up, we realize it's not the crowing that made the sun come up. But when we take a pill and then we feel better, we assume it was the pill that made us feel better. We don't stop to think that we might have felt better for some other reason. We jump to conclusions like in the story of a man who trained a flea to dance when it heard music, then he cut the flea's legs off one by one until it could no longer dance and concluded that the flea's organ of hearing must be in its legs!

So, there are a lot of ways we can get it wrong. Luckily, there's a way we can eventually get it right: by scientific testing. There's nothing mysterious or complicated about science: it's just a toolkit of commonsense ways to test beliefs against reality. If you believe you've lost weight and you step on the scale to test your belief, that's science. If you think you have a better way to grow carrots and you test your idea by planting two rows side by side, one with the old method and one with the new method, and see which row produces better carrots, that's science. To test medicines, we can sort a large number of patients into two equal groups and give one group the treatment we're testing and give the other group an inert placebo, such as a sugar pill. If the group that got the active treatment does significantly better, the treatment probably really works.

Jacqueline Jones was a fifty-year-old woman who had suffered from asthma since the age of two. She read about a miraculous herbal treatment that cured a host of ailments, including asthma. She assumed the information was true, because it included a lot of testimonials from people who had used it and were able to stop taking their asthma medications. They *knew* it worked; they had *seen* it work. Sick of the side-effects of conventional drugs, Jones stopped using her three inhalers, steroids, and nebulizer and took the herbal supplement instead. Within two days she had a massive asthma attack with complications that kept her hospitalized for six weeks.

All those people who said that herbal remedy had cured their asthma got it wrong. Asthma symptoms fluctuate. Maybe their symptoms would have improved anyway. Whatever the reason, the remedy had not been tested scientifically and was not effective for treating asthma. Believing those testimonials almost cost Jacqueline her life.

The next time a friend enthusiastically recommends a new treatment, stop and remember that they could be wrong. Remember Jacqueline Jones. Remember George Washington. Sometimes we get it wrong.

Harriet Hall, MD, was a retired Air Force physician and flight surgeon, writing about pseudoscientific and so-called alternative medicine. She was a contributing editor and frequent contributor to the *Skeptical Inquirer* and the blog *Science-Based Medicine*. She was author of *Women Aren't Supposed to Fly: Memoirs of a Female Flight Surgeon* and coauthor of the 2012 textbook *Consumer Health: A Guide to Intelligent Decisions*.

CHAPTER 4

The God Engine

HOW BELIEF DEVELOPS

James Alcock

Vol. 42, No. 5
September/October 2018

Whatever its form, religion is powerful and pervasive and, for billions of people, obviously important. Yet, while major religions such as Islam, Christianity, Hinduism, and Buddhism have endured since ancient times, others, despite having enjoyed great appeal for centuries, have disappeared into the history books. No longer does anyone worship Zeus, the supreme god of the ancient Greeks; Marduk, the Babylonian god of creation; Bast, the Egyptian goddess of protection; Jupiter, the supreme god of the Romans; the Incan Apocatequil; or the Aztec Huehueteotl. Those bygone gods were central figures in highly developed theocracies and were as real to their devotees as are today's deities to contemporary worshippers.

The continuing power of religious belief in all its many contradictory forms suggests that it serves important functions. Indeed, some researchers consider religion to have become culturally important because fear of the deity promoted social solidarity, cooperation, trust, and self-sacrifice. Important behaviors were either mandated or declared taboo by religion, and believers had little choice but to accept that a powerful supernatural being had deemed them so. This social control in turn increased the likelihood of the survival and reproduction of individuals as well as the long-term survival of the group itself. As religion became deeply established within a group, the religious beliefs and rituals taught to young people contributed an important part of their social identities, and their corresponding roles and duties further contributed to the functioning and cohesiveness of the group.

However, the prevailing view in modern psychology is that religious belief developed not because of those functions but rather as the automatic byproduct of brain systems that evolved for everyday cognition. That is, belief in the supernatural is a natural consequence of the way our brains work, a product of a metaphorical "God Engine" that endows it both with significant power over the lives of people and the groups to which they belong and with strong resistance to change. In other words, a number of automatic processes and cognitive biases combine to make supernatural belief the automatic default.

Components of the God Engine

A number of elements of this "God Engine" are particularly important in the development of religious belief:

We are born magical thinkers. Magical thinking—assuming some magical causation between an assumed cause and its effect—plays such a significant role in religion that some researchers consider it to be the very foundation of religious belief. Suppose that the sky darkens with the approach of a frightening thunderstorm. Someone importunes Zeus to send the storm away, and, by coincidence, the storm abates. Fear is reduced, the action of praying has been reinforced, and Zeus is credited with having quieted the storm.

We are born agency detectors. From a very early age, we look for reasons and intentions, for agency, behind threatening or awe-inspiring events. Failure to identify a cause can lead to the conclusion that the agent is present but invisible, resulting in increased belief in undetectable agencies. And because people everywhere have similar experiences, beliefs in supernatural beings—gods, ghosts, angels, and demons—have developed in every culture and society. Interpreting events as divine interventions, messages, warnings, or punishments serves as further confirmation of the existence of the divine being. Indeed, in every society that anthropologists have examined, uncontrollable tragedies have been viewed by many people as having been deliberately caused by some supernatural agent. An unexpected and violent storm may be interpreted as an expression of God's wrath in response to some transgression. The epidemic of HIV/AIDS was taken by some fundamentalist Christians in the United States to be God's punishment for homosexuality.

There are only a limited number of attributes that can be reasonably associated with a hypothetical supernatural agent, and therefore it is not surprising that many similarities are apparent among the supernatural beings envisioned by peoples around the planet. For example, we might attribute to a supernatural being counterintuitive physical properties such as being a ghost, counterintuitive biological properties such as never aging or dying, or counterintuitive psychological properties such as prescience or extraordinary perception. However, not all possible combinations of such abilities are likely to persist in the belief system or be passed on to subsequent generations. The representations of supernatural beings that persist are recognizable, easily remembered, easily communicated, and useful in dealing with problems. Examination of a wide variety of mythologies, anecdotes, cartoons, religious writings, and science fiction bears testament to this.

We develop theory of mind. Infants become able to distinguish between animate and inanimate objects by five months of age. This is followed by the development of *theory of mind* sometime before the third year as children come to understand that humans and animals have internal mental processes similar to their own. This leads directly to *dualism*, the notion that mind is separate from matter, which in turn paves the way for belief in disembodied spirits and intelligent deities whose minds function in a similar manner to our own. As a result, it should not be surprising that about half of all four-year-olds have an imaginary friend. Some cleverly designed experiments have shown that children's concepts of gods are not simply extensions of their concepts of what people are like. Being godlike involves powers that go beyond human capability.

We develop promiscuous teleological intuition. As they reason about the world around them, children appear innately prone to consider objects and events as serving an intentional purpose, and this teleological bias to perceive that things happen for a reason operates promiscuously (in that it is applied to just about everything). Of course, such reasoning eventually leads to contemplation about some sort of extraordinary intelligence that guides the world and belief in supernatural creation (Kelemen 2004).

Reality testing. Children have to learn to distinguish fantasy from reality. They learn that nightmares are not real, that the tooth fairy and Easter bunny are fictions, and that Santa Claus does not exist. Yet, where religious beliefs are concerned, religious parents not only do not teach their children to reality test, they typically teach them *not* to reality test. They are taught that religious beliefs are justified by faith alone and are not to be subjected to reason. In the more fundamentalist religions, children are further taught that it is not only inappropriate but sinful to question religious teachings, and the resultant guilt aroused by any doubts about their religion is a powerful deterrent to future intellectual challenges. In consequence, these beliefs may remain forever insulated from reality testing and therefore may be very unlikely to change.

Compare belief in Santa Claus with belief in a deity (see figure below). Both beliefs are communicated to children at a time when their ability to reason is undeveloped. Both beliefs involve physically impossible feats and make no logical sense. Both beliefs are typically shared by the child's peers, but while all adults look askance at any twelve-year-old who still believes in Santa Claus, religious parents look askance at the twelve-year-old who no longer believes in the deity.

As children strive to make distinctions between reality and fantasy, making wishes comes easily to them. Note that praying and making a wish are very similar, for each involves a mental process intended to bring about some desired

	Santa Claus	Deity
Taught to young children at an age when critical faculties are underdeveloped	✓	✓
Involves physically impossible feats	Climbs down chimneys Flying reindeer Delivers gifts around the world in a single night	Hears all prayers simultaneously Visits Earth as human or animal Transcends space, time, and physical laws
Makes no logical sense	Child is too young to question	It is a matter of faith; Logic is inappropriate
Belief is shared by peers, providing social support	✓	✓
Outgrowing the belief is a sign of maturity	✓	✗

Belief in Santa Claus vs. Deity. *Illustration by Chris Fix.*

outcome without any physical effort on one's own part. While praying is part of a system of *institutionalized* magical beliefs (religion) taught by adults and shared with family, wish-making is something that one does independently of others; it is *non-institutionalized*. When family and cultural influences support and encourage religious beliefs while discouraging and disparaging other magical beliefs, children, at some time between ages four and eight years, come to view praying and wish-making quite differently. They typically show both increasing belief in the power of prayer and decreasing confidence in the effectiveness of personal idiosyncratic magical beliefs, including wishing (Woolley 2000).

We are taught religion. Children do not become Christians, Muslims, Buddhists, or Jews on their own. They learn religion, which is intertwined with a group's history and culture. The development of language ability not only gives the growing child a powerful tool for the symbolic analysis of events, it also makes possible the acquisition of massive amounts of information from others. Children learn much about the world through their own experiences, but when it comes to things they cannot observe directly—for example, their internal organs, the shape of the Earth, or the revolution of the Earth around the Sun—they uncritically accept the teachings of adults. It is therefore not surprising that they uncritically accept religious teaching about invisible entities that they cannot observe directly.

We are more likely to remember "ontological" violations. Children develop an intuitive ontology—an understanding about what exists, about what is real—early in life. An *ontological violation* is something that is considered impossible in the normal world, such as waving one's arms in order to fly. Of course, many religious beliefs involve ontological violations—such as an omniscient deity who can hear people's prayers wherever they may be in the world.

Violations of ontological expectations are better remembered than beliefs that intuitively make sense. For example, "a man who walked through a wall" is an ontological violation, while "a man with six fingers," although unusual, does not violate our beliefs about what is possible and what is not. Research has found that reference to a man who walked through a wall is recalled more readily over a period of months than reference to a six-fingered man. This effect has been found not only in the West but also in Tibet and with the Fang people, a Bantu tribe in Gabon (Boyer 2001).

Religious beliefs involving violations of our intuitive (ontological) under-standing of the world are both attention-grabbing and memorable. Because they are more memorable, they are also more likely to be communicated to others. The advantage that a memorable idea has in terms of being transmit-ted from generation to generation may be small, but over many generations its influence is amplified as it becomes an inherent part of the culture. Psy-chologist Pascal Boyer points out that although individuals and groups can be credulous at times, there are limits to what people will believe in the name of religion, and so only certain religious beliefs, those that provide solace or offer explanations for strange events, are likely to be transmitted over extended periods of time. While something too counterintuitive (e.g., a species of super-intelligent beetles that control television networks) may not take hold, the notion of a god who possesses some human qualities but also has omniscience and prescience may be just counterintuitive enough but not too much. Boyer also points out that it does not take much effort to maintain religious beliefs, and that people do not have to work very hard to persuade themselves that these beliefs are true.

Persistence of Religious Belief

While supernatural belief is a natural byproduct of normal cognitive devel-opment, such belief is supported through experiences that seem to offer verification:

To sleep, perhaps to dream. There is considerable evidence, both cross-cultural and historical, that dreams provide people with what seems to be verifi-cation of their beliefs in supernatural beings and life after death and contribute to the development of religious rituals, beliefs, and concepts of the divine.

Emotional experience. While each religion has its own set of beliefs and prac-tices, and while the major religions are guided by holy texts supposedly divinely dictated or inspired eons ago, religion is much more than a set of dictates and beliefs, for strong emotional experience is often involved. It is one thing to debate the appeal of various religious beliefs and it is quite another to "experience" the divine. Many religious people have reported having felt the power of a divine presence, often during prayer. Some religions deliberately encourage such experiences, and the resurgence of charismatic and Pentecostal Christianity, sometimes involving speaking in tongues and laying on of hands, attests to the appeal of a mixture of emotion and spirituality.

Religion answers existential questions. Why are we here, and what happens when we die? Where did the world come from, and when did it begin? Religion provides answers to many questions that otherwise would go unanswered. Consider these creation stories:

Creation Story #1 (King James Bible, Book of Genesis)

In the beginning, God created the heaven and the earth. And the earth was without form, and void; and darkness was upon the face of the deep. And the Spirit of God moved upon the face of the waters. And God said, Let there be light: and there was light. And God saw the light, that it was good: and God divided the light from the darkness. And God called the light Day, and the darkness he called Night. And the evening and the morning were the first day.

Creation Story #2 (Native North American, *Achomawi*)

In the beginning, all was water. Then a cloud formed in the clear sky, became lumpy and turned into Coyote. Then a fog developed and became lumpy and Silver Fox was formed, and then Coyote and Silver Fox both became people. And they began to think. And they thought a canoe, and it became real and they floated in it for many years. One time while Coyote was sleeping, Silver Fox combed his hair and then took the combings, flattened them and spread them out on the water until they covered the surface of the water. Then he thought, "There should be a tree," and now there was a tree. And then he did the same with shrubs and rocks and people and birds and fish.

Creation Story #3 (The big bang, modern science)

About 13.8 billion years ago, a singularity, an infinitesimally small point of infinite density, suddenly exploded and began to expand rapidly to form the universe as we now know it, with a diameter at present of some 93 billion light-years and still expanding. (That is, light emanating from one edge of the universe would take 93 billion years to reach the other edge.) The universe is estimated to contain between 100 and 200 billion galaxies, with each galaxy comprising hundreds of billions of stars.

All three stories may seem fantastical in their own way, and all three are held on trust by those who believe them. While the first involves trust in the validity of the Book of Genesis and the second in the oral traditions of North American First Nations peoples, the third requires trust in the conclusions of modern science. While scientists understand the logic and the data that support the big bang explanation, it is beyond the layperson's ability to do so.

Such explanations provide a feeling of understanding to people puzzled about how the world began, even though they leave other mysteries in their wake. Those who attribute the origins of the universe to their god or gods conveniently ignore the question about where their god or gods came from. On the other hand, the big bang explanation also leads to another mystery: the origin of that singularity.

Reinforcement of religious belief. Religious beliefs are reinforced in a number of ways. As mentioned above, many religious people believe that they have experienced a divine presence, often during prayer. In addition, no matter to which god one prays, subsequent events that would have occurred whether one prayed or not can often be interpreted as the desired answers to one's prayers. After all, even without divine intervention, people sometimes unexpectedly recover from severe illness; rains do come to end droughts; lost loved ones are found safe; and people do get promotions at work. However, when preceded by prayer, these events may appear to be miraculous. And the gods cannot lose, for any apparent failure of prayer is generally explained away: "Sometimes, God says no."

Second, religion provides a bulwark against existential anxiety and fear of annihilation. It offers comfort in times of threat and can help one to deal with the loss of loved ones, especially if one believes that our personalities survive death. Religious belief and prayer also offer an important shift in focus when faced with anxiety and uncertainty. When under stress, people with strong religious conviction show less reactivity in the part of the brain associated with anxiety. The framework that religion provides can help an individual to remain calm and to deal with difficult circumstances. Of course, an emotionally balanced atheist with a good sense of purpose might react similarly.

Third, religion can provide an increased sense of control. Religion not only serves to reduce anxiety, for belief in a benevolent god who watches over the world and answers prayers provides a sense of control when all around us is chaos and calamity. "God is in His heaven—All's right with the world." As a result, belief in God typically increases when people are under extreme levels of stress. Reflecting this, researchers have found that both in Western and non-Western contexts, pronounced government instability is associated with increased religious belief.

Fourth, religion can provide companionship that goes beyond being a member of a congregation. Many religious people form strong emotional attachments to their deity. Such attachment may either mirror strong feelings of childhood closeness to parents or compensate for unsatisfactory relationships with parents. Further, if one believes in a god who listens and watches over us, this belief provides a sense of never being alone. This is very important in a world where loneliness is a significant problem for many.

Fifth, religion is often a source of significant self-enhancement. "God sees the little sparrow fall . . . I know he loves me too." Perhaps you feel unimportant or ignored or rejected, but if you know that your God loves you, this gives you a reason to feel good about yourself.

Sixth, religion contributes to group identity. Religion does not occur in a social vacuum. Religious beliefs and practices are shared and reinforced by family, friends, and neighbors and provide an important basis for social identification. Adherents feel themselves to be part of a moral community with its sacred values, norms, and ethical expectations. Religious identity provides something that no other social identity can offer: a sense of eternal belonging and a set of sacred beliefs that provide certainty about the world and one's place in it. These unique characteristics, combined with powerful emotional experiences and strong bonds with other members, imbue it with more importance to many people than any other form of social identity.

In conclusion, belief in the supernatural is a natural consequence of normal cognitive development, and so it should be no surprise that religion is both pervasive and enduring. While for some, childhood religious belief ultimately succumbs to critical thinking, for many others it is excluded from critical analysis and is instead accepted on faith as being uncorrupted truth. And then no matter which presumed deity is worshipped—a god with the head of an elephant, a god who throws thunderbolts from the sky, or a god who comes to Earth in human form—many events and personal experiences will serve up apparent confirmation that the deity is real.

James E. Alcock, PhD, is professor of psychology at York University, Toronto, Canada. He is a fellow of the Canadian Psychological Association and a member of the Executive Council of the Committee for Skeptical Inquiry and the Editorial Board of the *Skeptical Inquirer*. Alcock has written extensively about parapsychology and anomalous experience and has for many decades taught a psychology course focusing on these topics.

Detailed citations are provided in the original book chapter.

CHAPTER 5

Why We Believe—Long after We Shouldn't

Carol Tavris and Elliot Aronson

Vol. 41, No. 2
March/April 2017

It's pretty clear nowadays that we are not the rational animals we'd like to believe we are; in fact, we are more accurately called the "rationalizing animal." Skeptics are often puzzled when we calmly provide evidence that a popular belief is wrong, that some group is holding onto a way of doing things that's long past its sell-by date, and recipients of this valuable information don't say, "Why, *thank you*! I had no idea!" Why would people prefer to justify mistaken beliefs, behavior, and practices rather than change them for better ones? Isn't it good to know you didn't cause your child's autism with vaccinations?

As skeptics we are faced constantly with what psychologists call "the motivated rejection of science." Take global warming, for example. It's easy to assume that climate-change deniers are less educated or informed than wise scientists, but it's not so simple. An article in *Psychological Science* by Stephan Lewandowsky and Klaus Oberauer found that attitudes about global warming are unrelated to levels of scientific literacy, numeracy, or education. They *are* associated with political partisanship; that is, among liberals, higher levels of scientific literacy and education are associated with *increased* acceptance of climate change, the importance of vaccination, and trust in science. But among conservatives, higher levels of scientific literacy and education are associated with *reduced* acceptance. That's motivated cognition; people are emotionally motivated to reject findings that threaten their core beliefs or worldview. At present, the researchers found, public rejection of scientific findings is more prevalent on the political right than the left, yet, they added, "the cognitive mechanisms driving rejection of science are found regardless of political orientation." Meaning: It depends *what* scientific finding it is. Whether your worldview comes from the left or right, you will be tempted to sacrifice skepticism even when your side is promoting some cockamamie belief without evidence.

Decades ago, the great social psychologist Gordon Allport, in his brilliant book *The Nature of Prejudice*, offered this exchange to illustrate the weasely way a person with a prejudice or other entrenched belief argues with you.

> MR. X: The trouble with Jews is that they only take care of their own group.
>
> MR. Y: But the record of the Community Chest campaign shows that they give more generously, in proportion to their numbers, to the general charities of the community, than do non-Jews.
>
> MR. X: That shows they are always trying to buy favor and intrude into Christian affairs. They think of nothing but money; that is why there are so many Jewish bankers.
>
> MR. Y: But a recent study shows that the percentage of Jews in the banking business is negligible, far smaller than the percentage of non-Jews.
>
> MR. X: That's just it; they don't go in for respectable business; they are only in the movie business or run night clubs.

Notice that people like Mr. X—which is all of us on occasion—don't actually argue or respond to the point; they slide off your evidence and raise an irrelevant digression rather than face, let alone change, their fundamental belief. "I believe it" becomes enough.

The key motivational mechanism that underlies the reluctance to change our minds, to admit mistakes, and to be unwilling to accept unwelcome scientific findings is cognitive dissonance—the discomfort we feel when two cognitions, or cognition and behavior, contradict each other. Leon Festinger, who developed this theory sixty years ago, showed that the key thing about dissonance is that, like extreme hunger, it is uncomfortable, and, like hunger, we are motivated to reduce it. For smokers, the dissonant cognitions are "Smoking is bad for me" versus "I'm a heavy smoker." To reduce that dissonance, smokers either have to quit or justify smoking. Before we make a decision (about a car, a candidate, or anything else), we are as open-minded as we are likely to be; but after we make a decision, we have to reduce dissonance. To do this, we will emphasize everything good about the car we bought or the candidate we are supporting or the belief we accepted and notice only the flaws in the alternatives.

Dissonance theory comprises three cognitive biases in particular:

1. The bias that we, personally, don't have any biases—the belief that we perceive objects and events clearly, as they really are. Any opinion I hold must be reasonable; if it weren't, I wouldn't hold it. If my opponents—or kids or friends or partner—don't agree with me, it is because *they* are biased.

2. The bias that we are better, kinder, smarter, more moral, and nicer than average. This bias is useful for plumping up our self-esteem, but it also blocks us from accepting information that we have been not-so-kind, not-so-smart, not-so-ethical, and not-so nice.

3. The *confirmation bias*, the fact that we notice and remember information that confirms what we believe and ignore, forget, or minimize information that disconfirms it. We might even call it the consonance bias, because it keeps our beliefs in harmony by eliminating dissonant information before we are even aware of it.

Dissonance is painful enough when you realize that you bought a lemon of a car and paid too much for it. But it's most painful when an important element of the self-concept is threatened; your post-car-purchase dissonance will be greater if you see yourself as a car expert and superb negotiator. We have two ways to reduce dissonance: either accept the evidence and change the self-concept ("Yes, that was a foolish/incompetent/unethical thing to do; was I ever wrong to believe that") or deny the evidence and preserve the self-concept ("That study was fatally flawed"). Guess which is the popular choice?

Understanding cognitive dissonance helps explain the astonishing obstinacy that some people reveal when they are shown to be wrong. Consider the conspiracy theorists who vehemently deny the horrifying evidence that Adam Lanza killed twenty children at Sandy Hook Elementary School. They maintain it was all a conspiracy of the gun-control lobby, and they persist in that delusion even when faced by grieving parents holding photos of their beloved children. But dissonance theory explains why people can hold crazy ideas without necessarily being crazy. If we start from where the disbelievers are, holding core beliefs in the importance of owning guns, that guns are safe, and that gun-control people want to take their guns away, then information that guns were used for a rampage that left twenty little children (and six school staff) dead is powerfully dissonant. By denying the evidence that this tragedy occurred, they get to retain their gun beliefs *and* their self-esteem: why, they were smart and right all along to oppose gun control of any kind. Indeed, dissonance theory would predict that their opposition would become even stronger—look at the effort those bastards put into creating the fiction of Sandy Hook. They must *really* want to take our guns away.

The greatest danger of dissonance reduction occurs not when a belief or action is a one-time thing like buying a car, but when it sets a person on a course of action. The metaphor that we use in our book is that of a pyramid. Imagine that two students are at the top of a pyramid, a millimeter apart in their attitudes toward cheating: it is not a good thing to do, but there are worse crimes in the world. Now they are both taking an important exam, when they draw a blank on a crucial question. Failure looms, at which point each one gets an easy opportunity to cheat by reading another student's answers. After a long moment of indecision, one spontaneously yields and the other resists. Each gains something important, but at a cost: one gives up integrity for a good grade; the other gives up a good grade to preserve his integrity.

As soon as they make a decision—to cheat or not—they will justify the action they took in order to reduce dissonance, that is, to keep their behavior consonant with their attitudes. They can't change the behavior, so they shift their attitude. The one who cheated will justify that action by deciding that cheating is not such a big deal: "Hey, everyone cheats. It's no big deal. And I needed to do this for my future career." But the one who resisted the temptation will justify that action by deciding that cheating is far more immoral than he originally thought: "In fact, cheating is disgraceful. People who cheat should be expelled." By the time they finish justifying their actions, they have slid to the bottom and now stand at opposite corners of its base, far apart from one another. The one who didn't cheat considers the other to be totally immoral, and the one who cheated thinks the other is hopelessly puritanical—and, come to think of it, why don't I just buy the services of a professional cheater to take the whole course for me? I really need the credits, and so what if I never learn what this class requires? I'll learn on the job. Hey, neurosurgery can't be that hard.

As we go through life we will find ourselves on the top of many such metaphorical pyramids, whenever we are called upon to make important decisions and moral choices: for example, whether to accept growing evidence that a decision we made is likely wrong; decide whether or not a sensational rape or murder case in the media is true; whether to blow the whistle at company corruption or decide not to rock the boat. As soon as we make a decision, we stop noticing or looking for disconfirming evidence, and we are on that path to the bottom, where certainty lies.

This process blurs the distinction that people like to draw between "us good guys" and "those bad guys," or, occasionally in the skeptic world, "us smart, reasonable guys and those ignorant, crazy guys." Often, when standing at the top of the pyramid we are faced not with a clear go-or-no-go decision but instead with ambiguous choices whose consequences are unknown or unknowable. We make an impulsive decision, and then we justify it to reduce the ambiguity of the choice. And soon we are trapped in a process of action, justification, and further action that increases our commitment to that first tentative decision. Taking the next step down the pyramid in that direction is almost inevitable, because otherwise we have to go back up and say, "I was wrong to take that first little step." How do you corrupt an innocent person? How does a company or a country get enmeshed in illegal or unethical decisions? They only have to take a small step off the pyramid, and self-justification will do the rest.

Dissonance reduction has benefits, including letting us sleep at night—and besides it's good to hold an informed opinion and not change it with every fad or every new study that comes along. But it is also essential to be able to let go of that opinion when the weight of the evidence dictates, even if we are far down that pyramid. Dissonance reduction may be built into our mental wiring, but how we think about our mistaken actions and beliefs is not.

Living with dissonance requires us to learn how to admit our mistakes and separate them from our self-esteem. Our brains may be wired for self-justification, but that is no justification for not overriding the impulse—and we can. That's what the skeptical movement is designed to help us do: show that people can remain committed to their country, political party, friends, and family, yet understand that it is not disloyal to disagree with actions or policies or candidates we find wrong or reprehensible. And when *we* are faced with evidence of our own mistaken beliefs, we can learn to say: "When I, a kind and smart person, make a mistake, I remain a kind and smart person; the mistake remains a mistake. Now, how do I remedy what I did and make sure I don't repeat it?"

Skeptics already have an immense challenge in debunking pseudoscience, con artists, and conspiracy theories; to this burden we'd add another: facing our own sources of dissonance—ambiguity, complexity, and compromise. For some on the left, "compromise" means selling out; for some on the right, "compromise" means consorting with the enemy. But no politician will do everything we want; no feminist or civil rights activist can achieve 100 percent ideological purity; no human being can be 100 percent free of bias. That may be the most dissonant message of all.

Carol Tavris, PhD, is a social psychologist, writer, and lecturer. She is author or coauthor of many books including *Mistakes Were Made (but Not by Me)* and several introductory psychology textbooks.

CHAPTER 6

Progressophobia

WHY THINGS ARE BETTER THAN YOU THINK THEY ARE

Steven Pinker

Vol. 42, No. 3
May/June 2018

Intellectuals hate progress. Intellectuals who call themselves "progressive" *really* hate progress. It's not that they hate the *fruits* of progress, mind you: most pundits, critics, and their *bien-pensant* readers use computers rather than quills and inkwells, and they prefer to have their surgery with anesthesia rather than without it. It's the *idea* of progress that rankles the chattering class—the Enlightenment belief that by understanding the world we can improve the human condition.

An entire lexicon of abuse has grown up to express their scorn. If you think knowledge can help solve problems, then you have a "blind faith" and a "quasi-religious belief" in the "outmoded superstition" and "false promise" of the "myth" of the "onward march" of "inevitable progress." You are a "cheerleader" for "vulgar American candoism" with the "rahrah" spirit of "boardroom ideology," "Silicon Valley," and the "Chamber of Commerce." You are a practitioner of "Whig history," a "naïve optimist," a "Pollyanna," and of course a "Pangloss," a modern-day version of the philosopher in Voltaire's *Candide* who asserts that "all is for the best in the best of all possible worlds."

Professor Pangloss, as it happens, is what we would now call a pessimist. A modern optimist believes that the world can be *much, much* better than it is today. Voltaire was satirizing not the Enlightenment hope for progress but its opposite, the religious rationalization for suffering called theodicy, according to which God had no choice but to allow epidemics and massacres because a world without them is metaphysically impossible.

Epithets aside, the idea that the world is better than it was and can get better still fell out of fashion among the clerisy long ago. In *The Idea of Decline in Western History*, Arthur Herman shows that prophets of doom are the all-stars of the liberal arts curriculum, including Nietzsche, Arthur Schopenhauer, Martin

29

Heidegger, Theodor Adorno, Walter Benjamin, Herbert Marcuse, Jean-Paul Sartre, Frantz Fanon, Michel Foucault, Edward Said, Cornel West, and a chorus of eco-pessimists.[1] Surveying the intellectual landscape at the end of the 20th century, Herman lamented a "grand recessional" of "the luminous exponents" of Enlightenment humanism, the ones who believed that "since people generate conflicts and problems in society, they can also resolve them." In *History of the Idea of Progress*, the sociologist Robert Nisbet agreed: "The skepticism regarding Western progress that was once confined to a very small number of intellectuals in the nineteenth century has grown and spread to not merely the large majority of intellectuals in this final quarter of the century, but to many millions of other people in the West."[2]

Yes, it's not just those who intellectualize for a living who think the world is going to hell in a handcart. It's ordinary people when they switch into intellectualizing mode. Psychologists have long known that people tend to see their own lives through rose-colored glasses: they think they're less likely than the average person to become the victim of a divorce, layoff, accident, illness, or crime. But change the question from the people's *lives* to their *society,* and they transform from Pollyanna to Eeyore.

Public opinion researchers call it the Optimism Gap.[3] For more than two decades, through good times and bad, when Europeans were asked by pollsters whether their *own* economic situation would get better or worse in the coming year, more of them said it would get better, but when they were asked about their *country's* economic situation, more of them said it would get worse.[4] A large majority of Britons think that immigration, teen pregnancy, litter, unemployment, crime, vandalism, and drugs are a problem in the United Kingdom as a whole, while few think they are problems in their area.[5] Environmental quality, too, is judged in most nations to be worse in the nation than in the community, and worse in the world than in the nation.[6] In almost every year from 1992 through 2015, an era in which the rate of violent crime plummeted, a majority of Americans told pollsters that crime was rising.[7] In late 2015, large majorities in eleven developed countries said that "the world is getting worse," and in most of the last forty years a solid majority of Americans have said that the country is "heading in the wrong direction."[8]

Are they right? Is pessimism correct? Could the state of the world, like the stripes on a barbershop pole, keep sinking lower and lower? It's easy to see why people feel that way: every day the news is filled with stories about war, terrorism, crime, pollution, inequality, drug abuse, and oppression. And it's not just the headlines we're talking about; it's the op-eds and long-form stories as well. Magazine covers warn us of coming anarchies, plagues, epidemics, collapses, and so many "crises" (farm, health, retirement, welfare, energy, deficit) that copywriters have had to escalate to the redundant "serious crisis."

Whether or not the world really is getting worse, the nature of news will interact with the nature of cognition to make us think that it is. News is about things that happen, not things that don't happen. We never see a journalist saying to the camera, "I'm reporting live from a country where a war has not broken out"—or a city that has not been bombed, or a school that has not been shot up. As long as bad things have not vanished from the face of the earth, there will always be enough incidents to fill the news, especially when billions of smartphones turn most of the world's population into crime reporters and war correspondents.

And among the things that do happen, the positive and negative ones unfold on different time lines. The news, far from being a "first draft of history," is closer to play-by-play sports commentary. It focuses on discrete events, generally those that took place since the last edition (in earlier times, the day before; now, seconds before).[9] Bad things can happen quickly, but good things aren't built in a day, and as they unfold, they will be out of sync with the news cycle. The peace researcher John Galtung pointed out that if a newspaper came out once every fifty years, it would not report half a century of celebrity gossip and political scandals. It would report momentous global changes such as the increase in life expectancy.[10]

The nature of news is likely to distort people's view of the world because of a mental bug that the psychologists Amos Tversky and Daniel Kahneman called the Availability heuristic: people estimate the probability of an event or the frequency of a kind of thing by the ease with which instances come to mind.[11] In many walks of life this is a serviceable rule of thumb. Frequent events leave stronger memory traces, so stronger memories generally indicate more-frequent events: you really are on solid ground in guessing that pigeons are more common in cities than orioles, even though you're drawing on your memory of encountering them rather than on a bird census. But whenever a memory turns up high in the result list of the mind's search engine for reasons other than frequency—because it is recent, vivid, gory, distinctive, or upsetting—people will overestimate how likely it is in the world. Which are more numerous in the English language, words that begin with *k* or words with *k* in the third position? Most people say the former. In fact, there are three times as many words with *k* in the third position (*ankle, ask, awkward, bake, cake, make, take* . . .), but we retrieve words by their initial sounds, so *keep, kind, kill, kid,* and *king* are likelier to pop into mind on demand.

Availability errors are a common source of folly in human reasoning. First-year medical students interpret every rash as a symptom of an exotic disease, and vacationers stay out of the water after they have read about a shark attack or if they have just seen *Jaws*.[12] Plane crashes always make the news, but car crashes, which kill far more people, almost never do. Not surprisingly, many people have a fear of flying, but almost no one has a fear of driving. People rank tornadoes (which kill about fifty Americans a year) as a more common cause of death than asthma (which kills more than four thousand Americans a year), presumably because tornadoes make for better television.

It's easy to see how the Availability heuristic, stoked by the news policy "If it bleeds, it leads," could induce a sense of gloom about the state of the world. Media scholars who tally news stories of different kinds, or present editors with a menu of possible stories and see which they pick and how they display them, have confirmed that the gatekeepers prefer negative to positive coverage, holding the events constant.[13] That in turn provides an easy formula for pessimists on the editorial page: make a list of all the worst things that are happening anywhere on the planet that week, and you have an impressive-sounding case that civilization has never faced greater peril.

The consequences of negative news are themselves negative. Far from being better informed, heavy newswatchers can become miscalibrated. They worry more about crime, even when rates are falling, and sometimes they part company with reality altogether: a 2016 poll found that a large majority of Americans follow news about ISIS closely, and 77 percent agreed that "Islamic militants operating in Syria and Iraq pose a serious threat to the existence or survival of the United States," a belief that is nothing short of delusional.[14] Consumers of negative news, not surprisingly, become glum: a recent literature review cited "misperception of risk, anxiety, lower mood levels, learned helplessness, contempt and hostility towards others, desensitization, and in some cases, . . . complete avoidance of the news."[15] And they become fatalistic, saying things like "Why should I vote? It's not gonna help," or "I could donate money, but there's just gonna be another kid who's starving next week."[16]

Seeing how journalistic habits and cognitive biases bring out the worst in each other, how can we soundly appraise the state of the world?

The answer is to *count*. How many people are victims of violence as a proportion of the number of people alive? How many are sick, how many starving, how many poor, how many oppressed, how many illiterate, how many unhappy? And are those numbers going up or down? A quantitative mindset, despite its nerdy aura, is in fact the morally enlightened one, because it treats every human life as having equal value rather than privileging the people who are closest to us or most photogenic. And it holds out the hope that we might identify the causes of suffering and thereby know which measures are most likely to reduce it.

That was the goal of my 2011 book *The Better Angels of Our Nature*, which presented a hundred graphs and maps showing how violence and the conditions that foster it have declined over the course of history. To emphasize that the declines took place at different times and had different causes, I gave them names. The Pacification Process was a fivefold reduction in the rate of death from tribal raiding and feuding, the consequence of effective states exerting control over a territory. The Civilizing Process was a fortyfold reduction in homicide and other violent crimes which followed upon the entrenchment of the rule of law and norms of self-control in early modern Europe. The Humanitarian Revolution is another name for the Enlightenment-era abolition of slavery, religious persecution, and cruel punishments. The Long Peace is the historians' term for the decline of great-power and interstate war after World War II. Following the end of the Cold War, the world has enjoyed a New Peace with fewer civil wars, genocides, and autocracies. And since the 1950s the world has been swept by a cascade of Rights Revolutions: civil rights, women's rights, gay rights, children's rights, and animal rights.

Few of these declines are contested among experts who are familiar with the numbers. Historical criminologists, for example, agree that homicide plummeted after the Middle Ages, and it's a commonplace among international-relations scholars that major wars tapered off after 1945. But they come as a surprise to most people in the wider world.[17]

I had thought that a parade of graphs with time on the horizontal axis, body counts or other measures of violence on the vertical, and a line that meandered from the top left to the bottom right would cure audiences of the Availability bias and persuade them that at least in this sphere of well-being the world has made progress. But I learned from their questions and objections that resistance to the idea of progress runs deeper than statistical fallacies. Of course, any dataset is an imperfect reflection of reality, so it is legitimate to question how accurate and representative the numbers truly are. But the objections revealed not just a skepticism about the data but also an unpreparedness for the *possibility* that the human condition has improved. Many people lack the conceptual tools to ascertain whether progress has taken place or not; the very idea that things can get better just doesn't compute. Here are stylized versions of dialogues I have often had with questioners:

> "So violence has declined linearly since the beginning of history! Awesome!"

No, not "linearly"—it would be astonishing if any measure of human behavior with all its vicissitudes ticked downward by a constant amount per unit of time, decade after decade and century after century. And not monotonically, either (which is probably what the questioners have in mind)—that would mean that it always decreased or stayed the same, never increased. Real historical curves have wiggles, upticks, spikes, and sometimes sickening lurches. Examples include the two world wars, a boom in crime in Western countries from the mid-1960s to the early 1990s, and a bulge of civil wars in the developing world following decolonization in the 1960s and 1970s. Progress consists of trends in violence on which these fluctuations are superimposed—a downward swoop or drift, a return from a temporary swelling to a low baseline. Progress cannot always be monotonic because solutions to problems create new problems.[18]

But progress can resume when the new problems are solved in their turn.

By the way, the nonmonotonicity of social data provides an easy formula for news outlets to accentuate the negative. If you ignore all the years in which an indicator of some problem declines, and report every uptick (since, after all, it's "news"), readers will come away with the impression that life is getting worse and worse even as it gets better and better. In the first six months of 2016 the *New York Times* pulled this trick three times, with figures for suicide, longevity, and automobile fatalities.

> "Well, if levels of violence don't always go down, that means they're cyclical, so even if they're low right now it's only a matter of time before they go back up."

No, changes over time may be *statistical,* with unpredictable fluctuations, without being *cyclical,* namely oscillating like a pendulum between two extremes. That is, even if a reversal is possible at any time, that does not mean it becomes more likely as time passes. (Many investors have lost their shirts betting on a misnamed "business cycle" that in fact consists of unpredictable swings.) Progress can take place when the reversals in a positive trend become less frequent, become less severe, or, in some cases, cease altogether.

> "How can you say that violence has decreased? Didn't you read about the school shooting (or terrorist bombing, or artillery shelling, or soccer riot, or barroom stabbing) in the news this morning?"

A decline is not the same thing as a disappearance. (The statement "$x > y$" is different from the statement "$y = 0$.") Something can decrease a lot without vanishing altogether. That means that the level of violence today is *completely irrelevant* to the question of whether violence has declined over the course of history. The only way to answer that question is to compare the level of violence now with the level of violence in the past. And whenever you look at the level of violence in the past, you find a lot of it, even if it isn't as fresh in memory as the morning's headlines.

> "All your fancy statistics about violence going down don't mean anything if you're one of the victims."

True, but they do mean that you're less likely to *be* a victim. For that reason they mean the world to the millions of people who are not victims but would have been if rates of violence had stayed the same.

"So you're saying that we can all sit back and relax, that violence will just take care of itself."

Illogical, Captain. If you see that a pile of laundry has gone down, it does not mean the clothes washed themselves; it means someone washed the clothes. If a type of violence has gone down, then some change in the social, cultural, or material milieu has caused it to go down. If the conditions persist, violence could remain low or decline even further; if they don't, it won't. That makes it important to find out what the causes are, so we can try to intensify them and apply them more widely to ensure that the decline of violence continues.

"To say that violence has gone down is to be naïve, sentimental, idealistic, romantic, starry-eyed, Whiggish, utopian, a Pollyanna, a Pangloss."

No, to look at data showing that violence has gone down and say "Violence has gone down" is to describe a fact. To look at data showing that violence has gone down and say "Violence has gone up" is to be delusional. To ignore data on violence and say "Violence has gone up" is to be a know-nothing.

As for accusations of romanticism, I can reply with some confidence. I am also the author of the staunchly unromantic, anti-utopian *The Blank Slate: The Modern Denial of Human Nature*, in which I argued that human beings are fitted by evolution with a number of destructive motives such as greed, lust, dominance, vengeance, and self-deception. But I believe that people are also fitted with a sense of sympathy, an ability to reflect on their predicament, and faculties to think up and share new ideas—the better angels of our nature, in the words of Abraham Lincoln. Only by looking at the facts can we tell to what extent our better angels have prevailed over our inner demons at a given time and place.

"How can you predict that violence will keep going down? Your theory could be refuted by a war breaking out tomorrow."

A statement that some measure of violence has gone down is not a "theory" but an observation of a fact. And yes, the fact that a measure has changed over time is not the same as a prediction that it will continue to change in that way at all times forever. As the investment ads are required to say, past performance is no guarantee of future results.

"In that case, what good are all those graphs and analyses? Isn't a scientific theory supposed to make testable predictions?"

A scientific theory makes predictions in *experiments* in which the causal influences are controlled. No theory can make a prediction about the world at large, with its seven billion people spreading viral ideas in global networks and interacting with chaotic cycles of weather and resources. To declare what the future holds in an uncontrollable world, and without an explanation of why events unfold as they do, is not prediction but *prophecy*, and as David Deutsch observes, "The most important of all limitations on knowledge-creation is that we cannot prophesy: we cannot predict the content of ideas yet to be created, or their effects. This limitation is not only consistent with the unlimited growth of knowledge, it is entailed by it."[19]

Our inability to prophesy is not, of course, a license to ignore the facts. An improvement in some measure of human well-being suggests that, overall, more things have pushed in the right direction than in the wrong direction. Whether we should expect progress to continue depends on whether we know what those forces are and how long they will remain in place. That will vary from trend to trend. Some may turn out to be like Moore's Law (the number of transistors per computer chip doubles every two years) and give grounds for confidence (though not certainty) that the fruits of human ingenuity will accumulate and progress will continue. Some may be like the stock market and foretell short-term fluctuations but long-term gains. Some of these may reel in a statistical distribution with a "thick tail," in which extreme events, even if less likely, cannot be ruled out.[20] Still others may be cyclical or chaotic. For now we should keep in mind that a positive trend suggests (but does not prove) that we have been doing something right, and that we should seek to identify what it is and do more of it.

When all these objections are exhausted, I often see people racking their brains to find *some* way in which the news cannot be as good as the data suggest. In desperation, they turn to semantics.

> "Isn't Internet trolling a form of violence? Isn't strip-mining a form of violence? Isn't inequality a form of violence? Isn't pollution a form of violence? Isn't poverty a form of violence? Isn't consumerism a form of violence? Isn't divorce a form of violence? Isn't advertising a form of violence? Isn't keeping statistics on violence a form of violence?"

As wonderful as metaphor is as a rhetorical device, it is a poor way to assess the state of humanity. Moral reasoning requires proportionality. It may be upsetting when someone says mean things on Twitter, but it is not the same as the slave trade or the Holocaust. It also requires distinguishing rhetoric from reality. Marching into a rape crisis center and demanding to know what they have done about the rape of the environment does nothing for rape victims and nothing for the environment. Finally, improving the world requires an understanding of cause and effect. Though primitive moral intuitions tend to lump bad things together and find a villain to blame them on, there is no coherent phenomenon of "bad things" that we can seek to understand and eliminate. (Entropy and evolution will generate them in profusion.) War, crime, pollution, poverty, disease, and incivility are evils that may have little in common, and if we want to reduce them, we can't play word games that make it impossible even to discuss them individually.

I have run through these objections to prepare the way for my presentation of other measures of human progress. The incredulous reaction to *Better Angels* convinced me that it isn't just the Availability heuristic that makes people fatalistic about progress. Nor can the media's fondness for bad news be blamed entirely on a cynical chase for eyeballs and clicks. No, the psychological roots of progressophobia run deeper.

The deepest is a bias that has been summarized in the slogan "Bad is stronger than good."[21] The idea can be captured in a set of thought experiments suggested by Tversky.[22] How much better can you imagine yourself feeling than you are feeling right now? How much *worse* can you imagine yourself feeling? In answering the first hypothetical, most of us can imagine a bit more of a spring in our step or a twinkle in our eye, but the answer to the second one is: it's bottomless. This asymmetry in mood can be explained by an asymmetry in life (a corollary of the Law of Entropy). How many things could happen to you today that would leave you much better off? How many things could happen that would leave you much *worse* off? Once again, to answer the first question, we can all come up with the odd windfall or stroke of good luck, but the answer to the second one is: it's endless. But we needn't rely on our imaginations. The psychological literature confirms that people dread losses more than they look forward to gains, that they dwell on setbacks more than they savor good fortune, and that they are more stung by criticism than they are heartened by praise. (As a psycholinguist I am compelled to add that the English language has far more words for negative emotions than for positive ones.[23])

One exception to the Negativity bias is found in autobiographical memory. Though we tend to remember bad events as well as we remember good ones, the negative coloring of the misfortunes fades with time, particularly the ones that happened to us.[24] We are wired for nostalgia: in human memory, time heals most wounds. Two other illusions mislead us into thinking that things ain't what they used to be: we mistake the growing burdens of maturity and parenthood for a less innocent world, and we mistake a decline in our own faculties for a decline in the times.[25] As the columnist Franklin Pierce Adams pointed out, "Nothing is more responsible for the good old days than a bad memory."

Intellectual culture should strive to counteract our cognitive biases, but all too often it reinforces them. The cure for the Availability bias is quantitative thinking, but the literary scholar Steven Connor has noted that "there is in the arts and humanities an exceptionless consensus about the encroaching horror of the domain of number."[26] This "ideological rather than accidental innumeracy" leads writers to notice, for example, that wars take place today and wars took place in the past and to conclude that "nothing has changed"—failing to acknowledge the difference between an era with a handful of wars that collectively kill in the thousands and an era with dozens of wars that collectively killed in the millions. And it leaves them unappreciative of systemic processes that eke out incremental improvements over the long term.

Nor is intellectual culture equipped to treat the Negativity bias. Indeed, our vigilance for bad things around us opens up a market for professional curmudgeons who call our attention to bad things we may have missed. Experiments have shown that a critic who pans a book is perceived as more competent than a critic who praises it, and the same may be true of critics of society.[27] "Always predict the worst, and you'll be hailed as a prophet," the musical humorist Tom Lehrer once advised. At least since the time of the Hebrew prophets, who blended their social criticism with forewarnings of disaster, pessimism has been equated with moral seriousness. Journalists believe that by accentuating the negative they are discharging their duty as watchdogs, muckrakers, whistleblowers, and afflicters of the comfortable. And intellectuals know they can attain instant gravitas by pointing to an unsolved problem and theorizing that it is a symptom of a sick society.

The converse is true as well. The financial writer Morgan Housel has observed that while pessimists sound like they're trying to help you, optimists sound like they're trying to sell you something.[28] Whenever someone offers a solution to a problem, critics will be quick to point out that it is not a panacea, a silver bullet, a magic bullet, or a one-size-fits-all solution; it's just a Band-Aid or a quick technological fix that fails to get at the root causes and will blow back with side effects and unintended consequences. Of course, since nothing is a panacea and everything has side effects (you can't do just one thing), these common tropes are little more than a refusal to entertain the possibility that anything can ever be improved.[29]

Pessimism among the intelligentsia can also be a form of one-upmanship. A modern society is a league of political, industrial, financial, technological, military, and intellectual elites, all competing for prestige and influence, and with differing responsibilities for making the society run. Complaining about modern society can be a backhanded way of putting down one's rivals—for academics to feel superior to businesspeople, businesspeople to feel superior to politicians, and so on. As Thomas Hobbes noted in 1651, "Competition of praise inclineth to a reverence of antiquity. For men contend with the living, not with the dead." Pessimism, to be sure, has a bright side. The expanding circle of sympathy makes us concerned about harms that would have passed unnoticed in more callous times. Today we recognize the Syrian civil war as a humanitarian tragedy. The wars of earlier decades, such as the Chinese Civil War, the partition of India, and the Korean War, are seldom remembered that way, though they killed and displaced more people. When I grew up, bullying was considered a natural part of boyhood. It would have strained belief to think that someday the president of the United States would deliver a speech about its evils, as Barack Obama did in 2011. As we care about more of humanity, we're apt to mistake the harms around us for signs of how low the world has sunk rather than how high our standards have risen.

But relentless negativity can itself have unintended consequences, and recently a few journalists have begun to point them out. In the wake of the 2016 American election, the *New York Times* writers David Bornstein and Tina Rosenberg reflected on the media's role in its shocking outcome:

> Trump was the beneficiary of a belief—near universal in American journalism—that "serious news" can essentially be defined as "what's going wrong." . . . For decades, journalism's steady focus on problems and seemingly incurable pathologies was preparing the soil that allowed Trump's seeds of discontent and despair to take root. . . . One consequence is that many Americans today have difficulty imagining, valuing or even believing in the promise of incremental system change, which leads to a greater appetite for revolutionary, smash-the-machine change.[30]

Bornstein and Rosenberg don't blame the usual culprits (cable TV, social media, late-night comedians) but instead trace it to the shift during the Vietnam and Watergate eras from glorifying leaders to checking their power—with an overshoot toward indiscriminate cynicism, in which everything about America's civic actors invites an aggressive takedown.

If the roots of progressophobia lie in human nature, is my suggestion that it is on the rise itself an illusion of the Availability bias? Anticipating the methods I will use in the rest of the book, let's look at an objective measure. The data scientist Kalev Leetaru applied a technique called sentiment mining to every article published in the *New York Times* between 1945 and 2005, and to an archive of translated articles and broadcasts from 130 countries between 1979 and 2010. Sentiment mining assesses the emotional tone of a text by tallying the number and contexts of words with positive and negative connotations, like *good*, *nice*, *terrible*, and *horrific*. The following figure shows the results. Putting aside the wiggles and waves that reflect the crises of the day, we see that the impression that the news has become more negative over time is real. The *New York Times* got steadily more morose from the early 1960s to the early 1970s, lightened up a bit (but just a bit) in the 1980s and 1990s, and then sank into a progressively worse mood in the first decade of the new century. News outlets in the rest of the world, too, became gloomier and gloomier from the late 1970s to the present day.

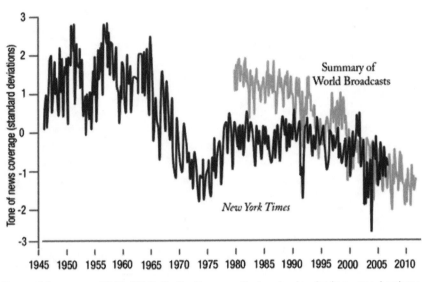

Tone of the news, 1945–2010. Plotted by month, beginning in January. *Leetaru, 2011.*

What is progress? You might think that the question is so subjective and culturally relative as to be forever unanswerable. In fact, it's one of the easier questions to answer.

Most people agree that life is better than death. Health is better than sickness. Sustenance is better than hunger. Abundance is better than poverty. Peace is better than war. Safety is better than danger. Freedom is better than tyranny. Equal rights are better than bigotry and discrimination. Literacy is better than illiteracy. Knowledge is better than ignorance. Intelligence is better than dull-wittedness. Happiness is better than misery. Opportunities to enjoy family, friends, culture, and nature are better than drudgery and monotony.

All these things can be measured. If they have increased over time, that is progress.

Granted, not everyone would agree on the exact list. The values are avowedly humanistic, and leave out religious, romantic, and aristocratic virtues like salvation, grace, sacredness, heroism, honor, glory, and authenticity. But most would agree that it's a necessary start. It's easy to extoll transcendent values in the abstract, but most people prioritize life, health, safety, literacy, sustenance, and stimulation for the obvious reason that these goods are a prerequisite to everything else. If you're reading this, you are not dead, starving, destitute, moribund, terrified, enslaved, or illiterate, which means that you're in no position to turn your nose up at these values—or to deny that other people should share your good fortune.

As it happens, the world does agree on these values. In the year 2000, all 189 members of the United Nations, together with two dozen international organizations, agreed on eight Millennium Development Goals for the year 2015 that blend right into this list.[31]

And here is a shocker: *The world has made spectacular progress in every single measure of human well-being.* Here is a second shocker: *Almost no one knows about it.*

Information about human progress, though absent from major news outlets and intellectual forums, is easy enough to find. The data are not entombed in dry reports but are displayed in gorgeous Web sites, particularly Max Roser's *Our World in Data*, Marian Tupy's *HumanProgress*, and Hans Rosling's *Gapminder*. (Rosling learned that not even swallowing a sword during a 2007 TED talk was enough to get the world's attention.) The case has been made in beautifully written books, some by Nobel laureates, which flaunt the news in their titles—*Progress, The Progress Paradox, Infinite Progress, The Infinite Resource, The Rational Optimist, The Case for Rational Optimism, Utopia for Realists, Mass Flourishing, Abundance, The Improving State of the World, Getting Better, The End of Doom, The Moral Arc, The Big Ratchet, The Great Escape, The Great Surge, The Great Convergence.*[32] (None was recognized with a major prize, but over the period in which they appeared, Pulitzers in nonfiction were given to four books on genocide, three on terrorism, two on cancer, two on racism, and one on extinction.)

Learning about human progress is not an exercise in optimism, cheeriness, or looking on the bright side: it's a matter of accuracy, of understanding the world as it really is.

Steven Pinker is Harvard College Professor and Johnstone Family Professor in the Department of Psychology at Harvard University. Until 2003, he taught in the Department of Brain and Cognitive Sciences at MIT. He conducts research on language and cognition; writes for such publications as the *New York Times*, *Time*, and the *New Republic*; and is the author of many books, including *The Language Instinct, How the Mind Works, Words and Rules, The Blank Slate*, and *The Stuff of Thought: Language as a Window into Human Nature*.

Notes

1. Herman 1997, p. 7, also cites Joseph Campbell, Noam Chomsky, Joan Didion, E. L. Doctorow, Paul Goodman, Michael Harrington, Robert Heilbroner, Jonathan Kozol, Christopher Lasch, Norman Mailer, Thomas Pynchon, Kirkpatrick Sale, Jonathan Schell, Richard Sennett, Susan Sontag, Gore Vidal, and Gary Wills.

2. Nisbet 1980/2009, p. 317.

3. The optimism gap: McNaughton-Cassill and Smith 2002; Nagdy and Roser 2016; Veenhoven 2010; Whitman 1998.

4. EU Eurobarometer survey results, reproduced in Nagdy and Roser 2016.

5. Survey results from Ipsos 2016, "Perils of Perception (Topline Results)," 2013, https://www.ipsos.com/sites/default/files/migrations/en-uk/files/Assets/Docs/Polls/ipsos-mori-rss-kings-perils-of-perception-topline.pdf, graphed in Nagdy and Roser 2016.

6. Dunlap, Gallup, and Gallup 1993, graphed in Nagdy and Roser 2016.

7. J. McCarthy, "More Americans say crime is rising in U.S.," *Gallup.com*, October 22, 2015, http://www.gallup.com/poll/186308/americans-say-crime-rising.aspx.

8. World is getting worse: Majorities in Australia, Denmark, Finland, France, Germany, Great Britain, Hong Kong, Norway, Singapore, Sweden, and the United States; also Malaysia, Thailand, and the United Arab Emirates. China was the only country in which more respondents said the world was getting better than said it was getting worse. YouGov poll, January 5, 2016, https://yougov.co.uk/news/2016/01/05/chinese-people-are-most-optimistic-world/. The United States on the wrong track: Dean Obeidallah, "We've been on the wrong track since 1972," *Daily Beast*, November 7, 2014, http://www.pollingreport.com/right.htm.

9. Source of the expression: B. Popik, "First draft of history (journalism)," *BarryPopik.com*, http://www.barrypopik.com/index.php/new_york_city/entry/first_draft_of_history_journalism/.

10. Frequency and nature of news: Galtung and Ruge 1965.

11. Availability heuristic: Kahneman 2011; Slovic 1987; Slovic, Fischof, and Lichtenstein 1982; Tversky and Kahneman 1973.

12. Misperceptions of risk: Ropeik and Gray 2002; Slovic 1987. Post-*Jaws* avoidance of swimming: Sutherland 1992, p. 11.

13. If it bleeds, it leads (and vice versa): Bohle 1986; Combs and Slovic 1979; Galtung and Ruge 1965; Miller & Albert 2015.

14. ISIS as "existential threat": Poll conducted for *Investor's Business Daily* by TIPP, March 28–April 2, 2016, http://www.investors.com/politics/ibdtipp-poll-distrust-on-what-obama-does-and-says-on-isis-terror/.

15. Effects of newsreading: Jackson 2016. See also Johnston and Davey 1997; McNaughton-Cassill 2001; Otieno, Spada, and Renkl 2013; Ridout, Grosse, and Appleton 2008; Unz, Schwab, and Winterhoff-Spurk 2008.

16. Quoted in J. Singal, "What all this bad news is doing to us," *New York*, August 8, 2014.

17. Decline of violence: Eisner 2003; Goldstein 2011; Gurr 1981; Human Security Centre 2005; Human Security Report Project 2009; Mueller 1989, 2004; Payne 2004.

18. Solutions create new problems: Berlin 1988/2013, p. 15; Deutsch 2011, pp. 64, 76, 350.

19. Deutsch 2011, p. 193.

20. Thick-tailed distributions: See chapter 19 and, for more detail, Pinker 2011, pp. 210–22.

21. Negativity bias: Baumeister, Bratslavsky, et al. 2001; Rozin and Royzman 2001.

22. Personal communication, 1982.

23. More negative words: Baumeister, Bratslavsky, et al. 2001; Schrauf and Sanchez 2004.

24. Rose-tinting of memory: Baumeister, Bratslavsky, et al. 2001.

25. Illusion of the good old days: Eibach and Libby 2009.

26. Connor 2014; see also Connor 2016.

27. Snarky book reviewers sound smarter: Amabile 1983.

28. M. Housel, "Why does pessimism sound so smart?" *Motley Fool*, January 21, 2016.

29. Similar points have been made by the economist Albert Hirschman (1991) and the journalist Gregg Easterbrook (2003).

30. D. Bornstein and T. Rosenberg, "When reportage turns to cynicism," *New York Times*, November 15, 2016. For more on the "constructive journalism" movement, see Gyldensted 2015, Jackson 2016, and the magazine *Positive News* (www.positive.news).

31. The UN Millennium Development Goals are: 1. To eradicate extreme poverty and hunger. 2. To achieve universal primary education. 3. To promote gender equality and empower women. 4. To reduce child mortality. 5. To improve maternal health. 6. To combat HIV/AIDS, malaria, and other diseases. 7. To ensure environmental sustainability. 8. To develop a global partnership for [economic] development.

32. Books on progress (in order of mention): Norberg 2016, Easterbrook 2003, Reese 2013, Naam 2013, Ridley 2010, Robinson 2009, Bregman 2017, Phelps 2013, Diamandis and Kotler 2012, Kenny 2011, Bailey 2015, Shermer 2015, DeFries 2014, Deaton 2013, Radelet 2015, Mahbubani 2013.

References

Bailey, R. 2015. *The end of doom: Environmental renewal in the 21st century.* New York: St. Martin's Press.

Baumeister, R., Bratslavsky, E., Finkenauer, C., & Vohs, K. D. 2001. Bad is stronger than good. *Review of General Psychology*, 5, 323–70.

Berlin, I. 1988/2013. The pursuit of the ideal. In I. Berlin, ed., *The crooked timber of humanity.* Princeton, NJ: Princeton University Press.

Bohle, R. H. 1986. Negativism as news selection predictor. *Journalism Quarterly*, 63, 789–96.

Bregman, R. 2017. *Utopia for realists: The case for a universal basic income, open borders, and a 15-hour workweek.* Boston: Little, Brown.

Combs, B., & Slovic, P. 1979. Newspaper coverage of causes of death. *Journalism Quarterly*, 56, 837–43.

Connor, S. 2014. The horror of number: Can humans learn to count? Paper presented at the Alexander Lecture. http://stevenconnor.com/horror.html.

Connor, S. 2016. *Living by numbers: In defence of quantity.* London: Reaktion Books.

Deaton, A. 2013. *The great escape: Health, wealth, and the origins of inequality.* Princeton, NJ: Princeton University Press.

DeFries, R. 2014. *The big ratchet: How humanity thrives in the face of natural crisis.* New York: Basic Books.

Deutsch, D. 2011. *The beginning of infinity: Explanations that transform the world.* New York: Viking.

Diamandis, P., & Kotler, S. 2012. *Abundance: The future is better than you think.* New York: Free Press.

Dunlap, R. E., Gallup, G. H., & Gallup, A. M. 1993. Of global concern. *Environment: Science and Policy for Sustainable Development,* 35, 7–39.

Easterbrook, G. 2003. *The progress paradox: How life gets better while people feel worse.* New York: Random House.

Eibach, R. P., & Libby, L. K. 2009. Ideology of the good old days: Exaggerated perceptions of moral decline and conservative politics. In J. T. Jost, A. Kay, & H. Thorisdottir, eds., *Social and psychological bases of ideology and system justification.* New York: Oxford University Press.

Eisner, M. 2003. Long-term historical trends in violent crime. *Crime and Justice,* 30, 83–142.

Galtung, J., & Ruge, M. H. 1965. The structure of foreign news. *Journal of Peace Research,* 2, 64–91.

Goldstein, J. S. 2011. *Winning the war on war: The surprising decline in armed conflict worldwide.* New York: Penguin.

Gurr, T. R. 1981. *Historical trends in violent crime: A critical review of the evidence* (vol. 3). Chicago: University of Chicago Press.

Herman, A. 1997. *The idea of decline in Western history.* New York: Free Press.

Human Security Centre. 2005. *Human Security Report 2005: War and peace in the 21st century.* New York: Oxford University Press.

Human Security Report Project. 2009. *Human Security Report 2009: The shrinking costs of war.* New York: Oxford University Press.

Ipsos. 2016. The perils of perception 2016. https://perils.ipsos.com/.

Jackson, J. 2016. Publishing the positive: Exploring the perceived motivations for and the consequences of reading solutions-focused journalism. https://www.constructive journalism.org/wp-content/uploads/2016/11/Publishing-the-Positive_MA-thesis -research-2016_Jodie-Jackson.pdf.

Johnston, W. M., & Davey, G. C. L. 1997. The psychological impact of negative TV news bulletins: The catastrophizing of personal worries. *British Journal of Psychology,* 88.

Kahneman, D. 2011. *Thinking, fast and slow.* New York: Farrar, Straus & Giroux.

Kenny, C. 2011. *Getting better: How global development is succeeding—and how we can improve the world even more.* New York: Basic Books.

Mahbubani, K. 2013. *The great convergence: Asia, the West, and the logic of one world.* New York: PublicAffairs.

McNaughton-Cassill, M. E. 2001. The news media and psychological distress. *Anxiety, Stress, and Coping,* 14, 191–211.

McNaughton-Cassill, M. E., & Smith, T. 2002. My world is OK, but yours is not: Television news, the optimism gap, and stress. *Stress and Health,* 18, 27–33.

Miller, R. A., & Albert, K. 2015. If it leads, it bleeds (and if it bleeds, it leads): Media coverage and fatalities in militarized interstate disputes. *Political Communication*, 32, 61–82.

Mueller, J. 1989. *Retreat from doomsday: The obsolescence of major war.* New York: Basic Books.

Mueller, J. 2004. *The remnants of war.* Ithaca, NY: Cornell University Press.

Naam, R. 2013. *The infinite resource: The power of ideas on a finite planet.* Lebanon, NH: University Press of New England.

Nagdy, M., & Roser, M. 2016. Optimism and pessimism. Our World in Data. https://ourworldindata.org/optimism-pessimism/.

Nisbet, R. 1980/2009. *History of the idea of progress.* New Brunswick, NJ: Transaction.

Norberg, J. 2016. *Progress: Ten reasons to look forward to the future.* London: Oneworld.

Otieno, C., Spada, H., & Renkl, A. 2013. Effects of news frames on perceived risk, emotions, and learning. *PLOS ONE*, 8, 1–12.

Payne, J. L. 2004. *A history of force: Exploring the worldwide movement against habits of coercion, bloodshed, and mayhem.* Sandpoint, ID: Lytton Publishing.

Phelps, E. A. 2013. *Mass flourishing: How grassroots innovation created jobs, challenge, and change.* Princeton, NJ: Princeton University Press.

Pinker, S. 2011. *The better angels of our nature: Why violence has declined.* New York: Penguin.

Radelet, S. 2015. *The great surge: The ascent of the developing world.* New York: Simon & Schuster.

Reese, B. 2013. *Infinite progress: How the internet and technology will end ignorance, disease, poverty, hunger, and war.* Austin, TX: Greenleaf Book Group Press.

Ridley, M. 2010. *The rational optimist: How prosperity evolves.* New York: HarperCollins.

Ridout, T. N., Grosse, A. C., & Appleton, A. M. 2008. News media use and Americans' perceptions of global threat. *British Journal of Political Science*, 38, 575–93.

Robinson, F. R. 2009. *The case for rational optimism.* New Brunswick, NJ: Transaction.

Ropeik, D., & Gray, G. 2002. *Risk: A practical guide for deciding what's really safe and what's really dangerous in the world around you.* Boston: Houghton Mifflin.

Rozin, P., & Royzman, E. B. 2001. Negativity bias, negativity dominance, and contagion. *Personality and Social Psychology Review*, 5, 296–320.

Schrauf, R. W., & Sanchez, J. 2004. The preponderance of negative emotion words in the emotion lexicon: A cross-generational and cross-linguistic study. *Journal of Multilingual and Multicultural Development*, 25, 266–84.

Shermer, M. 2015. *The moral arc: How science and reason lead humanity toward truth, justice, and freedom.* New York: Henry Holt.

Slovic, P. 1987. Perception of risk. *Science*, 236, 280–85.

Slovic, P., Fischof, B., & Lichtenstein, S. 1982. Facts versus fears: Understanding perceived risk. In D. Kahneman, P. Slovic, & A. Tversky, eds., *Judgment under uncertainty: Heuristics and biases.* New York: Cambridge University Press.

Sutherland, S. 1992. *Irrationality: The enemy within.* London: Penguin.

Tversky, A., & Kahneman, D. 1973. Availability: A heuristic for judging frequency and probability. *Cognitive Psychology*, 4, 207–32.

Unz, D., Schwab, F., & Winterhoff-Spurk, P. 2008. TV news—the daily horror? Emotional effects of violent television news. *Journal of Media Psychology*, 20, 141–55.

Veenhoven, R. 2010. Life is getting better: Societal evolution and fit with human nature. *Social Indicators Research* 97, 105–22.

Whitman, D. 1998. *The optimism gap: The I'm OK—they're not syndrome and the myth of American decline*. New York: Bloomsbury USA.

Part II

SCIENCE, MISINFORMATION, AND THE PUBLIC

CHAPTER 7

The War on Science, Anti-Intellectualism, and "Alternative Ways of Knowing" in 21st-Century America

H. Sidky

Vol. 42, No. 2
March/April 2018

At the start of the twenty-first century, over 40 percent of Americans did not know that the Earth orbits the sun in a year-long cycle (Otto 2016, 224). Another 52 percent did not know that dinosaurs died before the appearance of humans, and 45 percent were unaware that the world is older than 10,000 years. It is unnecessary to mention the equally alarming numbers of people who believe in ghosts, space aliens, paranormal monsters, devil possession, angels, demons, miracles, and so forth (Smith 2010, 22–23).

This mostly scientifically illiterate public seems to lack the necessary skills to distinguish between contending claims to knowledge or differentiate between fact and opinion. We now live in a scary and confusing "post-truth" era of disinformation, "fake news," "counterknowledge," "weaponized lies," conspiracy theories, magical thinking, and irrationalism (see Andersen 2017; Levitin 2016).

Bogus and irrational ideas (beliefs that have been falsified or are unfalsifiable) are thriving and seem to be widely received and accepted. However, tolerating irrationalism and scientific illiteracy poses many dangers. It is dangerous to individual well-being. Numerous people have died because of their trust in sham alternative medical cures, and many others have lost their life savings by believing in psychics and miracle workers (see Bridgstock 2009, 1–3; Coyne 2015, 229–239; Gilovich 1993, 5–6; Hines 2003, 38–41; Schick and Vaughn 2014, 12–13). More than that, acquiescence to irrationalism threatens the well-being of our society (see Mooney and Kirshenbaum 2009; Sharlet 2010). As philosophers Theodor Schick and Lewis Vaughn (2014, 13) have put it:

> A democratic society depends on the ability of its members to make rational choices. But rational choices must be based on rational beliefs. If we can't tell the difference between reasonable and unreasonable claims, we become susceptible to the claims of charlatans, scoundrels, and mountebanks.

Purveyors of supernaturalism, anti-intellectual dogmas, medieval credulities, and "alternative forms of knowledge" that are daily an affront to our intelligence and sensibilities are swarming with bluster and hubris that science is now defunct and offer their own "truths" and "ways of knowing" as better substitutes. However, before we submit to the assertions of religious ideologues, miracle workers, and quacks and make their "truths" the basis of our worldview, we need to ask the following question: Are the assertions that science is defunct based on compelling evidence? To answer this question, let's look at the circumstances that brought us here.

Explanations for the rise of anti-intellectualism and antiscience perspectives in this country would no doubt include many complex interconnected factors, such as globalization, demographic shifts, changes in the socioeconomic infrastructure, disparities in wealth and power, the disenchantment of the world by science and technology, and so forth. However, as the science writer Shawn Otto discusses in his recent book *The War on Science: Who Is Waging It, Why It Matters, What We Can Do About It* (2016), the decades-long systematic academic assault on science and rationalism stands above many other factors. Paul Gross and Norman Levitt's *Higher Superstition: The Academic Left and Its Quarrels with Science* (1994) also provides an insightful account of the academic war on science. More specifically, philosopher and historian of science Noretta Koertge (1998) explores the direct influence this assault on science has had on scientific literacy in various fields of study in the United States. Embarrassingly, cultural anthropology, the discipline to which I belong, was instrumental in this untoward enterprise (see Otto 2016, 175–176).

From the late 1960s cultural anthropologists—in concert with their counterparts in departments of English, education, journalism, political science, cultural studies, science studies, and humanities—collectively engaged in a seemingly well-intentioned intellectual enterprise to "speak truth to power." Their objective was to promote epistemological egalitarianism open to diverse viewpoints and create a more tolerant, multicultural society free of all the evils of modernity. They argued that modernity's hegemonic power and authoritarianism had to be exposed, and these savants claimed to possess the intellectual tools to accomplish this task. In their discourse, science and scientific truths (deceptively misconstrued as "absolute truths") were cast as the embodiment of that hegemonic power and its evils, such as racism, sexism, imperialism, colonialism, militarism, oppression, slavery, white supremacy, the atomic bomb, and the destruction of the biosphere.

This intellectual movement was known variously as *social constructivism*, *deconstructionism*, *post-structuralism*, and *postmodernism*. Here I shall use the label *postmodernism* (Sokal 2008, 269). The paragons of this movement consisted of a handful of French philosophers, including Michel Foucault, Jacques Derrida, Jean-François Lyotard, and Bruno Latour (see Sidky 2004, 394–412). Although their works differed in various respects, they shared general features, such as a disdain for the rationalist tradition of the Enlightenment, a disregard for empirical data and logic, the idea of "the cultural construction of knowledge," and subjective and intuitive approaches to knowledge. Moreover, although pontificating about science became their forte, none of these scholars were trained as professional scientists.

Irrationalist philosophers in the United States, such as Thomas Kuhn and Paul Feyerabend, also contributed to the postmodern antiscience program. In his highly acclaimed book *The Structure of Scientific Revolutions* (1970), Kuhn asserted that scientific truths depend upon agreement among scientists operating under a guiding intellectual umbrella, or paradigm, built around a core of ideas based on irrational cultural and sociopolitical factors. A paradigm persists for a while until mounting anomalies it cannot address result in a "scientific revolution" and the establishment of a new paradigm built upon new conventions linked to a different set of sociopolitical factors. According to Kuhn, the solution to questions are *relative* to a paradigm rather than empirical evidence. For this reason, paradigms are incommensurable, and there is no real growth of scientific knowledge. If true, this would mean that our knowledge of the world and universe today has not increased beyond the state of knowledge four hundred years ago, a view that verges on the ludicrous and is a misrepresentation of the history of science. Solutions under previous paradigms do not become "un-solutions" after paradigm shifts (Kuznar 2008, 57; Stove 2001). Hence scientists still use Newton's law of gravity to calculate the orbits of spacecraft (Stenger 2008, 114–115). As the philosopher David Stove (2001, 21–50) points out in his devastating critique, Kuhn's irrationalist view of science appears plausible because he relies on evocation, ambiguity, false equivalencies, and clever inconsistencies. There are only a few instances of Kuhnian type revolutions in the history of science. For this reason, the Nobel laureate physicist Steven Weinberg (1998) refers to Kuhn's ideas as "the revolution that didn't happen."

Paul Feyerabend, the author of *Against Method* (1975), similarly advocated the idea of the cultural construction of knowledge, asserting that scientific research protocols and methodology are merely ornamentations that legitimize truths established through irrational means subject to sociopolitical and historical factors. Starting from a reasonable observation that "all methodologies have limitations," Feyerabend (1975, 296) reached the erroneous conclusion, and a true *non sequitur*, that in the pursuit of knowledge "anything goes" both in the context of discovery and in the context of justification (Sokal 2008, 199). Therefore, there are no epistemological distinctions between science and religion or mythology. This is a gross misrepresentation of the scientific enterprise (Gross and Levitt 1994, 47).

Despite such epistemological problems, postmodernists were able to launch an all-encompassing disinformation campaign to delegitimize science and rationality. The distressing effects of this campaign were painfully brought to light for many after the 2016 U.S. presidential election. The assault on science centered on the idea of epistemological relativism. This entails the premise that conditions of knowledge are such that the truth and falsity of assertions are context-dependent, situated, and always relative to cultural and social backgrounds, political position, class, gender, ethnicity, race, and religion. Thus, the idea that scientific knowledge depends upon objective empirical evidence is false. Excluding the empirical dimension of the scientific enterprise, these writers misrepresented science as merely a "story" or narrative like any other that relies on rhetorical ornamentation and language games to persuade people of its legitimacy and authority. Epistemological relativism dictates that no representations of reality or story can be privileged because there are multiple and equally valid realities and truths. Moreover, because all truths are relative, postmodernists asserted, whose truth prevails is a coefficient of power and coercion (Foucault 1984, 75). The West is dominant and hegemonic, and hence its "truths" (i.e., science) are privileged.

To expose the exact nature of power relations, postmodern thinkers believed, one had to look at the linguistic context of truth claims because nothing exists apart from the discourse that constitutes them. In other words, apprehension of a reality outside the linguistic webs that entangle us is not possible, which is an assertion that goes against anthropological evidence, science, common sense, and everyday epistemology. We survive and act successfully in the world during our day-to-day interactions by assuming that reality "out there" exists (Abel 1976, 33). Despite many factors that bias our perceptions in various ways, our senses do not systematically deceive us all the time. That is why we do not intentionally bump into walls, walk off cliffs, or step into traffic. Hence, we are not hopeless prisoners of language. We navigate the world using the same principles encapsulated in the scientific method by continuously making decisions about our perceptions according to the rules of inductive/deductive hypothesis testing and refutation (Fox 1997, 341). Science does this more systematically and with greater rigor. As the philosopher Karl Popper (1972) put it, science is enlightened common sense. Science is *a human enterprise* that generates *approximate* understandings *in human terms* of "something" (call it "reality," the empirical world, or whatever) that seems to exist apart from our perceptual and cognitive apparatus rather than being generated by it (see Lett 1986; 1997). Postmodernists left that "something" out of their epistemological equation. For them, everything was about language and linguistic webs that form perceptual prisons from which there is no escape. However, these writers believed that they possessed the skills to reveal the occult codes of power by looking at seemingly inconsequential aspects of language, such as "tropes," rhetorical strategies, and figurative devices that are invisible to conventional analysts and even the authors of those texts.

Ironically, given that this enterprise was about epistemological egalitarianism and human dignity, those who did not accept postmodern premises were labeled racists, sexists, right-wing oppressors, colonialists, and the instruments of a defunct materialist worldview (the terms of opprobrium were endless). In the halls of American academia, postmodernism acquired a frightening authoritarianism similar to religion, complete with self-styled messiahs, infatuated acolytes, sacred texts, secret mantras, taboo words, and moral injunctions.

The postmodern antiscience perspective had several inherent flaws that both ensured its ultimate failure and sadly rendered its proponents entirely irrelevant as a political force today. First, it confused the authority of science with that of the person conveying scientific knowledge. In science, the ultimate arbiter is the evidence; it is gravity—not the scientist asserting that an apple will plummet to the ground—that is the defining condition of knowledge in the end. Sadly, relativist antiscience writers remain befuddled about this issue (e.g., Herzfeld 2017). It is an epistemological blunder to confuse the assertions of facts (e.g., the words used to describe gravity) with the facts themselves (that apples fall from trees) as aspects of the external world that exist irrespective of how we know or which words we use to write about them. By taking this stance, postmodernists transformed the reasonable position that "facts do not speak for themselves" into the absurd conclusion that "there are no facts," and that no knowledge of the empirical world is possible, which is a gross *non sequitur* (Spaulding 1988, 264).

Second, the postmodern perspective was self-contradictory because it claimed that all truths were relative to class, gender, ethnicity, and cultural background but excluded itself from the constraints of culture, history, and context (Sidky 2004, 399). As Schick and Vaughn (2014, 311–312) have put it:

> To say that everything is relative is to say that no unrestricted universal generalizations are true (an unrestricted universal generalization is a statement to the effect that something holds for all individuals, societies, or conceptual schemes). But the statement that "No unrestricted universal generalizations are true" is itself an unrestricted universal generalization. So if relativism in any of its forms is true, it's false.

Third, there were no specified rules for extracting the codes of power and encrypted significations from texts. Careful reading does not accomplish this task. So how does one proceed? Remarkably, the answer was through subjective means, using the postmodern scholars' personal and often oversimplified moral categories of exploitation versus resistance, with truth conflated with "good" provided by and suited to the analyst's moralistic sensibilities (Sahlins 1999). This enterprise was not about the discovery of new knowledge because the analyst herself or himself provided the "truth" (Salzman 2001, 136). These writers professed a self-righteous desire to "speak truth to evil" (Scheper-Hughes 1995), but it was their own "truth" arrived at using extraordinary capacities and hermeneutic ingenuities with which they credited themselves and denied everyone else (Sidky 2007, 68). However, their colossal blunder was to assume that the political and moral values they were promoting would be embraced by others in society at large along with their antiscience message.

Fourth, the postmodern discourse was characterized by strategic ambiguity. It was full of obscure literary allusions, baroque rhetorical forms, and contrived scientific-sounding jargon, such as "non-Euclidian space," "chaos theory," "reversal of cause and effect," and "endorphin of culture" that sounded erudite but made for incomprehensible texts. Somehow being abstruse was equated with being profound (Carneiro 1995, 14). However, most of their brilliant insights about knowledge and science were pure nonsense. It turned out that even the leading icons of the movement did not understand much of what was said. The New York University physicist Alan Sokal brought this to light by submitting a parody article full of absurdities and blatant *non sequiturs* to *Social Text*, one of the prestigious postmodern journals. The paper, with the lovely title "Transgressing the Boundaries: Towards a Transformative Hermeneutics of Quantum Gravity," was accepted and published in a special issue, called "Science Wars," devoted to refuting the critics of postmodernism. When Sokal revealed the hoax, the embarrassed postmodern savants reacted with indignation, hostility, and the only weapons they had: special pleading, specious rationalizations, and *ad hominin* attacks (e.g., Robbins and Ross 2000). The hoax revealed the true obscurantist and nonsensical nature of postmodern discourse. In their book *Fashionable Nonsense*, Sokal and Bricmont (1998, 207) made the following observation concerning the effects of postmodernism: "The deliberately obscure discourses of postmodernism, and the intellectual dishonesty they engender, poison a part of intellectual life and strengthen the facile anti-intellectualism that is already too widespread in the general public."

For forty years, the postmodern savants in universities across the country indoctrinated students with their antiscience message (Otto 2016, 198). The substitute they offered was epistemological relativism as the avenue to establish a genuinely just and tolerant society open to diverse viewpoints. At the time, few of these scholars considered the actual implications of their effort to disqualify objective empirical evidence as the basis for evaluating claims to knowledge and public policy. Science-minded scholars, however, were not so oblivious. As Sokal and Bricmont (1998, 209) pointed out: "If all discourses are merely 'stories' or 'narrations,' and none is more objective or truthful than another, then one must concede that the worst sexist or racist prejudices and the most reactionary socio-economic theories are 'equally valid.'"

Many of those indoctrinated in postmodern antiscience went on to become conservative political and religious leaders, policymakers, journalists, journal editors, judges, lawyers, and members of city councils and school boards. Sadly, they forgot the lofty ideals of their teachers, except that science is bogus (Otto 2016, 199). Thus, vast cadres of people with little interest in the message of multiculturalism and epistemological egalitarianism coopted the central lesson of postmodernism that truth is what one wants it to be to assert the legitimacy of their authoritarian dogmas, irrationalism, and bunkum. Even some hardcore antiscience philosophers now acknowledge the unpleasant consequences of their imprudent intellectual enterprise. As the noted postmodern "sociologist of science" Bruno Latour (2004, 227) has put it:

> . . . entire Ph.D. programs are still running to make sure that good American kids are learning . . . that facts are made up, that there is no such thing as natural, unmediated, unbiased access to truth, that we are always prisoners of language, that we always speak from a particular standpoint, and so on, while dangerous extremists are using the very same argument of social construction to destroy hard-won evidence that could save our lives.

Latour, to his credit, now goes even further in acknowledging the damage the science critique has caused. As noted in the March/April 2018 *Skeptical Inquirer,* Latour, in a recent interview for *Science* (de Vrieze 2017), said the science criticisms created the basis for antiscientific thinking and he now wants to help rebuild trust in science.

The effects of all this on today's media and politics are startling. Gonzo journalism has become widespread, and few in the profession consider speaking truth to power or even objectivity in reporting as part of their responsibilities (Otto 2016, 23, 129, 200). In this intellectual climate, pretentious and utterly unqualified politicians are flagrantly flaunting opinions on issues ranging from vaccines, human reproduction, stem cell research, the origins of the Earth, and human evolution, to the state of the biosphere, that are contrary to overwhelming historical and scientific evidence. In these cultural circumstances, institutions of higher learning have become beleaguered citadels in a vast ocean of irrationality expressed with bravado and pride, a blowback effect partly the creation of postmodern academics themselves. Emboldened xenophobia, scapegoating of ethnic minorities for social ills, and outright racism and bigotry have replaced political correctness, civility, and cultural sensitivity. In the same context, the nonacademic counterparts of postmodernism—pseudoscience, fortunetelling, astrology, and paranormal religions—are flourishing (see Sokal 2008, 263–370).

There are also the climate change deniers, oxymoronic scientific creationists, intelligent design exponents, and hordes of emboldened and intolerant religious fundamentalists. These ideologues along with their white supremacist allies bent on making America white again have taken over the political arena with a vengeance and are seeking to establish the theocracy of Jesus (see Blaker 2003; Hedges 2006; Sharlet 2010; Stenger 2003, 10). Taking advantage of these circumstances, profit-hungry energy extraction and agrochemical industries, seeking to dodge environmental and safety regulations and undermine policymaking based on scientific evidence, have formed an "unholy alliance" with fundamentalist churches presenting a unified front against science and rationality (Sokal 2008, xv).

Thus, during the second decade of the twenty-first century, it is not the Enlightenment view based on rationality and science but supernaturalism, anti-intellectualism, and obscurantism that compose the most potent forces in the private and national life of the United States. These developments are astonishing in a country historically known for secularism, the separation of church and state, science-driven technological innovations, and the exulted ideal that public policy must look to scientific evidence instead of appealing to emotion, religious dogma, or authority. The latter was the view cherished and espoused by this nation's founding figures such as Thomas Paine, Thomas Jefferson, and Benjamin Franklin.

The postmodern assault on science and its relativism has left us vulnerable to the absurdities of the defenders of supernaturalism, the deception of quacks, and the fanaticism of religious fascists and would-be dictators. History teaches that whenever and wherever irrationalism and relativism have acquired political force, human suffering, violence, oppression, and loss of life have inevitably followed. The example of Nazi Germany will suffice here. Welcome to the postmodern world(?).

Did the postmodern view of knowledge represent "the rearrangement of the very principles of intellectual perspective" (Herzfeld 2001, x, 2, 5, 9, 22), as one infatuated anthropologist put it? No. Was its case against science compelling and based on evidence? No. What the postmodern savants offered was disinformation and an intellectually dishonest enterprise that accomplished nothing aside from bewildering the American public about the role and function of science. Are we justified to abandon science in favor of the alternatives proposed by the purveyors of supernaturalism and other obscurantisms? No.

As Albert Einstein put it, science is one of the most precious things we have. It is valuable not because it guarantees absolute truths free of bias, error, and deception but because it is a unique self-correcting method for reducing bias, mistakes, and fraud to advance our understanding of the social and natural worlds and the universe. Science "is a language that all can use and share in and learn," as anthropologist Robin Fox (1992, 49) noted, and "the wretched of the earth want science and the benefits of science. To deny them this is another kind of racism." Among all the ways of knowing ever devised, only science strives to combat our confirmation biases by demanding that practitioners question their premises and to systematically expose their conclusions to the inspection of unsympathetic nonbelievers (Harris 1979, 27). The hallmark of science is the question "What is the evidence?" The hallmark of the alternative perspectives touted by our "home-grown ayatollahs" and obscurantist gurus is "I wish to believe" (Harris 1987, 14). Science remains our only path toward "thinking straight about the world," which is something urgently needed at this critical historical juncture as irrationalism and fanaticism are "bubbling up around us" (Gilovich 1993, 6; Sagan 1996, 27).

H. Sidky, PhD, is professor of anthropology and chief departmental advisor in the Department of Anthropology at Miami University (Ohio). He refers to himself as a scientific anthropologist. He is working on a book on the topic of this essay.

References

Abel, Reuben. 1976. *Man Is the Measure: A Cordial Invitation to the Central Problems of Philosophy*. New York: The Free Press.
Andersen, Kurt. 2017. *Fantasyland: How America Went Haywire: A 500-Year History*. New York: Random House.
Blaker, Kimberly (ed.). 2003. *The Fundamentals of Extremism: The Christian Right in America*. Michigan: New Boston Books.
Bridgstock, Martin. 2009. *Beyond Belief: Skepticism, Science and the Paranormal*. Cambridge: Cambridge University Press.
Carneiro, Robert. 1995. Godzilla meets new age anthropology: Facing the post-modernist challenge to a science of culture. *Europaea* (Italy) 1: 3–22.
Coyne, Jerry. 2015. *Faith vs. Fact: Why Science and Religion Are Incompatible*. New York: Penguin Books.
De Vrieze, Jop. 2017. 'Science Wars' veteran has a new mission. *Science* 358: 159.
Feyerabend, Paul. 1975. *Against Method*. London: New Left Books.
Foucault, Michel. 1984. *The Foucault Reader*. New York: Pantheon.
Fox, Robin. 1992. Anthropology and the 'teddy bear' picnic. *Society* 30(2): 47–55.
———. 1997. State of the art/science in anthropology. In *The Flight from Science and Reason*. Paul Gross, Norman Levitt, and Martin Lewis (eds.). New York: New York Academy of Science, pp. 327–345.

Gilovich, Thomas. 1993. *How We Know What Isn't So: The Fallibility of Human Reason in Everyday Life.* New York: The Free Press.

Gross, Paul, and Norman Levitt. 1994. *Higher Superstition: The Academic Left and Its Quarrels with Science.* Baltimore, MD: The Johns Hopkins University Press.

Harris, Marvin. 1979. *Cultural Materialism: The Struggle for a Science of Culture.* New York: Vintage Books.

———. 1987. *Why Nothing Works: The Anthropology of Daily Life.* New York: Simon & Schuster.

Hedges, Chris. 2006. *American Fascists: The Christian Right and the War on America.* New York: Free Press.

Herzfeld, Michael. 2001. *Anthropology: Theoretical Practice in Culture and Society.* Oxford: Blackwell.

———. 2017. Anthropological realism in a scientistic age. *Anthropological Theory* (August 13):1–22.

Hines, Terence. 2003. *Pseudoscience and the Paranormal.* Amherst, NY: Prometheus Books.

Koertge, Noretta. 1998. Postmodernism and the problem of scientific literacy. In *A House Built on Sand: Exposing Postmodern Myths about Science.* Noretta Koertge (ed.). Oxford: Oxford University Press, pp. 257–271.

Kuhn, Thomas. 1970. *The Structure of Scientific Revolutions.* Chicago: University of Chicago Press.

Kuznar, Lawrence. 2008: *Reclaiming a Scientific Anthropology.* Walnut Creek, CA: AltaMira.

Latour, Bruno. 2004. Why has critique run out of steam? From matters of fact to matters of concern. *Critical Inquiry* 30: 225–248.

Lett, James. 1986. *The Human Enterprise: A Critical Introduction to Anthropological Theory.* Boulder, CO: Westview Press.

———. 1997. *Science, Reason, and Anthropology: The Principles of Rational Inquiry.* Lanham, MD: Rowman & Littlefield.

Levitin, Daniel. 2016. *Weaponized Lies: How to Think Critically in the Post-Truth Era.* New York: Dutton.

Mooney, Chris, and Sheril Kirshenbaum. 2009. *Unscientific America: How American Scientific Illiteracy Threatens Our Future.* New York: Basic Books.

Otto, Shaw. 2016. *The War on Science: Who Is Waging It, Why It Matters, What We Can Do About It.* Minneapolis: Milkweed.

Popper, Karl. 1972. *Objective Knowledge: An Evolutionary Approach.* Oxford: Oxford University Press.

Robbins, Bruce, and Andrew Ross. 2000. Response: Mystery science theater. In *The Sokal Hoax: The Sham that Shook the Academy.* Editors of *Lingua Franca* (ed.). Lincoln, NE: University of Nebraska Press, pp. 54–58.

Sagan, Carl. 1996. *The Demon-Haunted World: Science as a Candle in the Dark.* New York: Ballantine Books.

Sahlins, Marshall. 1999. What is anthropological enlightenment? *Annual Review of Anthropology* 18: i–xxiii.

Salzman, Philip. 2001. *Understanding Culture: An Introduction to Anthropological Theory.* Prospect Heights, IL: Waveland.

Scheper-Hughes, Nancy. 1995. The Primacy of the Ethical: Propositions for a Militant Anthropology. *Current Anthropology* 36: 409–420.

Schick, Theodore, and Lewis Vaughn. 2014. *How to Think About Weird Things: Critical Thinking for a New Age.* New York: McGraw-Hill.

Sharlet, Jeff. 2010. *C Street: The Fundamentalist Threat to American Democracy.* New York: Little Brown.

Sidky, H. 2004. *Perspectives on Culture: A Critical Introduction to Theory in Cultural Anthropology.* Upper Saddle River, NJ: Prentice Hall.

———. 2007. Cultural materialism, scientific anthropology, epistemology, and "narrative ethnographies of the particular." In *Studying Societies and Cultures: Marvin Harris's Cultural Materialism and Its Legacy.* Lawrence Kuznar and Stephen Sanderson (eds.). Boulder, CO: Paradigm Publishers, pp. 66–77.

Smith, Jonathan. 2010. *Pseudoscience and Extraordinary Claims of the Paranormal.* West Sussex: Wiley-Blackwell.

Sokal, Alan. 2008. *Beyond the Hoax: Science, Philosophy and Culture.* Oxford: Oxford University Press.

Sokal, Alan, and Jean Bricmont. 1998. *Fashionable Nonsense: Postmodern Intellectuals' Abuse of Science.* New York: Picador.

Spaulding, Albert. 1988. Distinguished lecture: Archaeology and anthropology. *American Anthropologist* 90 (2): 263–271.

Stenger, Victor. 2003. *Has Science Found God? The Latest Results in the Search for Purpose in the Universe.* Amherst, NY: Prometheus Books.

———. 2008. *God: The Failed Hypothesis: How Science Shows That God Does Not Exist.* Amherst, NY: Prometheus Books.

Stove, David. 2001. *Scientific Irrationalism: Origins of a Postmodern Cult.* New Brunswick, NJ: Transaction Publishers.

Weinberg, Steven. 1998. The revolution that didn't happen. *New York Review of Books* (October 8).

CHAPTER 8

A Life Preserver for Staying Afloat in a Sea of Misinformation

Melanie Trecek-King

Vol. 46, No. 2
March/April 2022

My goals as a science educator are to teach students the essential skills of science literacy and critical thinking. Helping them understand the process of science and how to draw reasonable conclusions from the available evidence can empower them to make better decisions and protect them from being fooled or harmed.

Yet while educators agree that these skills are important, the stubborn persistence of pseudoscientific and irrational beliefs demonstrates that we have plenty of room for improvement. To help address this problem, I developed a general-education science course[1] that, instead of teaching science as a collection of facts to memorize, teaches students how to evaluate the evidence for claims to determine how we know something and recognize the characteristics of good science by evaluating bad science, pseudoscience, and science denial.

In my experience, science literacy and critical thinking skills are difficult to master. Therefore, it helps to provide students with a structured toolkit to systematically evaluate claims and allow for ample opportunities to practice. In previous semesters, I've had excellent results with James Lett's "A Field Guide to Critical Thinking"[2] in which he summarized the scientific method with the acronym FiLCHeRS (Falsifiability, Logic, Comprehensiveness of evidence, Honesty, Replicability, and Sufficiency of evidence).

While FiLCHeRS has served my students well, I've found myself adding rules and updating examples to help my students navigate today's misinformation landscape. The result is this guide to evaluating claims, summarized by the (hopefully memorable) acronym FLOATER, which stands for Falsifiability, Logic, Objectivity, Alternative explanations, Tentative conclusions, Evidence, and Replicability.

Think of FLOATER as a life-saving device. By using the seven rules in the toolkit we can protect ourselves from drowning in a sea of bad claims.

The foundation of FLOATER is skepticism. While skepticism has taken on a variety of connotations, from cynicism to denialism, scientific skepticism is simply insisting on evidence before accepting a claim and proportioning the strength of our belief to the strength and quality of the evidence.

Before using this guide, *clearly identify the claim* and define any potentially ambiguous terms. And remember, *the person making the claim bears the burden of proof* and must provide enough positive evidence to establish the claim's truth.

Rule 1: Falsifiability

It must be possible to think of evidence that would prove the claim false. It seems counterintuitive, but the first step in determining if a claim is true is to determine if you can prove it wrong.

Falsifiable claims can be proven false with evidence. If a claim is false, the evidence will disprove it. If it's true, the evidence won't be able to disprove it. Scientific claims must be falsifiable. Indeed, the process of science involves trying to disprove falsifiable claims. If the claim withstands attempts at disproof, we are more justified in tentatively accepting it.

Unfalsifiable claims cannot be proven false with evidence. They could be true, but because there is no way to use evidence to test the claim, any "evidence" that appears to support the claim is useless. Unfalsifiable claims are essentially immune to evidence. Four types of claims are unfalsifiable.

- **Subjective claims:** Claims based on personal preferences, opinions, values, ethics, morals, feelings, and judgments. For example, I may believe that cats make the best pets and that healthcare is a basic human right, but neither of these beliefs is falsifiable no matter how many facts or pieces of evidence I use to justify them.
- **Supernatural claims:** Claims that invoke entities such as gods and spirits, vague energies and forces, and magical human abilities such as psychic powers. By definition, the supernatural is above and beyond what is natural and observable and therefore isn't falsifiable. This doesn't mean these claims are necessarily false (or true!) but that there is no way to collect evidence to test them. For example, so-called "energy medicine," such as reiki and acupuncture, is based on the claim that illnesses are caused by out-of-balance energy fields that can be adjusted to restore health. However, these energy fields cannot be detected and do not correspond to any known forms of energy.

There are, however, cases where supernatural claims can be falsifiable. First, if a psychic claims to be able to impact the natural world in some way, such as moving/bending objects or reading minds, we can test the psychic's abilities under controlled conditions. And second, claims of supernatural events that leave physical evidence can be tested. For example, young-earth creationists claim that the Grand Canyon was formed during Noah's flood approximately 4,000 years ago. A global flood would leave behind geological evidence, such as massive erosional features and deposits of sediment. Unsurprisingly, the lack of such evidence disproves this claim. However, even if the evidence pointed to a global flood only a few thousand years ago, we still couldn't falsify the claim that a god was the cause.

- **Vague claims:** Claims that are undefined, indefinite, or unclear. Your horoscope for today says, "Today is a good day to dream. Avoid making any important decisions. The energy of the day might bring new people into your life." Because this horoscope uses ambiguous and vague terms, such as *dream*, *important*, and *might*, it doesn't make any specific, measurable predictions. Even more, because it's open to interpretation, you could convince yourself that it matches what happened to you during the day, especially if you spent the day searching for "evidence." Due to legal restrictions, many alternative medicine claims are purposefully vague. For example, a supplement bottle says it "strengthens the immune system" or a chiropractic advertisement claims it "reduces fatigue." While these sweeping claims are essentially meaningless because of their ambiguity, consumers often misinterpret them and wrongly conclude that the products are efficacious.
- ***Ad hoc* excuses:** These entail rationalizing and making excuses to explain away observations that might disprove the claim. While the three types of claims described thus far are inherently unfalsifiable, sometimes we protect false beliefs by finding ways to make them unfalsifiable. We do this by making excuses, moving the goalposts, discounting sources or denying evidence, or proclaiming that it's our "opinion."

For example, a psychic may dismiss an inaccurate reading by proclaiming her energy levels were low, or an acupuncturist might excuse an ineffective treatment by claiming the needles weren't placed properly along the patient's meridians. Conspiracy theorists are masters at immunizing their beliefs against falsification by claiming that supportive evidence was covered up and contradictory evidence was planted.

The rule of falsifiability essentially boils down to this: Evidence matters. And never assume a claim is true because it can't be proven wrong.

Rule 2: Logic

Arguments for the claim must be logical. Arguments consist of a conclusion, or claim, and one or more premises that provide evidence, or support, for the claim. In effect, the conclusion is a belief, and the premises are the reasons we hold that belief. Many arguments also contain hidden premises, or unstated assumptions that are required for the conclusion to be true, and therefore must be identified when evaluating arguments.

There are two types of arguments, which differ in the level of support they provide for the conclusion.

Deductive arguments provide conclusive support for the conclusion. Deductive arguments are valid if the conclusion must follow from the premises, and they are sound if the argument is valid *and* the premises are true. For the conclusion to be considered true, the argument must be both valid and sound. For example: "Cats are mammals. Dmitri is a cat. Therefore, Dmitri is a mammal." The conclusion has to follow from the premises, and the premises are true. Because this argument is both valid and sound, we must accept the conclusion.

In everyday language, the word *valid* generally means true. However, in argumentation, *valid* means the conclusion follows from the premises, regardless of whether the premises are true or not. The following example is valid but unsound: "Cats are trees. Dmitri is a cat. Therefore, Dmitri is a tree." The conclusion is valid because it follows from the premises, but the conclusion is wrong because of an untrue premise; cats aren't trees.

Inductive arguments provide *probable* support for the conclusion. Unlike deductive arguments, in which a conclusion is guaranteed if the argument is both valid and sound, inductive arguments provide only varying degrees of support for a conclusion. Inductive arguments whose premises are true and provide reasonable support are considered to be strong, while those that do not provide reasonable support for the conclusion are weak. For example: "Dmitri is a cat. Dmitri is orange. Therefore, all cats are orange." Even if the premises are true (and they are), a sample size of one does not provide reasonable support to generalize to all cats, making this argument weak.

Logical fallacies are flaws in reasoning that weaken or invalidate an argument. While there are more logical fallacies than can be covered in this guide, some of the more common fallacies include:

- **Ad hominem**: Attempts to discredit an argument by attacking the source.
- **Appeal to (false) authority**: Claims that something is true based on the position of an assumed authority.
- **Appeal to emotions**: Attempts to persuade with emotions, such as anger, fear, happiness, or pity in place of reason or facts.

- **Appeal to the masses**: Asserts that a claim is true because many people believe it.
- **Appeal to nature**: Argues that something is good or better because it's natural.
- **Appeal to tradition**: Argues that something is good or true because it's been around for a long time.
- **False choice**: Presents only two options when many more likely exist.
- **Hasty generalization**: Draws a broad conclusion based on a small sample size.
- **Mistaking correlation for causation**: Assumes that because events occurred together, there must be a causal connection.
- **Red herring**: Attempts to mislead or distract by referencing irrelevant information.
- **Single cause**: Oversimplifies a complex issue to a single cause.
- **Slippery slope**: Suggests an action will set off a chain of events leading to an extreme, undesirable outcome.
- **Straw man**: Misrepresents someone's argument to make it easier to dismiss.

Consider the following example: "GMO foods are unhealthy because they aren't natural." The conclusion is "GMO foods are unhealthy," and the stated premise is "They aren't natural." This argument has a hidden premise, "Things that aren't natural are unhealthy," which commits the appeal to nature fallacy. We can't assume that something is healthy or unhealthy based on its presumed naturalness. (Arsenic and botulinum are natural, but neither is good for us!) By explicitly stating the hidden premise and recognizing the flaw in reasoning, we see that we should reject this argument.

Rule 3: Objectivity

The evidence for a claim must be evaluated honestly.

Richard Feynman famously said, "The first principle is that you must not fool yourself, and you are the easiest person to fool."

Most of us think we're objective; it's those who disagree with us who are biased, right?

Unfortunately, every single one of us is prone to flawed thinking that can lead us to draw incorrect conclusions. While there are numerous ways we deceive ourselves, three of the most common errors are:

- **Motivated reasoning:** Emotionally biased search for justifications that support what we want to be true.

- **Confirmation bias**: Tendency to search for, favor, and remember information that confirms our beliefs.
- **Overconfidence effect**: Tendency to overestimate our knowledge and/or abilities.

The rule of objectivity is probably the most challenging rule of all, because the human brain's capacity to reason is matched only by its ability to deceive itself. We don't set out to fool ourselves, of course. But our beliefs are important to us; they become part of who we are and bind us to others in our social groups. So when we're faced with evidence that threatens a deeply held belief, especially one that's central to our identity or worldview, we engage in motivated reasoning and confirmation bias to search for evidence that supports the conclusion we want to believe and discount evidence that doesn't. If you're looking for evidence you're right, you will find it. You'll be wrong, but you'll be confident you're right.

Ultimately the rule of objectivity requires us to be honest with ourselves—which is why it's so difficult. The problem is, we're blind to our own biases.

The poster children for violating the rule of objectivity are pseudoscience and science denial, both of which start from a desired conclusion and work backward, cherry-picking evidence to support the belief while ignoring or discounting evidence that doesn't. There are, however, key differences:

- **Pseudoscience** is a collection of beliefs or practices that are portrayed as scientific but aren't. Pseudoscientific beliefs are motivated by the desire to believe something is true, especially if it conforms to an individual's existing beliefs, sense of identity, or even wishful thinking. Because of this, the standard of evidence is very low. Examples of pseudoscience include various forms of alternative medicine, cryptozoology, many New Age beliefs, and the paranormal.
- **Science denial** is the refusal to accept well-established science. Denial is motivated by the desire not to believe a scientific conclusion, often because it conflicts with existing beliefs, personal identity, or vested interests. As such, the standard of evidence is set impossibly high. Examples include denying human-caused climate change, evolution, the safety and efficacy of vaccines, and GMO safety.

In both these cases, believers are so sure they're right, and their desire to protect their cherished beliefs is so strong, they are unable to see the errors in their thinking.

To objectively evaluate evidence for a claim, pay attention to your thinking process. Look at *all* the evidence—even (especially) evidence that contradicts

FLATER

Thinking Is Power

Evaluate claims with this life-saving toolkit

FALSIFIABILITY
It must be possible to think of evidence that would disprove the claim.

LOGIC
Arguments for the claim must be logical and not commit fallacies.

OBJECTIVITY
Evidence for the claim must be evaluated honestly, without bias or self-deception.

ALTERNATIVE EXPLANATIONS
Other ways of explaining the observation must be considered.

TENTATIVE CONCLUSIONS
A conclusion can change with new evidence.

EVIDENCE
Evidence for a claim must be reliable, comprehensive, and sufficient.

REPLICABILITY
Evidence for a claim should be able to be repeated.

"FLOATER." *Illustration by Wendy Cook.*

what you want to believe. No denial or rationalization. No cherry-picking or *ad hoc* excuse-making. If the evidence suggests you should change your mind, then that's what you must do.

It also helps to separate your identity from the belief, or evidence that the belief is wrong will feel like a personal attack. And don't play on a team; be the referee. If defending your beliefs is more important to you than understanding reality, you will likely fool yourself.

Rule 4: Alternative Explanations

Other ways of explaining the observation must be considered. It's human nature to get attached to a single explanation, often because it came from someone we trust or it fits with our existing beliefs. But if the goal is to know the real explanation, we should keep in mind that we might be wrong and consider alternative explanations.

Start by brainstorming other ways to explain your observation. (The more the better!) Ask yourself: What else could be the cause? Could there be more than one cause? Or could it be a coincidence? In short, propose as many (falsifiable) explanations as your creativity allows. Then try to disprove each of the explanations by comprehensively and objectively evaluating the evidence.

Next, determine which of the remaining explanations is the most likely. One helpful tool is Occam's razor, which states that the explanation that requires the fewest new assumptions has the highest probability of being the right one. Basically, identify and evaluate the assumptions needed for each explanation to be correct, keeping in mind that the explanation requiring the fewest assumptions is most likely to be correct and that extraordinary claims require extraordinary evidence.

For example, one morning you wake up to find a broken glass on the floor. Naturally, you want to know how it got there! Maybe it was a burglar? Could it have been a ghost? Or maybe it was the cat? You look for other signs that someone was in your house, such as a broken window or missing items; without other evidence, the burglar explanation seems unlikely. The ghost explanation requires a massive new assumption for which we currently don't have proof: the existence of spirits. So while it's possible that a specter was in your house during the night, a ghost breaking the glass seems even less likely than the burglar explanation, because it requires additional, unproven assumptions for which there is no extraordinary evidence. Finally, you look up to see your cat watching you clean shards of glass off the floor and remember seeing him push objects off tables and counters. You don't have definitive proof it was the cat, but it was *probably* the cat.

Rule 5: Tentative Conclusions

In science, any conclusion can change based on new evidence. A popular misconception about science is that it results in proof, but scientific conclusions are always tentative. Each study is a piece of a larger picture that becomes more clear as the pieces are put together. However, because there is always more to learn (more pieces of the puzzle yet to be discovered), science doesn't provide absolute certainty; instead, uncertainty is reduced as evidence accumulates. There's always the possibility that we're wrong, so we have to leave ourselves open to changing our minds with new evidence.

Some scientific conclusions are significantly more robust than others. Explanations that are supported by a vast amount of evidence are called theories. Because the evidence for many theories is so overwhelming, and from many different independent lines of research, they are very unlikely to be overturned—although they may be modified to account for new evidence.

Importantly, this doesn't mean scientific knowledge is untrustworthy. Quite the opposite: science is predicated on the humility of scientists and their willingness and ability to learn. If scientific ideas were set in stone, knowledge couldn't progress.

Part of critical thinking is learning to be comfortable with ambiguity and uncertainty. Evidence matters, and the more and better our evidence, the more justified we are in accepting a claim. But knowledge is not black or white. It's a spectrum with many shades of gray. Because we can never be 100 percent certain, we shouldn't be overly confident!

Therefore, the goal of evaluating claims and explanations isn't to prove them true. Disprove those you can, then tentatively accept those left standing proportional to the evidence available and adjust your confidence accordingly. Be open to changing your mind with new evidence and consider that you might never know for sure.

Rule 6: Evidence

The evidence for a claim must be reliable, comprehensive, and sufficient. Evidence gives us reasons to believe (or not believe) a claim. In general, the more and better the evidence, the more justified we are in accepting a claim. This requires that we assess the quality of the evidence based on the following considerations:

THE EVIDENCE MUST BE RELIABLE

Not all evidence is created equal. To determine if the evidence is reliable, we must look at two factors:

- **How the evidence was collected.** A major reason science is so reliable is that it uses a systematic method of collecting and evaluating evidence.

  However, scientific studies vary in the quality of evidence they provide. Anecdotes and testimonials are the least reliable and are never considered sufficient to establish the truth of a claim. Observational studies collect real-world data and can provide correlational evidence, while controlled studies provide causational evidence. At the top of the hierarchy of evidence are meta-analyses and systematic reviews, as they are a combination of other studies and therefore look at the big picture.
- **The source of the information.** Sources matter; unreliable sources do not provide reliable evidence. In general, the most reliable sources are peer-reviewed journals, because as the name suggests, the information had to be approved by other experts before being published. Reputable science organizations and government institutions are also very reliable. The next most reliable sources are high-quality journalistic outlets that have a track record of accurate reporting. Be skeptical of websites or YouTube channels that are known to publish low-quality information and be very wary of unsourced material on social media. In addition, experts are more reliable than nonexperts, because they have the qualifications, background knowledge, and experience necessary to understand their field's body of evidence. Experts can be wrong, of course, but they're much less likely to be wrong than nonexperts. If the experts have reached consensus, it is the most reliable knowledge.

THE EVIDENCE MUST BE COMPREHENSIVE

Imagine the evidence for a claim is like a puzzle, with each puzzle piece representing a piece of evidence. If we stand back and look at the whole puzzle, or *body of evidence*, we can see how the pieces of evidence fit together and the larger picture they create.

You could, either accidentally or purposefully, cherry-pick any one piece of the puzzle and miss the bigger picture. For example, everything that's alive needs liquid water. The typical person can live for only three or four days without water. In fact, water is so essential to life that, when looking for life outside of Earth, we look for evidence of water. But what if I told you that all serial killers have admitted to drinking water? Or that it's the primary ingredient in many toxic pesticides? Or that drinking too much water can lead to death?

By selectively choosing these facts (or pieces of the puzzle), we can wind up with a distorted, inaccurate view of water's importance for life. So if we want to better understand the true nature of reality, it behooves us to look at all the evidence—including (especially!) evidence that doesn't support the claim. And be wary of those who use single studies as evidence; they may want to give their position legitimacy, but in science you don't get to pick and choose. You have to look at all the relevant evidence. If independent lines of evidence are in agreement, or what scientists call consilience of evidence, the conclusion is considered very strong.

THE EVIDENCE MUST BE SUFFICIENT

To establish the truth of a claim, the evidence must be sufficient. Claims made without evidence provide no reason to believe and can be dismissed. In general:

- **Extraordinary claims require extraordinary evidence.** Essentially, the more implausible or unusual the claim, the more evidence that's required to accept it.
- **Claims based on authority are never sufficient.** Expertise matters, of course, but experts should provide evidence. "Because I said so," is never enough.
- **Anecdotes are never sufficient.** Personal stories can be very powerful. But they can also be unreliable. People can misperceive their experiences, and, unfortunately, they can also lie.

As an example, let's say you own a company, and Jamie works for you. She is an excellent employee, always on time, and always does great work. One day, Jamie is late for work. If Jamie tells you her car broke down, you most likely will believe her. You have no reason not to—although if you're really strict you may ask for a receipt from the tow truck driver or mechanic. But what if Jamie tells you she's late because she was abducted by aliens? I don't know about you, but my standard of evidence just shot through the roof. That's an extraordinary claim, and she bears the burden of proof. If she tells you that one of the aliens took her to another dimension and forced her to bear offspring but then reversed time to bring her back without physical changes . . . Again, just speaking for myself, I'm either going to assume she's lying or suggest she see a professional.

Rule 7: Replicability

Evidence for a claim should be able to be repeated. *Replicability* (and its related terms) can refer to a range of definitions, but for the purpose of this guide it means the ability to arrive at a similar conclusion no matter who is doing the research or what methodology they use. The rule of replicability is foundational to the self-correcting nature of science, because it helps to safeguard against coincidence, error, or fraud.

The goal of science is to understand nature, and nature is consistent; therefore, experimental results should be too. But it's also true that science is a human endeavor, and humans are imperfect. This can lead to fraud or error. For example, in 1998, Andrew Wakefield published a study claiming to have found a link between the MMR (measles, mumps, rubella) vaccine and autism. After scientists all over the world tried unsuccessfully to replicate Wakefield's findings—with some studies involving millions of children—it was discovered that Wakefield had forged his data as part of a scheme to profit off a new vaccine. The inability to replicate Wakefield's study highlights the importance of not relying on any single study.

Conversely, we can be significantly more confident in results that are successfully replicated independently with multiple studies. And we can be the most confident in conclusions that are supported by multiple independent lines of evidence, especially those from completely different fields of science. For example, because evidence for the theory of evolution comes from many diverse lines, including anatomical similarities, shared developmental pathways, vestigial structures, imperfect adaptations, DNA and protein similarities, biogeography, fossils, etc., scientists have great confidence in accepting that all living things share a common ancestor.

Conclusion

Using FLOATER's seven rules to evaluate claims can help us make better decisions and protect us from being fooled (or even harmed) by false or misleading claims. Evaluating claims this way will likely take practice. But don't get discouraged; it's worth it!

Melanie Trecek-King is an associate professor of biology at Massasoit Community College. Her website is www.ThinkingIsPower.com.

Notes

1. See *Skeptical Inquirer* issue 46:1, 2022.
2. See *Skeptical Inquirer* issue 14:2, 1990.

CHAPTER 9

Surviving the Misinformation Age

David J. Helfand

Vol. 41, No. 3
May/June 2017

In 2016 many in the mainstream media portentously declared we had entered the age of "post-truth politics" (Drezner 2016) and now live in a "post-factual democracy" (Barret 2016). With monetized "fake news" sites proliferating, tweets inconsistent with reality dominating political debate, and most citizens busily constructing echo chambers of their personal beliefs through their social media accounts, the hysteria may seem warranted. But as Alexios Mantzarlis of the Poynter Institute reminds us (Mantzarlis 2016), politicians, media commentators, and your next-door neighbor have been playing fast and loose with the "truth" for a long time.

Indeed, the classical scholar Edward M. Harris noted in his paper dissecting "Demosthenes Speech Against Medias" (Harris 1989) that 2,400 years ago in Athens, "although a witness who perjured himself could be prosecuted . . . an orator who spoke in court could indulge in as much fabrication as he wished without fear of punishment." Harris went on to state: "In short, nothing aside from the knowledge of the audience and the limits of plausibility restrained the orator from inventing falsehoods and distorting the truth."

Public prevarication, then, is nothing new. What is novel is the technology-saturated environment in which it is now embedded. It is the "knowledge of the audience" and the "limits of plausibility"—not the falsehoods and distortions—that have changed.

How has the "knowledge of the audience" evolved over the tenure of *Homo sapiens* on this Earth? For more than 95 percent of our history, knowledge was limited but was tested daily against reality. The hunter-gatherer who picked a basket of poisonous berries was soon eliminated from the gene pool, as was the youth who led his kin toward the hungry lions instead of the grazing gazelles. Those few who parsed the patterns of the stars and so could predict the wildebeests' migration were accorded special veneration (we used to call them "experts"). There was also, no doubt, much misinformation abroad in those halcyon days—lightning evinced the anger of the gods, and neighboring kin groups were largely shunned as the hostile "other," whether they were hostile or not. But with simple survival as the foremost concern, the "knowledge of the audience" in general comported well with reality.

The average citizen today lives in a very different world. As Arthur C. Clarke's celebrated third law has it, "Any sufficiently advanced technology is indistinguishable from magic," and such magic permeates and defines the world of the typical American adult. From self-parking cars and GPS to iPads, airplanes, and LASIK surgery, most people have no clue how the technology that envelops them works or what physical principles underlie its operation—it is, truly, "indistinguishable from magic." And, living in this magical world, the "limits of plausibility" are easily expanded. If the talking box on your dashboard knows exactly where you are and can tell you how to get where you are going, why should talking to dead relatives not be plausible? If shining a light in your eye can eliminate your need for glasses, why shouldn't wearing magnets cure your arthritis?

If a scientific "expert" tells you that magnet therapy is nonsense, he's just exemplifying Clarke's first law: "When a distinguished elderly scientist . . . states that something is impossible, he is very probably wrong." And since the magnets worked wonders for your sister-in-law's best friend, they will probably work for you. If most of one's world is indistinguishable from magic, it is both reasonable and practical to adopt magic as an operating principle. And since only wizards understand magic, consulting them (the homeopaths, the astrologers, the mediums, and the mystics) makes perfect sense.

Thus, the "limits of plausibility" have vanished, and the "knowledge of the audience" is constructed from Facebook feeds, personal experience, and anecdote. The average American is largely insulated from the physical reality his ancestors were forced to confront daily and, as such, resides in a world of self-reinforcing magical thinking.

What we have entered, then, is not the "post-factual" or "post-truth" era but the Misinformation Age. Facts still exist. Good approximations of the truth can still be found. And information has never been more plentiful: IBM calculated a few years ago that we generate 2.5 quintillion bytes of information per day, enough to fill a bookcase half a kilometer tall and stretching around the Earth at the equator—every day. How much of that information is nonsense is anyone's guess. The problem is that everyone feels equally well-qualified to make such a guess and then post it on their blog where it becomes their personal version of the truth that can be easily shared and propagated. And that's how misinformation begins.

There was a time when most people writing on a particular topic did so because they had acquired some degree of specialized knowledge. They had read what was already known about the subject, had conducted some observations or even experiments of their own, and had concluded they might have something to contribute to the advancement of our understanding of the topic at hand. They might even be said to be an "expert" on the subject. The Internet has exploded this model. While the democratization of both access to knowledge and the ability to contribute to it provided by the Internet has obvious benefits, it also has a very serious downside.

Tom Nichols, writing in the *Federalist* (Nichols 2014), describes this downside as "The Death of Expertise," which he characterizes as "a Google-fueled, Wikipedia-based, blog-sodden collapse of any division between professionals and laymen, students and teachers, knowers and wonderers—in other words, between those with any achievement in an area and those with none at all."

This, he says, creates a culture in which "everyone's opinion about anything is as good as anyone else's." Thus, Jenny McCarthy can state that her "mommy instinct" is far superior to scientific medical evidence on vaccination safety, and millions of Google-fed zombies nod in agreement and back up their foundationless opinions from a treasure trove of misinformation and related nonsense on the Internet.

This cornucopia of misinformation feeds another great American pastime: conspiracy theories. A recent survey (Poppy 2017) of 1,511 American adults found that 54 percent believe the 9/11 attacks involved a U.S. government conspiracy, while 42 percent believe global warming is a conspiracy or a hoax—the same percentage who believe in alien encounters. Thirty percent believe President Obama was born in Kenya. I was a little surprised to see that only 24 percent believe the Moon landings were a hoax, but perhaps the Apollo program has faded so far from the collective memory that no one cares anymore. Most interesting was the result that 32 percent believe the North Dakota crash was a government cover up—despite the fact that the researchers completely fabricated an incident they called the "North Dakota crash" and inserted it into the survey to see how many respondents view everything as a conspiracy.

For those of us still convinced that facts about the physical world can be discovered and that a rational analysis of those facts can be useful in creating predictive models of that world, a counterinsurgency seems in order. Where should we begin?

Science is the most powerful intellectual tool humankind has yet invented. Unlike the comforting certainty other worldviews provide, science recognizes its facts as contingent and its models as limited in their application. It is important to realize, however, that science is, at once, both a system for fact discovery and a set of values—skepticism, a reliance on evidence, interpretation using deductive and inductive reasoning, etc. Scientists hold that both the fact-discovery system and these values are crucially important. But, since values are a touchy subject with most people and are at best indirectly testable against reality, it is likely wisest to defer to Jonathan Swift's dictum that "Reasoning will never make a man correct an ill opinion, which by reasoning he never acquired" (Swift 1721), and leave the values part aside for now. My recommendation for an opening gambit in a counterattack on the Misinformation Age is to stick to the facts.

For these purposes, my definition of a simple fact is a measurement of some physical quantity, performed with the best available instruments according to a precisely defined procedure, quoted with an associated uncertainty, and passed through a skeptical review, preferably one that repeats and verifies the measurement. A compound fact can be deduced from a number of simple facts.

A good example of a compound fact is the statement that the dominant component of the CO_2 currently being added to the atmosphere arises from the burning of fossil fuel. I have given many classes and public talks on the subject of climate change, and while my audiences have not all agreed with my conclusions about the gravity of the situation or my proposals for mitigation, I have yet to encounter objection to this fact once I take the time to carefully lay out the evidence. I proceed as follows:

Step 1

I describe how we can count atoms and molecules, one by one, and show a table that lists the numbers of each kind in a sample of a million particles of air. This counting process is, of course, quite remarkable (bordering on magic?), but since most audiences have no concept of the size of an atom (and thus how remarkable it is that we can count them one by one), they can accept the atmospheric concentrations as facts, since counting is a straightforward process everyone understands.

Step 2

I show the first two years of the Keeling curve of CO_2 concentration from 1958 and 1959. This plot shows the number of CO_2 molecules rising steadily from October to May, and then falling symmetrically from May through September. A discussion of how plants breathe in CO_2 and breathe out oxygen during the growing season, and then how bacteria break down the plant tissue and release CO_2 in winter, is also a plausible story that is readily accepted. When a keenly thoughtful person objects because the southern hemisphere has opposite seasons, I reward the thoughtfulness and then show a map of the world pointing out how much more plant-covered land area there is in the temperate zone of the northern hemisphere versus that same zone in the southern hemisphere.

Step 3

I show the entire fifty-eight years of the Keeling curve in which the monotonic trend upward dwarfs the seasonal fluctuations. I then pose the question: How can we know where this additional CO_2 is coming from?

Step 4

I show the O_2 concentration in the atmosphere as a function of time. The decline in O_2 is a surprise to almost everyone but is unmistakable in the data, as are the seasonal fluctuations that are perfectly out of phase with the CO_2 annual pattern. Here I reiterate the plants' CO_2 to O_2 respiration so everyone is comfortable with it—it is always good to tell people something they know, even if they just learned it five minutes ago, as it keeps them engaged with the line of argument. The interesting point to make here is that the amount of O_2 that has disappeared is just equal to the amount needed to explain the CO_2 increase if the CO_2 comes from the combustion (combining with oxygen) of carbon-containing material: C plus O_2 equals CO_2. It is important to note that this is just a correlation and cannot be interpreted as causation—emphasizing the care with which we are accumulating facts and not jumping to conclusions.

Step 5

A digression into isotopes is now required—in particular, that carbon has three common isotopes (C-12, C-13, and C-14). I emphasize how these isotopes are chemically identical but that the heavier ones move more slowly and are thus discriminated against in chemical reactions. This explains why plants have less C-13 and C-14 than does the air they breathe. I also provide a brief introduction to radioactive decay to explain how C-14 gradually converts to plain old nitrogen, the dominant constituent of the atmosphere, on a timescale of 5,730 years.

Step 6

The penultimate data points are the ratios of C-13 and C-14 to C-12, both from direct measurements of the atmosphere over the past forty years and from tree rings going back many centuries. These data show a gradual decline in the C-13/C-12 ratio beginning round 1800 at the start of the industrial revolution that accelerates rapidly over the past few decades, just as the total CO_2 skyrockets. The C-14/C-12 ratio has also been declining rapidly over the past thirty years. By providing pictures of the various sources of carbon (CO_2 from volcanoes and ocean-air exchange, as well as C from living plants, nuclear bomb tests in the atmosphere, and that long-dead plants equal fossil fuels), I am ready for the inescapable conclusion.

Step 7

The declining ratios rule out volcanoes and ocean-atmosphere exchange, since both have higher C-13/C-12 ratios. The falling C-13 values means plants must be involved. The plunging C-14 values mean we must be adding CO_2 to the air that is highly deficient in C-14 and that can't come from modern plants whose C-14 was enriched by the bomb tests in the 1950s. It must come from long-dead plants in which the C-14 has all decayed away. Thus, the dominant fraction of the new CO_2 in the atmosphere must come from burning fossil fuels. QED.

In my experience, this approach has two virtues. First, for all but the most committed science deniers, it establishes the unequivocal role of humans in changing the composition of the atmosphere. Second, it illustrates the process of uncovering facts about the world. I do *not* tout dire predictions about the future of the planet, nor do I suggest policy prescriptions to solve this problem. The former are far too uncertain to constitute "facts," and the latter involves values about which reasonable people may differ. But starting with a fact on which we can agree establishes both a point of connection and a reality-based platform for further discussion.

The Misinformation Age provides poor support for individual decision making and poses a potential disaster for the formation of rational public policy. A counterinsurgency is definitely called for. But our actions will be ineffective if they are politicized (Foster 2017) and unpersuasive unless we scrupulously abide by the principles of a scientific mind. The reproducibility problems in biomedical research (Begley and Ellis 2012) and, more recently, in psychology (Nousek et al. 2015) undermine our credibility. Participation in—even cultivation of—media hype over scientific findings is likewise extremely unhelpful. Assertions of authority will (and perhaps should) be ignored. The power of science lies in its skeptical, rational, evidence-based approach to understanding the world. This power begins with facts, and, in my experience, these facts are the best tools with which to start the revolution.

David J. Helfand is a professor of astronomy at Columbia University and a CSI fellow, as well as past-president of the American Astronomical Society. His book *A Survival Guide to the Misinformation Age* enumerates the scientific habits of mind needed for the counter-revolution.

References

Barret, Nicholas. 2016. Brexit has locked us millennials out of the union we voted for. *Financial Times* (June 26). Available online at https://www.ft.com/content/82a1a548 -3b93-11e6-8716-a4a71e8140b0.

Begley, G.C., and L.M. Ellis. 2012. Drug development: Raise standards for preclinical cancer research. *Nature* 483: 531–533. doi: 10.1038/483531a.

Drezner, Daniel W. 2016. Why the post-truth political era might be around for a while. *Washington Post* (June 16). Available online at https://www.washingtonpost.com/ posteverything/wp/2016/06/16/why-the-post-truth-political-era-might-be-around-for -a-while/?utm_term=.a0da80e33c16.

Foster, Craig A. 2017. Skepticism, at heart, is not partisan. *Skeptical Inquirer* 41(1) (January/February): 14–15.

Harris, Edward M. 1989. Demosthenes speech against medias. *Harvard Studies in Classical Philology* 92: 117–136.

Mantzarlis, Alexios. 2016. No, we're not in a "post-fact" era. Available online at http:// www.poynter.org/2016/no-were-not-in-a-post-fact-era/421582/.

Nichols, Thomas. 2014. The death of expertise. Available online at http://thefederalist .com/2014/01/17/the-death-of-expertise.

Nousek, Brian, and the Open Science Collaboration. 2015. Estimating the reproducibility of psychological science. *Science* 349 doi: 10.1126/science.aac471.

Poppy, Carrie. 2017. Survey shows Americans fear ghosts, the government, and each other. *Skeptical Inquirer* 41(1) (January/February): 6–18, quoting The Chapman University Survey of American Fears 2016 found online at https://blogs.chapman.edu/ wilkinson/2016/10/11/americas-top-fears-2016/.

Swift, Jonathan. 1721. A letter to a young gentleman lately entered into holy orders. *In The Works of Jonathan Swift*. London: H.G. Bohn, 1856.

CHAPTER 10

How to Repair the American Mind

SOLVING AMERICA'S COGNITIVE CRISIS

Guy P. Harrison

Vol. 45, No. 3
May/June 2021

> *Every person we save is one less zombie to fight.*
>
> —*World War Z*, 2013 film

What is the great lesson of 2020? A pandemic killed hundreds of thousands of people and ravaged economies while people disagreed on basic facts. Conspiracy beliefs ran amok. Unscientific racism surged on social media. Medical quackery enjoyed a boom year. What was the common thread that ran through all of it? What should we have learned from such an extraordinarily eventful year?

The crucial ever-present factor in 2020 was *critical thinking*. Those who thought well were less likely to tumble into the rabbit holes of thinking QAnon is true, COVID-19 is a hoax, 5G towers help spread the virus, racism is scientific, hydroxychloroquine cures COVID-19, demon sperm is a problem, tracking devices are in vaccines, there is mass election fraud, etc. The ability and willingness to lean toward evidence and logic rather than side with blind trust and emotion was the key metric behind the madness. We may view the current year, 2021, as the test to see if we were paying attention in 2020. So far, it doesn't look good.

The January 6 invasion of the U.S. Capitol Building by rioters and vandals was more evidence that political propaganda and irrational beliefs have reached crisis levels. Millions of Americans now seem hypnotized by dishonest news sources, medical quackery claims, social media manipulation, and preposterous conspiracy beliefs. Some unknown number of them are willing to break laws and threaten others with violence. We can passively wait and hope that this dangerous offshoot from comparatively harmless believers shrinks, or our society can make intelligent proactive efforts toward ensuring that it does not become a greater problem.

Gullibility, fanaticism, and political trickery are not new, of course. America has always suffered a costly love affair with fraud and fantasy (Anderson 2017). But it all feels faster, louder, and more dangerous these days. One no longer needs to be a charismatic apocalyptic preacher with a brick-and-mortar church or a well-funded politician to pollute minds at a steady clip. Today anyone with a Facebook or Twitter account has the potential power to ignite wildfires of public lunacy.

The Capitol insurrection was a tornado strike made of swirling irrational beliefs, a national self-inflicted wound that never should have happened. A sitting U.S. president and his allies used television, radio, and social media to weaponize some of the most gullible people in America. Key instruments of modern communication could be informing, entertaining, and uplifting us almost exclusively. Increasingly, however, they are maliciously exploited to infect vulnerable people with mental viruses that transform them into either tragic fools or dangerously deranged mission-focused zombies (Harrison 2017). But it may only get worse. The next level of synthetic video/audio media, called deepfakes, is nearly impossible to identify as false, and an avalanche of it is about to drop on our heads (Westerlund 2019).

Our present course may be unsustainable. The synergy of increasingly sophisticated deception aimed at unthinking masses promises more crippling confusion, disruption, and chaos, perhaps more than America can endure. Every minute worrying about nefarious microchips in vaccines is time not spent intelligently evaluating risk and assessing evidence. Every day sacrificed at the altar of a conspiracy belief or at the feet of a hollow demagogue is another day lost to possible social and political progress for all.

Our best hope is to attack the source of the problem. But what is it? What is the real root of this crisis, and what is the solution most likely to work? Attempting to corral or mute those who promote fraud and bad beliefs will not work. Another con artist will always be waiting in the shadows, and disturbed minds will keep conjuring up baseless beliefs. The reason our problem of mass delusions and rampant disinformation can exist to the degree it currently does is because too many American minds are incapable of handling close encounters of the irrational kind. The key problem is that America is a nation of believers more than a nation of thinkers. Therefore, our primary target should not be the few who sell lies and fantasies but the many who so eagerly buy them.

Vast numbers of people do not know how to think critically and are insufficiently aware of how easy it can be for anyone, regardless of general education or intelligence, to be lured into a bogus belief. This abundance of unprotected minds provides the necessary foundation for our growing crisis.

Minus many millions of people in such a vulnerable state, empty claims and ridiculous beliefs could not rage across the land collecting converts with the ease they do now.

I detail the problem of conspiracy theory belief in one of my books, *Good Thinking*, and explain how gossip naturally appeals to us and makes us feel special when we know something others do not (Harrison 2013a). Confirmation bias, a troublesome process even the best of minds contend with, helps cement these beliefs by making them seem logical and evidence based. Familiarity with this and other such information makes people significantly less vulnerable. Bad beliefs can take root in any mind, even one that may be gifted and filled with knowledge. Some conspiracy believers demonstrate remarkable mental sharpness and vigor in protecting the nonsense inhabiting their heads. Picture a world-class lawyer skillfully and effectively defending a guilty client. General ignorance, or what some might describe as innate stupidity, are not the key problems. This crisis grows not from *no thinking* but from *bad thinking*.

If you are skeptical of the claim that America is in the midst of a critical thinking crisis, consider that many (possibly most) of the January 6 Capitol rioters were QAnon believers. Going all in with QAnon means believing that a long list of celebrities and political elites—including Lady Gaga, Bill Gates, Joe Biden, and Tom Hanks—operate an international satanic cannibalistic child sex-abuse organization. It is a claim so extraordinarily vacuous that it can almost serve as the perfect litmus test for the pathological absence of critical thinking.

QAnon is a movement with no clearly identified founders or leaders and no formal doctrine. It is a big-tent/grab-bag of many old and new conspiracy theories. It accommodates a long list of suspicions, fears, resentments, and prejudices (Blaskiewicz 2018). QAnon is the Costco of conspiracy theories, the Walmart of weird beliefs: "Whatever you want, we got it." It also has a *Da Vinci Code* element that seems to appeal to many believers who excitedly search for clues and follow online crumbs toward big "revelations." Despite the absence of any credible evidence and the overwhelming unlikeliness of it all, the combination of fuzziness, inclusiveness, and flexibility is working. QAnon has sucked up millions of unprepared minds in recent years.

According to a recent NPR/IPSOS national poll, less than half (47 percent) of Americans say QAnon's core claims are false (Newall 2020). Seventeen percent of U.S. adults—millions of voting-age citizens—admit to believing. Also alarming: 37 percent are "unsure." Imagine being on the fence about whether Tom Hanks and Beyoncé are working with Satan to traffic child sex-slaves around the world.

As if it were not interesting enough already, some people include lizard aliens in the QAnon recipe (Wallis 2021). These infiltrators from outer space supposedly include Queen Elizabeth, George W. Bush, and Barack Obama. The usual claim is that they are either shapeshifters or merely hiding beneath human skinsuits. This is not how the twenty-first century was supposed to be going for us. As a child nurtured on *Star Trek* reruns, I imagined our species solving poverty, ending war, and colonizing other worlds by now. Silly me. Here I am today discussing a popular belief that reptilian extraterrestrials reside in Buckingham Palace.

How can we prevent QAnon and other such beliefs from corroding our nation's collective sanity to the point of no return? Unfortunately, there is no quick fix available. But there is a preventive treatment. Most won't like it because it's slow and involves a lot of work. But it is a solution, perhaps the only one with a fair chance of success. Playing the long game of critical thinking education is the only way to deny the irrational-belief beast and the steady supply of victims it depends on.

Extreme political manipulation, social media idiocy, QAnon, and other cognitive disasters likely would dry up and shrink to insignificance if robbed of their current deep pool of unquestioning targets ready for assimilation. Those who understand the need to stop America's slide into ever-deepening irrationality must push our society to raise up new generations of *thinking citizens* who are capable of identifying and shrugging off unproven claims. The American mind can be repaired in the long term by teaching the skills and principles of critical thinking to every child. I am aware of the grandiose and cliché-like feel that comes with citing education as the only salvation from a big problem. But in this case, it really is the way.

Making critical thinking a national educational norm is the cognitive vaccine America needs to have a fighting chance of maintaining sufficient sanity. Good thinking prevents and alleviates bad thinking. Young students can be taught reason and skepticism as basic life skills. This would not be the kind of education that involves learning a bunch of facts for later regurgitation. Critical thinking is more like learning a trade. As one might train to weld or build furniture, one can learn how to think well out in the world.

Critical thinking courses for all elementary, middle school, and high school students might include age-appropriate lessons on how to ask the right questions when confronted with an unusual or important claim; a review of common logical fallacies (with an emphasis on relevance to everyday experiences); how to select reliable information sources; a basic survey of the surprising but normal workings of a human brain (how the brain processes visual input, seeks patterns, why memory is unreliable, subconscious influence on conscious thinking, etc.); review how the "critical thinking" concept can be abused and misrepresented (Many QAnon believers, for example, urge people to "think critically" and often say "do your own research." But this means little when poor information sources, flawed logic, and bogus evidence are attached to such advice.); historical review of past mass delusions, frauds, and costly mistakes rooted in poor thinking; and discussions about the many positive benefits of good thinking (increased odds for a safer, more efficient, and productive life).

Given its importance to individual and national health, why not teach critical thinking every day in every school? Why not give it the same attention and emphasis as reading, mathematics, the Pledge of Allegiance, or anything else? Doing this would not preclude addressing the social and health needs of struggling Americans. It would not stand in the way of the need for intelligent social media regulation, vigilance against domestic terrorism, or general science and history education.

As a parent and former teacher, I have seen how easily young students can pick this up. With guidance and encouragement, children can become highly proficient at thinking their way through spurious claims, recognizing potential problems of perception, and spotting bias and lies. From the earliest ages possible, children in the United States, and all countries, should be taught thinking skills because it will serve them well throughout their entire lives. It also just may save our civilization from implosion one day.

Having written several books and given many lectures and interviews on this topic, I know from experience that some people cringe, if not shriek in horror, at the thought of teaching critical thinking to ten-year-olds. But these concerns seem rooted in a misunderstanding of what critical thinking is. To be clear, it is not a list of approved ideas and taboo beliefs. Critical thinking is the means of figuring out if something makes sense and is likely to be true or not. Nothing is threatened apart from lies and errors in reasoning. Hopefully, one would want these exposed to avoid wasting time, energy, or money or risking good health. Thomas Paine put it well: "It is error only, and not truth, that shrinks from inquiry."

Anyone who opposes critical thinking education is effectively taking a position against reason and reality.

Considering the current American landscape, this is inexcusable stark negligence (Uscinski 2019). We cannot continue to fail so many children. They need the necessary tools to be able to navigate the increasingly complex and foreboding information jungles of twenty-first-century civilization. We owe students better than leaving them to exit schools on graduation day as soft targets for con artists and victims-in-waiting for delusion peddlers.

"Critical thinking" is viewed by some as code for "condescension and elitism." This is another unfair judgment. No one owns a patent on thinking well. No one can keep this from you if you want it, regardless of age, income, education, or social status. Relying on critical thinking as a matter of daily routine is a personal choice. It is the attempt to get most things right most of the time. This is too useful, too vital now to be left to university philosophy classes or spurned as a sign of intellectual snobbery. Critical thinking is a collection of down-in-the-trenches people skills that are available to everyone (Harrison 2013b). It's about doing the work to figure out important things based on reason more than emotion and on analysis more than trust or tradition. It also means reevaluating conclusions and changing your mind when it makes sense to do so. Put simply, this is the conscious attempt to dodge lies and false beliefs while moving in the general direction of truth and reality. There is no more reliable safeguard against becoming someone's fool or the sad pawn of an empty fantasy.

This is on us, the grownups. Adults who see the value and need for teaching reason-based, independent thinking to all children must act. Push politicians, school boards, and parents to prepare children for the countless lies, irrational temptations, and cognitive landmines they will encounter in life. What else can we do? The U.S. government cannot outlaw the inclination to believe nonsense. Regulations won't purge the internet of every lie. Our brains are not going to suddenly evolve beyond their natural tendencies to lead us astray when it comes to perceiving and calculating reality. The answer lies with us. Teach our children thinking skills so that they can be their own editors and fact checkers. Children who grow up in this century must be their own guardians of truth. But they will fall short unless someone cares enough to teach them how.

Guy Harrison is an award-winning journalist and the author of eight books that promote science and reason. His books include *Think: Why You Should Question Everything*, an introduction to critical thinking appropriate for all ages. His most recent book is *At Least Know This: Essential Science to Enhance Your Life*. Follow Harrison on Twitter at @harrisonauthor.

References

Anderson, Kurt. 2017. *Fantasyland: How America Went Haywire: A 500-Year History.* New York, NY: Random House.

Blaskiewicz, Robert. 2018. Of course, Qanon. *Skeptical Inquirer* online (August 9). Available online at https://skepticalinquirer.org/exclusive/of-course-qanon1/.

Harrison, Guy P. 2013a. *Good Thinking: What You Need to Know to be Smarter, Safer, Wealthier, and Wiser.* Amherst, NY: Prometheus Books, 215–229.

———. 2013b. *Think: Why You Should Question Everything.* Amherst, NY: Prometheus Books.

———. 2017. *Think Before You Like: Social Media's Effect on the Brain and the Tools You Need to Navigate Your Newsfeed.* Amherst, NY: Prometheus Books.

Newall, Mallory. 2020. More than 1 in 3 Americans believe a "deep state" is working to undermine Trump. IPSOS (December 30). Available online at https://www.ipsos.com/en-us/news-polls/npr-misinformation-123020.

Uscinski, Joseph E. 2019. Conspiring for the common good. *Skeptical Inquirer* 43(4) (July/August): 40–44.

Wallis, Paul. 2021. Op-Ed: Biden is JFK Jr in a mask and an intergalactic being, says QAnon. *Digital Journal* (January 16). Available online at http://www.digital journal.com/news/world/op-ed-biden-is-jfk-jr-in-a-mask-and-an-intergalactic-being -says-qanon/article/584035#ixzz6l4GkXuOo.

Westerlund, Mika. 2019. The emergence of deepfake technology: A review. *Technology Innovation Management Review* 9: 39–52. DOI: 10.22215/timreview/1282.

CHAPTER 11

The Politicalization of Scientific Issues

Jeanne Goldberg

Vol. 41, No. 5
September/October 2017

> *My dear Kepler, what would you say of the learned here, who,
> replete with the pertinacity of the asp, have steadfastly refused
> to cast a glance through the telescope? What shall we make of
> this? Shall we laugh, or shall we cry?*

These words of Galileo, written in a letter to his friend Johannes Kepler, expressed his frustration related to the fact that evidence clearly supportive of heliocentrism was not respected and was in fact rejected as being heretical, in direct opposition to biblical scripture. Galileo was hopeful that if people who believed in the ancient theory of geocentrism would, to paraphrase him, "just look through the lens" of his telescope, they would see evidence to support the theory of heliocentrism (in which the Earth and its planets revolve around the Sun), first contemplated in Hellenistic times and then later supported by Polish astronomer Nicolaus Copernicus's work *On the Revolutions of the Celestial Orbs*, published in 1543.

Aristotle's work in physics and astronomy was largely respected among astronomers at the time Copernicus's book was published, and they had difficulty accepting Copernicus's work. In addition, biblical views were prevalent among the population. Galileo was well aware of this fact but stated that "the Bible is written in the language of the common person who is not an expert in astronomy." He argued that "Scripture teaches us how to go to heaven, not how the heavens go" (Van Helden 1995). His discoveries, published in 1632 in *Dialogue Concerning the Two Chief World Systems*, and those of Kepler further supported the scientific foundation of Copernicus's work, ensuring that most serious astronomers subsequently were Copernicans.

Galileo's book, however, was incendiary by espousing a worldview that contradicted one long accepted. There are striking similarities of Galileo's world with ours today in the twenty-first century. Since his time, however, scientific research has furthered our understanding of the world and led to advances that have transformed the lives of billions of global citizens.

Why, then, have partisan politics permeated the discussions and decisions related to science-based issues such as climate change, evolution, vaccination, GMO technology, stem cell research, and other topics not only here in the United States but globally? Is a lack of understanding, disinterest, or ignorance of scientific facts to blame? Is scientific literacy and research not prioritized in our nation? What threats to people's lives are posed by accepting—or at least considering—scientific evidence? If citizens would "just look through" (Galileo's) telescope rather than the proverbial looking glass, would they understand the importance of science for themselves and be more accepting of the findings of scientific experts?

It is illuminating to step back in history again and consider the important role that philosophy played in the ancient world. Philosophy, the study of the fundamental nature of knowledge, reality, and existence, is regarded as a distinct academic subject today. Philosophy in the ancient world, however, represented the discipline of studying the natural world in a rational way, as a variety of scientific disciplines do today. Science and philosophy, considered to be such distinctly different disciplines today, were in effect one branch of knowledge in the ancient world.

Consider the poem that Lucretius wrote in 50 BCE, "On the Nature of Things." In *The Swerve: How the World Became Modern*, Stephen Greenblatt (2011) tells a fascinating story about a papal secretary who, in the Middle Ages, traipsed across Europe in search of a copy of this reportedly lost poem. The story of the adventures of this secretary is in itself intriguing, but the actual poem was earthshaking in its time and, interestingly, still is today!

Lucretius's poem portrayed religions as cruel and superstitious, fueled by ignorance and fear. In his poem, he proposed a scientific world vision in which all things, animate and inanimate, are composed of invisible particles, moving randomly and continuously in a void. There is no creator; living things have come into existence over eternity by random collisions of the particles and have evolved by a process of trial and error. Their purpose is only to survive, reproduce, and participate in a life of pleasure. Humans are not at the top privileged level of existence, and by understanding their own insignificance and the fact that there is no afterlife they will appreciate the wonder of life and be filled with pleasure (Greenblatt 2011, 185–201). The poem, which addressed Lucretius's natural ("scientific") worldview, was regarded as subversive and heretical, and those who openly supported it risked their lives. In fact, in 1600, the Roman Catholic Church Inquisition questioned Giordano Bruno, a defrocked Dominican monk, Italian philosopher, and scientist, and then burned him at the stake for openly supporting the views expressed by Lucretius in "On the Nature of Things."

The fusion of science and philosophy was also a cultural feature of pre–Revolutionary Era America, reflecting the values of The Enlightenment in Europe. In *Anti-Intellectualism in American Life*, Richard Hofstadter describes the fact that the early American Puritans, although criticized for serious, cruel actions, "came as close to being . . . a class of intellectuals intimately associated with a ruling power . . . as America has ever had" (Hofstadter 1962, 59).

A respect for science continued into the Revolutionary Era as the American Founding Fathers demonstrated their support of science and reason. Although Christianity was an important cultural feature of U.S. history, they embraced the secular values of Christianity in preference to its dogma. There is a plethora of quotations to support this fact. *In Poor Richard's Almanack*, Benjamin Franklin stated that "The way to see by faith is to shut the eye of reason." Thomas Paine wrote, "To argue with a man who has renounced the use and authority of reason, and whose philosophy consists in holding humanity in contempt, is like administering medicine to the dead" (Paine 1778). (Interestingly, George Washington, John Adams, James Madison, and Thomas Jefferson were citizen scientists who found, to quote Thomas Jefferson, "supreme delight" in pursuing scientific topics by conducting their own experiments.)

In a broader sense, the Founding Fathers understood the synergistic relationship of science and democracy. They transported the scientific method of testing of hypotheses and ideas, peer review, and free speech without fear of retaliation into a governmental context. John Adams referred to the "science of government" and applied the relationship of the scientific principle of equilibrium to the system of checks and balances, a critical component of democracy. The Founding Fathers also recognized the potential of scientific knowledge to solve problems and improve the lives of Americans in the future (Union of Concerned Scientists 2017).

The years following the Revolutionary Era in our country, however, marked an abrupt departure from the rational, secular orientation of the Founders. A current of anti-intellectualism, described in Richard Hofstadter's brilliant book *Anti-Intellectualism in American Life* (Hofstadter 1962), grew progressively stronger during the early nineteenth century, changing the public's attitudes toward science and other scholarly subjects. Hofstadter describes the early nineteenth century westward migration of the U.S. population away from the cultural centers in coastal New England as a critical factor in the development of anti-intellectualism. An upsurge of revivalistic religious sentiment accompanied this migration too, with a demand for preachers who could deliver emotional appeals to the uneducated lay people in the frontier regions. Dwight L. Moody, a prominent evangelist of the period, scorned reading books and described learning as "an encumbrance to the man of spirit." His scorn for science was widely shared among the public, amplified when he stated that students were taught that man was the "offspring of a monkey" (Hofstadter 1962).

As science became more complex and differentiated, the common man began to feel dependent upon "experts," many of whom were located in coastal eastern urban areas. Individuals' treasured feelings of self-sufficiency and independence were compromised, leading to resentment. Americans respected intelligence but not intellectuals, and they complained that the value of an idea was governed by its utility—it had no value in itself. This belief pervaded the business world as well, creating Americans' strong admiration for a self-made man, one whose path led to economic and social success without much education, often in spite of it.

So what has happened in our country? How have issues with scientific implications become politicized and analyzed through partisan rather than objective lenses? How can political candidates gain public support by professing views that contradict objective scientific evidence?

It is a paradox that in a nation that has been in the forefront of scientific developments, climate change was mentioned as an important issue by only one candidate, Bernie Sanders, in the 2016 presidential campaign. No questions related to climate change were posed by moderators in the televised debates. It is also remarkable that in the current 115th Congress, there are 222 members with law degrees while there are fourteen physicians. There is only one PhD scientist, Representative Bill Foster, who has a PhD in physics. Eighteen members of the House have no degree beyond their high school diploma (Manning 2017).

The War on Science by Shawn Otto is an authoritative new source of information concerning the questions raised in this article. He makes the point that "knowledge is power, and power is political" (Otto 2016). In view of its dynamic, ever-changing knowledge base, science may threaten vested interests and conflict with fixed worldviews, including those based on superstition and religion.

Another interesting paradox that Otto points out, however, is that the scientific method is inherently democratic since one's hypotheses and results are subject to evaluation and testing by other independent sources; nevertheless, the public often regards scientists as authoritarian, arrogant elitists (Otto 2016). Scientists in our country admittedly have often retreated to the laboratories and haven't actively engaged with the public, with the exception of a few stellar individuals such as Neil deGrasse Tyson, E.O. Wilson, Rachel Carson, and Carl Sagan. These scientists have performed an excellent service by making complex scientific issues understandable and interesting to the public. Unfortunately, their work is often undermined by the public's feelings of envy and alienation.

Anti-science views have been amplified by the political wave of populism that is sweeping not only America but also Europe. Key elements of populism are anti-elitism and nativism, which can translate into anti-immigrant views (Toker 2016). The role of globalization in the creation of not only economic but also knowledge inequality has amplified these feelings of resentment.

A critical driver of the politicization of science is the perception of threat to religious beliefs, and the school choice issue brings this into sharp focus. For example, many evangelicals feel that tax-supported vouchers should be used to promote anti-science religious dogmas such as creationism. Indeed, the recently appointed Education Secretary Betsy DeVos stated several years ago that one of the goals of our schools should be "to confront the culture in ways that will continue to advance God's kingdom" (Rizga 2017). Evangelicals feel threatened by evolution, now universally accepted by the scientific community. Interestingly, 81 percent of self-identified white, born-again evangelicals voted for the Republican presidential candidate Donald Trump in the 2016 presidential election, while only 16 percent voted for the Democrat Hillary Clinton (Smith and Martinez 2016).

In addition to posing a threat to one's religious beliefs, some individuals perceive scientific research as a threat to their business interests. This is especially true if the research results in governmental regulatory policies that are perceived to harm profits. Climate change is an obvious example of this dynamic. Shawn Otto points to the example of Exxon, which in the 1970s promoted climate science only to take the opposite stance as a climate denier in recent times (Otto 2016). The Republican Party, traditionally the party representing business interests, formulated a platform in the 2016 presidential election that supported cuts to scientific research, halting of funding for the U.N.'s Framework Convention on Climate Change, cancellation of the Clean Power Plan, and other deregulatory actions. The Trump Administration appointment to Secretary of the Environmental Protection Agency is an individual who has sued that very agency many times to weaken its environmental regulations and would like to see its power severely reduced.

Another facet of the politicization of science in our country is the effect of the postmodernism movement, which occurred in the latter part of the twentieth century. This movement represents a distrust of the Enlightenment principle of rationality. Although the term *postmodernism* has traditionally been applied to the humanities, it has broad implications for attitudes toward science, promoting the idea that truth is contextual, depending on one's culture, education, and life experiences. This attitude is misplaced when dealing with scientific facts such as evolution. Otto further describes the role of journalists in promoting the "other side" of disagreements regarding scientific issues, even when the weight of evidence overwhelmingly supports one conclusion. This mistakenly gives opinion the same weight as fact (Otto 2016).

One of the unfortunate results of postmodernism is that individuals see scientific issues in ways that fit their preconceptions and make them comfortable.

As individuals mature, they may gravitate toward the political party whose views they share on other nonscientific issues and then proceed to adopt unquestionably, almost in a tribal fashion, the views of that party on scientific issues. The public's gravitation to biased television reporting, social media, and Internet resources that fit their worldview as sources of information on scientific issues further calcifies their opinions.

Unfortunately, the public's respect for scientific developments can be modulated by fear. Progressive Democrats have traditionally been strong supporters of scientific research and have endorsed the validity of evolution and climate change; nevertheless, some regard GM foods, vaccination, fluoridation of drinking water, and a variety of chemicals as threatening developments. Even many scientifically literate progressives are skeptical about the safety of GM foods and are concerned that the food industry's vested interests may outweigh safety issues. Conservative Republicans, on the other hand, may or may not personally approve of GM foods, and business interests could override their safety concerns (Funk and Kennedy 2016).

The issue of vaccination is complex, involving strange bedfellows. This debate arose in recent times as a result of a *Lancet* medical journal article that contained fraudulent information indicating that autism could result from vaccination (General Medical Council 2010). Some progressive Democrats, usually supportive of regulations that they see as contributing to public welfare, object to vaccination on the (faulty) grounds that it may result in autism. Some conservatives, mainly Republicans, object to it because they feel that their personal freedom is threatened by school requirements for vaccination.

Another issue with strange bedfellows is food supplements. In his illuminating book *Do You Believe in Magic?*, Paul A. Offit outlines the steps by which Congressional members from both the Democrat and Republican parties enacted legislation in 1976 that effectively freed the entire supplement industry from the FDA requirement that products had to be shown to be both safe and effective. Later attempts to pass legislation requiring supplements to meet FDA requirements were made primarily by Democratic legislators but were defeated by Republicans who had political constituencies or personal financial investments in the industry (Offit 2013).

In *The War on Science*, Shawn Otto describes the "marriages" between different segments of our society, which by sharing common agendas became the two major modern political parties in our country. These marriages are as follows:

1. The anti-regulatory, pro-corporate business interests and the anti–reproductive-control religious interests found their representation in the Republican Party.

2. The pro-environment, pro-choice, anti-corporate elements of scientists and environmentalists found their interests best represented in the Democratic Party.

These marriages have catalyzed the extreme polarization regarding scientific issues that we witness today. In the current parlance, this is "identity politics" (Otto 2016).

It is beyond the scope of this article to propose potential solutions to this situation. Shawn Otto, however, has articulated a comprehensive strategy in his book to raise the awareness of the importance of science in our democracy. Scientists, teachers, and businessmen can engage in public outreach, and candidates running for public office must demonstrate their knowledge of and commitment to science in public debates. Otto lists ways in which concerned citizens can become effective activists and offers an exhaustive list of organizations that are engaged in this effort (Otto 2016). The hundreds of thousands of citizens in the United States and across the globe who participated in the March for Science on April 22, 2017, sent a strong, clear message to world leaders and to other citizens that science plays a vital role in our lives and is ignored only at our peril. After all, it's difficult to name any issue that isn't either directly or indirectly related to science!

An increasing number of businesses are endorsing policies and positions that are fact-based. Jeffrey Immelt, the CEO of GE, for example, stated recently, "We believe climate change is real and the science is well accepted. We hope that the United States continues to play a constructive role in furthering solutions to these challenges" (*Wall Street Journal* Business section, March 30, 2017).

As one of many nations that are intensely focused on the welfare of children and grandchildren, an emphasis on personal, corporate, and governmental responsibility could be powerful and universally appealing. Although a rational approach to scientific issues is essential, emotional appeals regarding specific issues incorporating the message "for the children's sake" might incentivize people with differing views and orientations to work together on controversial issues. Imagine the power of this approach with regard to environmental preservation, for example. A video could be produced showing a child enjoying a bird's song today in 2017 but not in 2050 because of environmental destruction. Or consider the alternative, a video of the child looking and listening to the same bird in 2050 as a result of environmentally responsible actions! This child-centered approach has been woefully underutilized.

In conclusion, there is no doubt that a threat to our democracy exists when there is scientific illiteracy, complacency, or extreme polarization regarding scientific issues among the general public. This is fertile ground for powerful vested interests to use baseless "information" (i.e., "fake news") to lobby for their positions on issues that threaten or support their views. This constitutes a form of authoritarianism that can be used to impede scientific progress and, in the long run, cause a government to fail. We have only to look at examples where that has occurred (e.g., China during the Cultural Revolution, Nazi Germany, and the Ottoman and Roman Empires) to see the catastrophic results. We must "look through Galileo's lens" rather than through an imaginary looking glass and respect the power of science to preserve our democracy in the United States and globally.

Jeanne Goldberg, MD, is a retired radiologist and a previous chair of the Florida Division of the American Cancer Society Breast Cancer Task Force. She is a science writer and active in several environmental organizations. She authored "Politicization of Scientific Issues: Looking through Galileo's Lens or through the Imaginary Looking Glass," the cover feature of our September/ October 2017 issue, and "From the Spectral to the Spectrum: Radiation in the Crosshairs" in our September/October 2018 issue.

References

Funk, Cary, and Brian Kennedy. 2016. The new food fights: U.S. public divides over food science. Pew Research Center (December). Available online at http://www.pew internet.org/2016/12/01/the-new-food-fights/.

General Medical Council. 2010. Andrew Wakefield: Determination of serious professional misconduct. Available online at http://www.bmj.com/content/342/bmj.c7452.

Greenblatt, Stephen. 2011. *The Swerve: How the World Became Modern.* New York and London: W.W. Norton and Company.

Hofstadter, Richard. 1962. *Anti-intellectualism in American Life.* New York: Vintage Books, a Division of Random House.

Manning, Jennifer E. 2017. Membership of the 115th Congress: A Profile. Congressional Research Service 7-5700. Available online at www.crs.gov.

Offit, Paul A. 2013. *Do You Believe in Magic?* New York: HarperCollins Publishers.

Otto, Shawn. 2016. *The War on Science.* Minneapolis: Milkweed Editions.

Paine, Thomas. 1778. The American Crisis: Lancaster, March 21, 1778. The Crisis.

Rizga, Kristina. 2017. Betsy DeVos wants to use America's schools to build "God's kingdom." *Mother Jones (March/April).* Available online at www.motherjones.com/politics/2017/01/betsy-devos-christian-schools-vouchers-charter-education-secretary.

Smith, Gregory A., and Jessica Martinez. 2016. How the faithful voted: A preliminary 2016 analysis. Pew Research Center (November 9). Available online at http://www.pew research.org/fact-tank/2016/11/09/how-the-faithful-voted-a-preliminary-2016 -analysis/.

Toker, Daniel. 2016. Is populism a threat to science? *The Humanist (August).* Available online at https://thehumanist.com/news/science/populism-threat-science.

Union of Concerned Scientists. 2017. Science and Democracy in the United States: A Rich History. Available online at http://www.ucsusa.org/center-for-science-and -democracy/science-and-democracy-in-the-US-history-html#.WJjE37GZOuU.
Van Helden, Al. 1995. Copernican system. The Galileo Project. Available online at http://galileo.rice.edu/sci/theories/copernican_system.html.

CHAPTER 12

Public Debate, Scientific Skepticism, and Science Denial

Harris L. Friedman, Michael Mann, Nicholas J. L. Brown, and Stephan Lewandowsky

Vol. 41, No. 1
January/February 2017

When scientists discover a distant planet that is made of diamonds (Bailes et al. 2011), public admiration is virtually assured. When the same scientific method yields findings that impinge on corporate interests or people's lifestyles, the public response can be anything but favorable. The controversy surrounding climate change is one example of a polarized public debate that is completely detached from the uncontested scientific fact that Earth is warming from greenhouse gas emissions (e.g., Cook et al. 2013). How can scientists navigate those contested waters, and how can the public's legitimate demand for involvement be accommodated without compromising the integrity of science?

Denial of Science

Public debate and skepticism are essential to a functioning democracy. There is evidence that skeptics can differentiate more accurately between true and false assertions (Lewandowsky et al. 2009). However, when tobacco researchers are accused of being a "cartel" that "manufactures alleged evidence" (Abt 1983, 127), or when a U.S. senator labels climate change a "hoax" that is ostensibly perpetrated by corrupt scientists (Inhofe 2012), such assertions are more indicative of the denial of inconvenient scientific facts than expressions of skepticism (Diethelm and McKee 2009). The dividing line between denial and skepticism may not always be apparent to the public, but existing research permits its identification because denial expresses itself in similar ways regardless of which scientific fact is being targeted (Diethelm and McKee 2009). For example, denial commonly invokes notions of conspiracies (Lewandowsky et al. 2015; 2013; Mann 2012). Conspiratorial content is widespread in anti-vaccination material on the Internet (Briones et al. 2012) as well as on blogs that deny the reality of climate change (Lewandowsky et al. 2015).

A second common feature of denial, which differentiates it further from legitimate debate, involves personal and professional attacks on scientists both in public and behind the scenes. To illustrate, the first two authors (Lewandowsky and Mann) have been variously accused of "mass murder and treason" or have received email from people who wanted to see them "six feet under." Such correspondence is not entirely random: Abusive mail tends to peak after the posting of scientists' email addresses on websites run by political operatives.

Those public attacks are paralleled by prolific complaints to scientists' host institutions with allegations of research misconduct. The format of such complaints ranges from brief enraged emails to the submission of detailed multipage dossiers, typically suffused with web links and richly adorned with formatting. In the tobacco arena, there is evidence that such complaints are highly organized (Landman and Glantz 2009). The triage between vexatious complaints and legitimate grievances causes considerable expenditure of public funds when university staff are tied up in phone calls, email exchanges, and responding to persistent approaches while also trying to examine the merit of complaints.

A further target for contrarian activity involves preliminary results or unpublished data. This modus operandi was also pioneered by the tobacco industry, which campaigned hard to gain unhindered access to epidemiological data (Baba et al. 2005). At first glance, it might appear paradoxical that an industry would sponsor laws ostensibly designed to ensure transparency of research. However, access to raw data is necessary for the re-"analyses" of data by entities sympathetic to corporate interests. In the case of tobacco, those analyses have repeatedly downplayed the link between smoking and lung cancer (see Proctor 2011).

A curious feature of all these lines of attack is that they tend to be accompanied by calls for "debate." Often the same individuals who launch complaints with institutions to silence a scientist also proclaim that they want to enter into a "debate" about the science that they so strenuously oppose.

Public Skepticism and the Scientific Process

Given that scientific issues can have far-reaching political, technological, or environmental consequences, greater involvement of the public in policy decisions can only be welcome and may lead to better outcomes. To illustrate, the town of Pickering in Yorkshire, England, recently revised its flood management plan as a result of a year-long collaboration between the local public and scientists (Whatmore and Landström 2011). The plan that was ultimately accepted differed considerably from the initial draft produced by scientists without local public input. Notably, Pickering escaped the flooding that gripped other parts of Yorkshire during the winter of 2015–2016 (Lean 2016).

Notwithstanding the public's entitlement to be involved, scientific debates must still be conducted according to the rules of science. Arguments must be evidence-based, and they are subject to peer review before they become provisionally accepted. Arguments or ideas that turn out to be false are eventually discarded—a process that sometimes seems to take too long but that arguably has served science and society well (Alberts et al. 2015).

Although these strictures are rigorous and may appear daunting to the layperson, they do not exclude the public from scientific debate. It is important to show that the public can participate in scientific debate, because otherwise denialist activities might acquire a sheen of legitimacy as the only avenues open to the public to question scientific findings.

Recently, two of us (Friedman and Brown) were coauthors of an article (Brown et al. 2013) that received much coverage for its criticism of a long-standing, much-cited finding in the field of positive psychology. Positive psychology studies the strengths that enable individuals to thrive and aims to aid in the achievement of a satisfactory and fulfilling life. At the time when the project that led to our article began, Brown (the first author of that paper) was essentially a stranger to academia, having only attended three weeks of a weekend master's program in psychology at the age of fifty-one while working full time as a civil servant.

When he doubted the validity of some of positive psychology's findings that were presented as fact in his classroom, he pursued the issue by contacting a researcher (Friedman) by email based only on the hope that Friedman might be sympathetic to his puzzlement. Once a dialog with the expert had been established—and once Brown had convinced his interlocutor of his sincerity—a fruitful scientific collaboration ensued that has thus far led to the publication of six articles. Notably, this collaboration differs from conventional student-professor interactions in that the parties initially were not known to each other and had no professional relationship prior to an unsolicited approach by email.

To be sure, the process of getting the first rebuttal article published was not easy, given the stature (e.g., more than 350 citations) of the article reporting the original, erroneous finding (Fredrickson and Losada 2005). Brown and Friedman encountered a certain amount of resistance—which would mostly qualify as bureaucratic rather than sinister, despite some apparent conflicts of interest—to the acceptance of both their initial rebuttal article (on the basis of some rather bureaucratic interpretations of customary publishing practices), and to their attempts to write a subsequent comment on the original author's reply (on the basis that the standard sequence of replies to a target article was now finished).

Ultimately, the system worked as it should: everyone remained calm and polite, and the various publishing and appeals processes were tested and observed to work. In the end, all articles appeared in print in the same journal, the scientific record was corrected, the field of positive psychology took stock, and nobody felt the need to publish home addresses or other personal details on the Internet (a harassing process known as "doxxing" that is popular not only with political operatives who oppose climate science but also with anti-vaccination activists and others). The contrast between the approach followed by Brown and the refusal to engage in the scientific process that is characteristic of denial as we described earlier in this article is striking.

The Need for Vigorous Debate

We underscore that there is plenty of room for honest and vigorous debate in science, even among collaborators: One of us (Brown) is an enthusiastic proponent of the widespread adoption of genetically modified organisms (GMOs) as a way to alleviate global food shortages, whereas two of us (Mann and Lewadowsky), while provisionally accepting the safety of GMOs, are concerned about their indirect consequences, such as the emergence of herbicide-resistant weeds that has been associated with GMO-related overuse of herbicides (Gilbert 2013). One of us (Friedman) is concerned about both their indirect consequences and their potential safety to individuals.

Two of us (Friedman and Brown) are not convinced beyond doubt that highly complex climate models are as yet sufficiently validated to be used as the basis of major public policy decisions that might have effects for many decades; the other two authors (Lewandowsky and Mann) acknowledge the uncertainty inherent in climate projections but note that, contrary to popular intuition, any uncertainty provides even greater impetus for climate mitigation (Lewandowsky et al. 2014). Notwithstanding those disagreements, the present authors found common ground for this article.

Although we believe that scientific evidence should inform political debate, we acknowledge that it is no substitute for it. To illustrate, the scientific evidence shows that the fallout from the Fukushima nuclear accident poses no discernible risk to people in North America (e.g., Fisher et al. 2013), but that finding should only guide, and not preclude, political debate about the safety of nuclear power. Whatever the science may say about the safety of nuclear power—for example, that it causes 100 times fewer fatalities than renewable biomass (Markandya and Wilkinson 2007)—those data might be legitimately overridden by the "dread" that nuclear power evokes in people. However, even dread does not justify harassment or threats of violence against scientists who measure nuclear fallout (Hume 2015).

Enhancing the Resilience of the Scientific Enterprise

Opinion surveys regularly and consistently show that public trust in scientists is very high (Pew Research Center 2015). However, the position of the scientist as a neutral, disinterested proponent of "the truth" should not be taken for granted. For example, when Brown and Friedman's first article on positive psychology (Brown et al. 2013) was published, it was cited on several forums and blogs dedicated to creationist ideas or to climate change denial. The argument typically ran thus: If psychologists can be as badly wrong as Brown et al. showed, and if psychologists are scientists, then how much confidence can we have in the pronouncements of other scientists? While such flawed logic is easily refuted in reasoned debate, it might be preferable if scientists refrained from giving provocateurs the opportunity to raise this kind of question in the first place. We suggest that the scientific community should respond to both legitimate skepticism and politically motivated denial with a three-pronged approach.

First, legitimate public concern about a lack of transparency and questionable research practices must be met by ensuring that research lives up to rigorous standards. We endorse most current efforts in this regard, and one of us (Lewandowsky) is a member of a relevant initiative involving the use of peer review to facilitate openness.

Second, we believe that daylight is the best protection against politically motivated maneuverings to undermine science. The first part of this article is one effort toward such transparency.

Finally, skeptical members of the public must be given the opportunity to engage in scientific debate. We have shown how two of the present authors—an academic and a member of the public who had been to three evening classes before his skepticism was aroused—teamed up to critique a widely cited finding and showed it to be unsupportable. None of their activities fell within the strategies and techniques of denial that we reviewed at the outset, clarifying that denial is not an "avenue of last resort" for members of the public who are desperate to contribute to science or even correct it but rather a politically motivated effort to undermine science.

References

Abt, C.C. 1983. The Anti-Smoking Industry (Philip Morris internal report). September. Available online at http://legacy.library.ucsf.edu/tid/vob81f00. Accessed May 6, 2012.

Alberts, B., R.J. Cicerone, S.E. Fienberg, et al. 2015. Self-correction in science at work. *Science* 348: 1420–1422. doi: 10.1126/science.aab3847.

Baba, A., D.M. Cook, T.O. McGarity, et al. 2005. Legislating "sound science": The role of the tobacco industry. *American Journal of Public Health* 95: S20–S27. doi: 10.2105/ AJPH.2004.050963.

Bailes, M., S. Bates, V. Bhalerao, et al. 2011. Transformation of a star into a planet in a millisecond pulsar binary. *Science* 333: 1717–1720.

Briones, R., X. Nan, K. Madden, et al. 2012. When vaccines go viral: An analysis of HPV vaccine coverage on YouTube. *Health Communication* 27: 478–485. doi: 10 .1080/10410236.2011.610258.

Brown, N.J.L., A.D. Sokal, and H.L. Friedman. 2013. The complex dynamics of wishful thinking: The critical positivity ratio. *American Psychologist* 68: 801–813. doi: 10 .1037/a0032850.

Cook, J., D. Nuccitelli, S.A. Green, et al. 2013. Quantifying the consensus on anthropogenic global warming in the scientific literature. *Environmental Research Letters* 8: 024024. doi: 10.1088/1748-9326/8/2/024024.

Diethelm, P., and M. McKee. 2009. Denialism: What is it and how should scientists respond? *European Journal of Public Health* : 2–4. doi: 10.1093/eurpub/ckn139.

Fisher, N.S., K. Beaugelin-Seiller, T.G. Hinton, et al. 2013. Evaluation of radiation doses and associated risk from the Fukushima nuclear accident to marine biota and human consumers of seafood. *Proceedings of the National Academy of Sciences* 110: 10670–10675. doi: 10.1073/pnas.1221834110.

Fredrickson, B.L., and M.F. Losada. 2005. Positive affect and the complex dynamics of human flourishing. *American Psychologist* 60: 678–686. doi: 10.1037/0003-066X .60.7.678.

Gilbert, N. 2013. Case studies: A hard look at GM crops. *Nature* 497: 24–26. doi: 10 .1038/497024a.

Hume, M. 2015. Canadian researcher targeted by hate campaign over Fukushima findings. *The Globe and Mail.* Available online at http://www.theglobeandmail.com/news/british -columbia/canadian-researcher-targeted-by-hate-campaign-over-fukushima-findings/ article27060613/.

Inhofe, J. 2012. *The Greatest Hoax: How the Global Warming Conspiracy Threatens Your Future.* Washington, DC: WND Books.

Landman, A., and S.A. Glantz. 2009. Tobacco industry efforts to undermine policy-relevant research. *American Journal of Public Health* 99: 45–58. doi: 10.2105/AJPH .2004.050963.

Lean, G. 2016. UK flooding: How a Yorkshire town worked with nature to stay dry. *The Independent.* Available online at http://www.independent.co.uk/news/uk/home -news/uk-flooding-how-a-yorkshire-flood-blackspot-worked-with-nature-to-stay-dry -a6794286.html.

Lewandowsky, S., J. Cook, K. Oberauer, et al. 2015. Recurrent fury: Conspiratorial discourse in the blogosphere triggered by research on the role of conspiracist ideation in climate denial. *Journal of Social and Political Psychology* 3: 142–178. doi: 10.5964/ jspp.v3i1.443.

Lewandowsky, S., G.E. Gignac, and K. Oberauer. 2013. The role of conspiracist ideation and worldviews in predicting rejection of science. *PLoS ONE* 8: e75637. doi: 10.1371/journal.pone.0075637.

Lewandowsky, S., J.S. Risbey, M. Smithson, et al. 2014. Scientific uncertainty and climate change: Part I. Uncertainty and unabated emissions. *Climatic Change* 124: 21–37. doi: 10.1007/s10584-014-1082-7.

Lewandowsky, S., W.G.K. Stritzke, K. Oberauer, et al. 2009. Misinformation and the war on terror: When memory turns fiction into fact. In W.G.K. Stritzke, S. Lewandowsky, D. Denemark, et al. (Eds.), *Terrorism and Torture: An Interdisciplinary Perspective* (pp. 179–203). Cambridge, UK: Cambridge University Press.

Mann, M.E. 2012. *The Hockey Stick and the Climate Wars: Dispatches from the Front Lines.* New York: Columbia University Press.

Markandya, A., and P. Wilkinson. 2007. Energy and health 2: Electricity generation and health. *The Lancet* 370: 979–990. doi: 10.1016/S0140-6736(07)61253-7.

Pew Research Center. 2015. Public Esteem for U.S. Military Highest, Scientific Achievements Second in Global Comparison. Available online at http://www.pewinternet.org/2015/01/29/public-and-scientists-views-on-science-and-society/.

Proctor, R.N. 2011. *Golden Holocaust: Origins of the Cigarette Catastrophe and the Case for Abolition.* Berkeley, CA: University of California Press.

Whatmore, S.J., and C. Landström. 2011. Flood apprentices: An exercise in making things public. *Economy and Society* 40: 582–610. doi: 10.1080/03085147.2011.602540.

CHAPTER 13

Clear Thinking about Conspiracy Theories in Troubled Times

Joseph E. Uscinski

Vol. 45, No. 1
January/February 2021

This article is based on a Skeptical Inquirer Presents live online presentation on July 30, 2020.

I'm going to discuss the latest polls, particularly those about COVID-19 conspiracy theories. I'm going to consider why these theories are popular or not, and then I'm going to go meta. I don't think I need to spend that much space addressing this audience about why conspiracy theories might be dangerous or why we should rely on authoritative information rather than on poorly sourced information.

I'll describe what's wrong with the popular discussions of conspiracy theories. What are the things that the journalists and perhaps some scholars are getting wrong? And I'm going to answer what I think is one of the more important questions nowadays: What are the dangers of believing the wrong things about beliefs in the wrong things?

To start, a conspiracy theory is an accusatory perception in which a small group of powerful people is acting in secret for their own benefit against the common good—and in a way that undermines our bedrock ground rules against the widespread use of force and fraud. In addition, this theory hasn't been found to be true by the appropriate experts, using data and evidence that is available for anyone to refute.

There are numerous conspiracy theories about COVID-19 out there. There are, in fact, too many to debunk them all. We've seen a deluge, whether they're about Bill Gates or George Soros being behind COVID-19 or that 5G technology is spreading the virus further. Some people have even responded by burning down cell towers. Some think that when a vaccine is released, we're going to be microchipped and tracked by the government. Some think that Big Pharma is behind the COVID-19 "scam," and they're going to make money by selling us a phony vaccine for a phony disease. And others are saying that doctors and hospitals are faking patients to make money.

So there are a lot of weird theories out there. Luckily, not one of these is believed that widely.

Most COVID-19 conspiracy theories fall into two broad categories, which I've been polling on since March. The first is that the disease has been exaggerated for political gain, usually to hurt President Donald Trump in an election year. The other is that it is some sort of bioweapon created or spread on purpose. When we poll on the idea that COVID-19 is being exaggerated to hurt Trump, we get about 29 percent of Americans agreeing. We get a similar number, 31 percent, agreeing with the idea that it's some form of bioweapon.

Some of the most-fringe ideas don't convince as many people, but these two ideas do each get about a third of Americans buying in. How come? Why do people believe conspiracy theories? A more general question is why do people believe anything really? And the answer to that question is, well, there are a lot of reasons. What's important to know is that there's not going to be any single reason people might believe in conspiracy theories, because there are so many conspiracy theories out there, all with their own idiosyncratic reasons people would come to believe them.

So *conspiracy theory* is a big bucket, and there could be a lot of reasons somebody could believe in any specific theory. I've also been polling on other ideas that have to do with COVID-19. And conspiracy theories aren't the only ones that are dangerous. For example, in my most recent poll of Americans, we got almost 15 percent believing that people who are right with God won't be injured by the coronavirus; almost 30 percent believed that prayer will protect them from COVID-19. Those ideas can be just as dangerous as believing that the disease is exaggerated.

So why do people believe conspiracy theories? In my research, the big explanatory factors are the latent dispositions that people have. The two I focus on are conspiracy thinking and denialism. Conspiracy thinking exists on a continuum; we all have it to one degree or another. Some people have it very strongly; others have it far less. But most people are somewhere in the middle. People who have very high levels of conspiracy thinking tend to think that everything is the product of a shadowy conspiracy. The people on the low end tend to be resistant to conspiracy theories.

Denialism works in a similar way. Some people have an antagonistic relationship with authoritative sources of information. When they hear something on the news, from scientists, or from government agencies, they say, "Well, I'm just not going to believe it." All of us have this to some degree in that we resist information we don't want to hear. But the people with elevated levels *just don't believe things*, whether it's from the news, government scientists, or other authoritative sources.

Another factor that drives conspiracy beliefs is our group attachments. The group attachment I am most concerned with is our partisan political attachments. Whether a Republican, Democrat, or something else, we engage in motivated reasoning. We believe in information from groups that we trust. When our group wins, we say, "That's how it should be, because our group is righteous and just." But when our group loses, we sometimes say we were cheated; *the other side* engaged in illicit practices, and that's why we lost.

Furthermore, we take our cues from the leaders we trust. If you follow the president and he engages in conspiracy theories, then you'll be more likely to believe those conspiracies. The next factor is information. If we're told information from sources that we trust that there is a conspiracy afoot, then we'll be likely to believe it. The final factor is that we can imagine our own conspiracy theories and make them up ourselves. We don't need someone else to share them with us. If we have high levels of conspiracy thinking, then it's not hard to imagine that everything we encounter during the day is part of some conspiracy.

So those are the broad reasons. We can put it into a very simple conceptual model: As information comes into our brains during the day, that information will be laid over the set of dispositions that we carry with us. And that information will be interpreted by those dispositions, which will then inform our particular beliefs about the world.

A person with elevated levels of conspiracy thinking will interpret the same information very differently from a person who has lower levels. The same information can lead two people to very different conclusions about the world. It is therefore our dispositions that divide us. To varying degrees, all of us have these dispositions operating within us. I imagine that most in this audience have low levels of conspiracy thinking. But for the mass public, this is a powerful force.

In sum, people believe these things for a lot of reasons. We have our motivated reasoning at play. We have conspiracy thinking at play. Those drive us to either engage with conspiracy theories or not and then to engage with particular conspiracy theories.

Why should we care if people believe in these theories?

If our beliefs are disconnected from our shared reality, then those beliefs are potentially harmful. If you believe that COVID-19 is a hoax, then you're not going to engage in best practices, such as frequent hand washing, mask wearing, or social distancing; you run the risk of further spreading the disease. But downstream from that, when people start to believe in a lot of conspiracy theories, it can make them distrust our institutions and our scientists, and that can lead them to believe in yet more conspiracy theories and detach themselves further from our shared reality. There are real reasons we need to fight against these beliefs.

But we need to get the causal locus right. Right now, a lot of the discussions about conspiracy theories get it wrong. With even the best intentions, journalists are saying the wrong things. The best-intentioned legislators believe the wrong things, and they may act the wrong way. And if they don't act the right way, then they could very well injure our rights. They could wind up censoring social media. And really, they may not have any impact on conspiracy theories at all.

What if it were the case that the people in Congress who want to legislate social media to tamp down on conspiracy theories are the people who are actually spreading the most conspiracy theories? For example, if you go back a few months, there was a hearing with tech moguls in front of Congress, with legislators complaining that there are so many conspiracy theories online. They asked: What happens if politicians want to share misinformation on Facebook? The question I had was, if it is indeed political elites who are sharing these ideas, is it really the fault of Facebook? Why can't Congress and the parties police their own if they are so concerned? Politicians would spread misinformation anyway without social media, and it would be spreading because of the people in Congress and the White House. We need to do things to dissuade politicians from sharing conspiracy theories, but social media may not be the problem.

We all have a role to play, but we should understand that the methods in which misinformation gets shared are still the old-style methods. And motivated reasoning and dispositions lead people to their beliefs. The internet may not be as much of a spreader as we think. To explain why, let's talk about some of those misconceptions a lot of people have. Much of the reason for regulating social media would be because people are supposedly becoming more conspiratorial now than they were in the past and that that effect is due to social media.

Here's the empirical question: Are Americans believing more conspiracy theories now than in the past? The headlines would make you think that they are. We have newspaper headlines saying we're now in the golden age of conspiracy theories. I would forgive you if you thought that were true. The problem is, if you go look through the headlines for the past sixty years, you will find journalists saying this almost every year, and it can't always be true, or else we would have fallen off the conspiracy cliff by now. But we haven't.

What does the data say? First, beliefs in many conspiracy theories have not increased. Belief in Birtherism, for example, has been flat. Belief in JFK-assassination conspiracy theories have almost been cut in half from where they were in the 1970s. In fact, those beliefs were almost 80 percent for decades. And it's only been since the introduction of the internet that they've come down. In my latest poll in March, they were around 44 percent. That's perhaps the first time they haven't been a majority belief in decades.

When we poll on conspiracy thinking over time, we find that it hasn't gone up. I've been polling this since 2012, and we're not finding that Americans are becoming more conspiratorial than in the past. Nor are we finding that they're believing in conspiracy theories more than they have in the past; the data just doesn't show that, at least thus far. If you have the impression that social media are turning everyone into raving conspiracy theorists, rest assured, we're not there yet.

I think the best example I could give is the QAnon conspiracy theory, a conspiracy theory that is rather fringe. And it's extreme, too, in the sense that the beliefs are extreme. It proclaims that President Trump is fighting the deep state, which is composed of satanic pedophiles and sex traffickers. A lot of the reporting lately claims that more people are believing this conspiracy theory now. Yes, it's true that some people who believe it are running for Congress and maybe one or two of them might win. But it's not getting bigger.

If you look at the headlines, QAnon is scary. It's big and getting bigger, and these beliefs are thriving on Facebook. Well, scary, yes, but big, relatively speaking? No. What do the data say? After a month of prolonged media coverage in 2018, we polled on it in Florida. We asked people what they thought about QAnon on a scale of zero to 100. And on average, people rated it poorly, at about a 24. To put that in comparison, we also asked about Fidel Castro, and if you know anything about Florida, you know Floridians don't like Castro. QAnon came out only a few points better than Castro—not well liked at all.

I've repeated these polls in the past few months, both in Florida and nationwide. QAnon has not increased in popularity. Other polls that ask about it find that most Americans still don't know what it is. And there are only about 6 percent in some polls who say that they support it or agree with it. It's not that big; its support is deeper than it is wide.

What is the role of the internet? We hear a lot that the internet spreads conspiracy theories. First, we have to be careful with the word *spreads*. When we say it is spreading, does that mean that it's changing minds? Or do we mean that it's just able to be accessed in other parts of the world? Because if we mean the latter, then, yes, obviously we can access things on the internet that we couldn't before. But is it changing minds? That's a very different matter.

Beliefs in conspiracy theories haven't gone up in recent decades. The forces that drive conspiracy theorizing exist regardless of the internet. And it may very well be the case that in previous decades or centuries, conspiracy theories either spread faster or had worse consequences. If you go back in this country 400 years, we were drowning and crushing "witches" for conspiring with Satan. There was an Illuminati panic 200 years ago. Shortly after that, there was a Freemason freak-out in the 1830s and 1840s. There were two Red Scares in the past century. It's not clear that people were immune to conspiracy theorizing in the good old days. Obviously, they exist today, but it's not clear that they exist more or have more of an impact now. And, even if they did, it's not clear that it would be due to the internet.

Another thing we must think about is that we have libraries that we carry with us everywhere in our pockets. Now accessible through our phone, we have the world's knowledge available to us at the touch of a button. But we seem to think that on the internet, only the conspiracy theories have an influence on us and that somehow when we go to the internet, it's only a swamp of conspiracy theories. That's just not true.

Furthermore, there's a hundred years of media-effects research that shows that news, campaigns, and political advertisements just don't have that much of an effect on people as we commonly think that they do. In fact, the more recent studies show that the net impact of campaigns is almost zero. I imagine many people, including this audience, are not going to be affected by campaign communications this year. You knew who you're going to vote for a long time ago. And that's the case in many elections. Many of our political choices are made long before we even know who the candidates are, because our dispositions, which are longstanding, drive this stuff. When we go on to the internet, we're not just lemmings who are getting tossed one direction or the other by different sets of information that changes our mind back and forth. We're picking and choosing. And we tend to seek out things that we already agree with because it makes us feel good; we don't change our minds very often. At the end of the day, it doesn't really matter what's on the internet. We still have to choose to access it and then we still have to accept it. It still has to comport with the things that we already believe. So, yes, it's true that there are problems with the information on the internet, and we need to clean it up. But it may not be quite as impactful as people say.

A lot of people thought that the 2016 election was decided by Russian bots or something like that. The studies coming out show that the impact just wasn't as big as some think. The fake news didn't have the impact that people originally thought it did.

Another claim that's popular is that Republicans and conservatives are more likely to believe in conspiracy theories. There are, of course, good reasons to believe this. You had a Republican Senator walk onto the Senate floor with a snowball and say because he was able to make a snowball climate change didn't exist. And you've got a fellow in the White House who makes all sorts of crazy conspiracy claims. I think the kookiest is that Ted Cruz's dad was behind the assassination of JFK. But here's the thing. Those are two elites, and they don't represent a lot of Republicans in the mass public.

What do the data say? First, many conspiracy theories are believed about equally by people on the left and the right. In my recent polls, I asked if Jeffrey Epstein was assassinated to cover up what he knows. I found near equal numbers of Republicans and Democrats answering affirmatively. Conspiracy theories about the JFK assassination are believed equally by people on the left and the right. Fringier conspiracy theories, whether it's the Freemasons or AIDS being created in a laboratory, are also believed equally by people on the left and the right. There are also conspiracy theories that are believed more by people on the left—whether it's ideas that Trump conspired with Russia or that the 1 percent and corporations control everything for some nefarious purpose. Those who believe those are more on the left than the right.

There are good reasons for this. The forces that drive people to believe in conspiracy theories, whether it's motivated reasoning or other mechanisms, operate on both the left and right. When we measure conspiracy thinking in the mass public, we find that it's near equal on the left and right.

Another claim we hear in the media is that conspiracy theories are for political extremists. Well, the fact is it depends on what we mean by *political* and what we mean by *extremist*. Some conspiracy theories are going to be believed by the "political extremist," strong partisans and strong conservatives, as long as that conspiracy is somehow attached to conservatism or it's being pushed by Republican and conservative elites, for example. We find the people who tend to be strong Republicans are very likely to believe in climate change conspiracy theories but only because they listen to what Republican elites tell them and because Republican elites keep saying that climate change is a hoax. But this is less about extremism and more about listening to party leaders. After every election, the losing side always thinks that the other side cheated. That has to do with motivated reasoning: no one likes to look in the mirror and say, "Well, gee, our ideas aren't that good or our candidate wasn't that good." Instead, it must be that the other side cheated. I also find this with COVID-19 being exaggerated, because the president and conservative media elites have said that COVID-19 is a deep state hoax or a Democrat hoax, or, as Rush Limbaugh said, Dr. Fauci isn't even a real doctor. Some personalities from Fox News were tweeting about how people should film the hospitals because there aren't real patients there. Well, that's going to drive conservatives and Republicans who are paying attention to believe in those theories.

But absent these theories having partisan or ideological content, cues, or circumstances, you're not going to find political extremists believing in them. In fact, you'll find people from both parties and independents buying in. There isn't really a strong ideological or partisan valence to Rothschild conspiracy theories or Freemason conspiracy theories or ideas about vaccines, GMOs, or the Holocaust being exaggerated because those don't really have much to do with mainstream politics—or Republicans or Democrats.

When we poll on QAnon, we find that very few people like it. But we also find that equal numbers of Republicans and Democrats claim to support it. And even though it's always called a far-right-wing conspiracy theory, there's nothing really *right wing* about it. There's nothing conservative about it. The people who support it come from all across the political spectrum. What binds those people together is a disdain for the political establishment and elevated levels of conspiracy thinking.

It's not like somebody got into researching George W. Bush and that led them down the path to reading Ronald Reagan's speech, and then they read Milton Friedman, and then all of a sudden it's *Satanic baby eaters*. It just doesn't work like that. If you are a strong Republican or you're a strong Democrat, you're ingrained in the political system. You're not going to believe a lot of these wacky conspiracy theories because you feel comfortable in your part of the system. People who feel disconnected from the political establishment are going to buy into these wacky conspiracy theories.

We tend to think that conspiracy theorizing is an us-versus-them dynamic, that we're the rational ones and everyone else is a conspiracy kook out there. Well it's not an us-versus-them. There are an infinite number of conspiracy theories out there. On any given poll, I can't ask about all of them because there are just too many. But what we find is this: the more conspiracy theories we ask about on any given survey, the fewer people we find believing in none of them. In March, I asked about twenty-three conspiracy theories; I had 91 percent believing at least one. Imagine if I was to ask about fifty or 100 conspiracy theories. I would probably have everyone buying into at least one if not a few.

Again, this is just part of the human condition, and we're all going to fall victim from time to time to a conspiracy theory. There's nothing really wrong or pathological about it. But we do want to make sure that our beliefs are tethered to evidence.

The second-to-final misconception I want to address is: Are these beliefs just an attempt by people to find a big cause for a big event? The answer is not really; this is just an optical illusion. There are conspiracy theories about big events, such as 9/11, the Kennedy assassination, and COVID-19. But all events attract conspiracy theories in varying degrees. There are conspiracy theories about everything, big and small.

What we need to think about is the fact that "big event" is a subjective idea to everybody, just as "big cause" is. Even if we all were looking to attach big causes to big events, it doesn't mean we'd get to any particular conspiracy theory about any particular event. For example, Kennedy-assassination conspiracy theories had 80 percent of people believing them for decades. But far more people believe in Kennedy conspiracy theories than in the 9/11 conspiracy theories. Was Kennedy really bigger than 9/11? Three times bigger? Or the moon landing conspiracy theories? When we poll on them, we get about 5 percent belief. And deep state conspiracy theories, when I polled on that in March, we got 50 percent believing in a deep state. What event is that about that people are trying to explain? Or when people say aliens landed in Roswell and it's being covered up by the government? Was that really a big event that somebody found tinfoil and sticks in the desert? And now 30 percent of Americans believe it. That didn't seem like that big of an event. And what are the events that are driving GMO conspiracy theories or vaccine conspiracy theories? I go back to the idea that it's really our dispositions that drive us to these beliefs and not some sort of search for a particular type of cause or explanation.

COVID-19 conspiracy theories are new. It does seem that these ideas are dangerous and scary because so many of them are wacky. But here's the thing: I've heard so many conspiracy theories that I'm just bored by all of them. Just as people say that COVID-19 is a bioweapon, people were saying Zika virus was a bioweapon, the swine flu was a bioweapon, and AIDS was a bioweapon. Every new disease is a bioweapon. It's the same theory using different nouns. Now people are saying that Bill Gates is behind it. But before him it was George Soros and the Koch brothers. Before them, it was the Rothschilds, the Freemasons, or the Kennedys. It's always some famous, rich person who's behind everything. There's really nothing new there. There is much more continuity than change in the conspiracy theorizing that Americans do.

Indeed, conspiracy theories can be very troublesome. We should work to keep our beliefs tethered to the truth and the best evidence as much as we can, because bad actions can spring from bad beliefs. This is especially true during a pandemic. We need to make sure that we're following the World Health Organization and the CDC and not Todd from Twitter. With that said, there are reasons to be somewhat hopeful and have some faith in humanity. There is more continuity than change. Our believing in conspiracy theories is nothing new, and it's not necessarily worse. We shouldn't be blaming old human problems on new technologies. We should be blaming ourselves! If anything, we should try to steer believers toward better beliefs with sympathy and empathy. But the mechanisms that lead to conspiracy theories are longstanding—and they're just part of being human.

Joseph Uscinski is associate professor of political science at the University of Miami, College of Arts & Sciences. He studies conspiracy theories, public opinion, and mass media. He is coauthor of *American Conspiracy Theories* (Oxford, 2014) and editor of *Conspiracy Theories and the People Who Believe Them* (Oxford, 2018).

CHAPTER 14

Life, the Quniverse, and Everything (Part I)

Stephanie Kemmerer

Vol. 45, No. 2
March/April 2021

This article was completed before the events of January 6 at the U.S. Capitol in which QAnon followers were among those leading the assault.

In the 1999 film *The Matrix*, the entire plot pivots on a choice that Neo must make. He must choose between a red pill and blue pill; the red pill will wake him up, and the blue pill will allow him to go back to life as he has always known it.

The film, with its clever mix of action and philosophy, is still as relevant today and may in fact be more relevant than when it premiered. Its pill metaphor has become inevitably enmeshed in the primordial stew of conspiratorial thinking and has been adopted by almost every fringe ideology.

Those "in the know" are considered the "red-pilled" or "woke," while the rest of us "blue-pilled sheeple" toil away in our lives, unaware or unwilling to accept the existence of the "Deep State." The blue-pilled stay asleep, while the red-pilled are schooled in the secret forbidden knowledge—that is somehow readily available for anyone to see online. Perhaps far more menacing than those who become red-pilled are those who give into "black-pilling," a sense of utter indoctrinated nihilism embraced by many mass shooters and incels (dangerous loners who often devolve to the point of violence).

Enter QAnon, the ultimate conspiracy—the mothership; the umbrella; the "Choose-Your-Own-Adventure" saucerful of secrets and codes. (To be clear, QAnon is not officially affiliated with the Guy Fawkes–mask-wearing hacktivists known as Anonymous.)

QAnon first appeared in October 2017 on the message board 4chan. This anonymous poster was not the first "Insider Anon" to attempt trolling the masses. There were several other Anon accounts, but QAnon stuck. "Q" claimed to have top secret Q Clearance, which is in fact an actual thing. Q began posting Nostradamus-esque riddles and codes and links, like an online scavenger hunt. Q's ultimate secret? That Hollywood, financial, and political elites were secretly controlling the world all the while engaging in Satanism, pedophilia, torture, and cannibalism to extract adrenachrome—a chemical said to prolong life—from children. This myth is not new and played a part in another topic I investigated.

While I was writing the three-part series on Nazi occultism for the podcast *Even the Podcast Is Afraid*, I dove into the origins of fascism prior to World War II. One of the most enduring and damning culprits that led to the Holocaust was a completely debunked, fictional—and plagiarized—book titled *The Protocols of the Meetings of the Learned Elders of Zion*, commonly known as *The Protocols of Zion*. This book emerged in 1903 and claimed to tell the "true" story of a meeting of a secret cabal of Jewish people who controlled the world.

The QAnon conspiracy takes heavily from this concept, and indeed at its core, almost every false conspiracy theory ends up somehow naming the evil culprits as those of Jewish descent.

Hollywood, financial, and political elites? It doesn't take a bulletin board filled with red strings and pushpins to decode what that means. Nor does it take much to understand the roots of the QAnon concept of Satanic cannibals who indulge in a sort of "blood sacrifice." This stems all the way back through the centuries to the Blood Libel Myth, which has somehow just been regurgitated through the years, being molded to fit the zeitgeist of any particular era.

Declaring one's enemies Satanists is an ancient trope as well. It would be difficult for anyone to support such evil. In actuality, true Satanism, as defined by Anton LaVey, the founder of the Church of Satan, is nothing more than atheism spiced up with a bit of mysterious "magick" and sexual ambiguity. A careful reading of LaVey's *The Satanic Bible* shows it is nothing more than a misinterpretation of the made-up text *Enochian Aethyrs*, with certain "angelic" words changed to "demonic" ones. Because most people who buy into Satanic Panic are oblivious to the historical, etymological, and mythological roots of this concept, Satan is just the ultimate bad guy stripped completely of his classical antihero roots and cast off into the darkness.

Clever too is the idea of adding children into this mix. While Satan represents the most evil force ever imagined, children represent the most innocent and fragile of humans. Maybe you can get away with giving Satan a pass, but how could anyone, ever, possibly condone the torture, rape, and murder of children?

Using the ultimate evil to inflict damage against the ultimate good is the most fallacious, conspiratorial basis for any misguided belief system. But it works, time and time again. And it persists like the stains on Lady MacBeth's hands.

QAnon is also a stale leftover slice of "PizzaGate," which emerged right before the 2016 election. It seemed to fade, and then it came back with a vengeance and morphed into QAnon and later on "WayfairGate," which emerged in early 2020 when people began speculating that expensive industrial cabinets with human-sounding names were really children being sold online.

The absurdity of QAnon might not have spread as far if it had remained where it had started. The Chan message boards are not for the casual internet user, so at the start, QAnon was not widespread or easily accessible. The problem began when the Chan users began sharing "Q Drops" (posts) on mainstream social media platforms. This made the Q "crumbs" easy for widespread "baking." (Q's codes are called "crumbs" or "breadcrumbs," and followers are encouraged to "bake the crumbs" into whatever nonsense they choose.)

There are even QAnon aggregator sites such as QAnon.pub, which contain all the Q Drops ever posted. While these aggregator sites usually attract QAnon followers, they are also utilized by reporters and podcasters for legitimate research.

Another cryptic and biblical tenet of QAnon is their belief in "The Storm." "The Storm Is Upon Us" is one of their catchphrases. So much energy has been put into the importance of this storm, yet no one—not even Q, who may in fact be several people—has defined what exactly "The Storm" even is. Anything can be "The Storm," and a lot of time and energy has been put into defining what it is. QAnon followers, once they have received their "red-pill" indoctrination, declare that they have experienced "The Great Awakening."

QAnon plays into the fantastical notions and concepts of heroic identity. Disgraced (and recently pardoned) General Michael Flynn, made a dog whistler reference to QAnon in one of his speeches, saying, "We have an Army of Digital Soldiers." QAnon followers rejoiced at this and adopted the moniker "Digital Soldiers." This is their ultimate fantasy; their ultimate LARP (Live Action Role Playing). QAnon is a never-ending LARP; these "digital soldiers" brave the battlefields of the internet with memes and hashtags as their weapons. ("Meme War" is another term used by QAnon followers. Nothing about QAnon is decidedly new or original, and even their main catchphrase, "Where We Go One, We Go All," was borrowed from a line in the film *White Squall*.)

Perhaps the most unapologetically comical aspect of the QAnon conspiracy is the "Jesus" figure; the force for ultimate good: Donald Trump, the man who in only four years reduced every core value of American democracy into a nonstop rally of contrarian beliefs, dog whistles to neofascist politics, and absolute pandemonium. Setting aside all the other ridiculous plot twists of QAnon, Donald Trump as the hero lies somewhere in between hyperbole and sheer lunacy. But this characterization of Trump as the savior has led to the spread of Q. While not every Trump supporter is a follower of QAnon, every QAnon follower is a supporter of Trump.

At its core, QAnon represents one of the most extreme examples of fundamentalist, dominionist Christian, right-wing belief systems. According to an internal FBI memo, QAnon—like many right-wing conspiracy organizations—has been classified as a potentially terroristic threat, and the conspiracists are often listed as hate groups by the Southern Poverty Law Center. Former "Moonie" cult member, deprogrammer, and author Steven Hassan has classified QAnon as a cult on his Freedom of Mind organization website. (His book *The Cult of Trump* is highly recommended. I suggest the Audible version, because it includes actual clips of Trump speaking to illustrate Hassan's points.)

QApostates

The most dangerous aspects of the QAnon conspiracy cult are its erosive properties. While they have hijacked and hidden behind hashtags such as #SaveThe-Children, QAnon has actually been the cause of destroying lives and tearing families apart. On the SubReddit "QAnon Casualties," innumerable posters discuss how either their former belief in QAnon or a loved one's current belief has led to everything from the loss of a job to divorce. Families have been torn apart by a cultic belief that claims to want to "save the children."

Jitarth Jadeja is perhaps the most well-known former QAnon follower and was in fact catapulted to a sort of reluctant fame due to his postings on this SubReddit. Jadeja, who currently resides in Australia, says he was slowly drawn into the dark world of QAnon after he experienced what can only be described as a sense of collective disillusionment and disbelief after Trump won the 2016 election. Jadeja began to question how such a thing could happen and began to seek alternative views of reality, because Trump's win represented the antithesis of reality as he had known it. He found himself gravitating toward Alex Jones, and the slippery slope finally landed him right into the gaping maw of QAnon.

I interviewed Jadeja by telephone. The further he slipped into the "rabbit hole," the more he became withdrawn from the real world. "Everyone who falls into QAnon is miserable," Jadeja says. Oftentimes these people already have a mental or emotional instability, and they likely possess some form of social dysfunction as well as low self-esteem. But Q is there to pick them up: "QAnon tells them they can be a hero. It's about being a hero and getting credit for it," adds Jadeja.

Despite the cries about saving the "Mole Children" being kept as sex slaves in underground tunnels and their fury at some imagined "Deep State," QAnon is ultimately about one thing: "It's not about kids. It's not about the cabal. It's about being heroes," Jadeja says. Jadeja spent quite some time in the Q Hole, as a follower from December 2017 until June 2019.

The tipping point came for him after a Q post predicted that Trump would use the phrase "Tippy top." (Q Drops that turn out to be "true" are called Q Proofs; sometimes the drops themselves are meant as a proof, such as the "Tippy

Top" post.) Jadeja discovered that while Trump did use the phrase as predicted, he also found that it is a phrase Trump has used repeatedly in the past. That was the moment Jadeja realized he had been lied to. It was a moment of harsh self-realization.

From his experiences, looking back, he was able to see all of the clues that it was fake, clues that he could not see at the time he was enmeshed in it. QAnon is all about the misuse of information and a clever PsyOp of disinformation. (PsyOp is short for Psychological Operations, which is used by the military and defense agencies. It is the usage of tactical psychology, akin to some of the concepts of Sun Tzu's *The Art of War*.) The way out of QAnon is through facts and knowledge. "The more people know about it, the more likely they are to leave it," Jadeja told me. The QAnon crowd misuses information to suit their needs, while disregarding actual information. "People use data the way a drunk uses a lamp post—for support rather than illumination."

Jadeja was thrust into the world of the media after his recovery, when he began posting his story in the SubReddit. He began receiving requests for interviews and at first declined. He later realized that his story of recovery could possibly help others find a similar path. While Jadeja is by no means the only outspoken former QAnon follower, he says, "I'm the most visible."

Conspiracy culture received a gift in the form of the coronavirus pandemic. With more people stuck at home with more time on their hands (mixed with dangerous algorithms used on social media platforms and YouTube), more people began sinking into the darkness. The pandemic itself became the center of many conspiracies, with many referring to it as a hoax or a "Plandemic."

One such casualty of the pandemic is Leila Hay from Northern England. She fell into QAnon right around the time the lockdowns began and emerged after only a short period of about six months. I was also able to interview her.

"I got out right before the point where I couldn't get out," says Hay, adding, "I have a network that's been very helpful to me." Hay points out two aspects of her personality that seem to apply to many QAnon followers: "I'm very obsessive. I'm also quite vulnerable." Her entry was by way of the algorithms used on YouTube, which has since wiped a lot of QAnon content from its platform. "It's a toxic environment; there's no real intelligence behind it."

Hay's escape from the Q Hole came by way of Twitter. She came across a page for a podcast titled *QAnon Anonymous*. Thinking it was a pro-QAnon podcast, she found herself drawn into the posts. She soon realized it was a decidedly anti-QAnon podcast, but she kept reading, emerging once more into the light.

The *QAnon Anonymous* podcast is inarguably one of the biggest thorns in the side of QAnon and Q "Adjacent" followers. The three hosts—Travis View, Julien Feeld, and Jake Rockatansky—present easily digestible and educated facts that debunk QAnon and other conspiracies. The episodes are well-researched, well-written, and often both frightening and hilarious at the same time. The hosts have even gone undercover to several QAnon rallies and conferences, most recently to one of the Arizona "Stop the Steal" post-election protests, where they encountered the "Frog Man" himself, Alex Jones.

Rockatansky expressed joy in hearing how they helped Hay escape the cult of Q. Diving so deeply into such a dark topic has its downsides for all three of the hosts. "We've had to figure out ways to keep ourselves grounded and not get taken along for the ride." He says the pandemic definitely played a very big role in the proliferation of Q but reiterates the importance of the 2016 election as well. "The whole world, their minds were broken by the event, and I think a lot of people were sure that one thing was going to happen and when it didn't happen, what they understand about the world was ripped away from them."

"There's a corporate interest in driving division; in not being able to empathize with the people you're instructed to hate," he says. There is no way around it. America has become more divided in the past four years, and at the helm of this divisive ship is Trump, steering us straight toward an iceberg at full speed.

It wasn't just Fox News that gave Trump his popularity. Rockatansky points to mainstream media as well, which have been on "a four-year 'hate tour' to make up for the fact that they kinda created this guy." He says, "Our brains have been collectively cooked by [the] media, especially post–Cold War American films from the '80s and '90s. People are still trying to fit reality into archetypes that were created then."

Rockatansky stresses that one of the most unique aspects of QAnon is its real-time malleability and ability to adjust to events as they occur. "I think they know that battle can never be won because it's not real. . . . You always get to be angry; you always get to be fighting. It's a comfortable place. . . . The fun of it is that it gets to keep going. There's never going to be any real answers. Once you have answers, you're never going to be satisfied with them."

Now, in the aftermath of the 2020 election, with the outcome being decidedly what Q did not predict, QAnon followers are in a state of doubled-down denial, sentiments that are disturbingly being echoed by other elected officials and newly elected officials such as Marjorie Taylor Greene, who will soon be our most "red-pilled" member of the House of Representatives.

Some QAnon followers have made threats against their own lives or those of their families if Trump does not maintain his stranglehold on the presidency. Despite the new conspiracies that arose in the wake of Joe Biden's historic win and Trump's refusal to concede, as of January 20, 2021, at noon, there will be a new president. But how far will QAnon followers go when that happens?

Fredrick Brennan (a.k.a. "Hot Wheels"), the original founder of the message board 8chan (now known as 8kun), has made some very frightening predictions of what may come along with the change of power on January 20.

Brennan is not only the original founder of 8chan; he also lived in the Philippines and worked alongside Jim Watkins. Watkins has been in the spotlight in the aftermath of the election, spouting conspiracies about the Dominion voting system. Watkins has a stake in these conspiracies. He has formed a Q Caucus and stands to profit from the proliferation of QAnon. Some speculate that Watkins may actually be Q. If he isn't, he has a direct line of communication to whoever is.

Brennan ultimately remanded ownership of 8chan to Watkins in 2015. (Q originally began posting on 4chan in October 2017 but migrated to 8chan on September 19, 2018.) 8chan ultimately turned into a cesspool of filth, with many users posting images of child pornography and racist rhetoric. 8chan had also included postings from persons who later went on to commit mass shootings.

Of QAnon followers, Brennan says, "They haven't had to reckon with the Biden administration, and that is the thing we were afraid would lead to violence. . . . I don't necessarily see bright times ahead at least for political violence. . . . When Trump is officially out of office, I just don't know where their narrative can go next."

Brennan points to this bizarre doubling-down and says, "As far as they're concerned there's going to be a second Trump administration. . . . I don't know how some of them move on" from Biden's win. He says there is a very real possibility that "some of them might be willing to commit violent acts." Of the QAnon followers who have been arrested in recent years, he says, "What we've seen are some of the dumb ones get[ting] caught."

Brennan has been unabashedly outspoken against Watkins and 8kun in recent years. His opinions on Watkins even led to a last-minute escape from the Philippines, where he had resided for several years. Brennan made a social media post that made reference to Watkins possibly having dementia. Due to the strict libel laws in the Philippines, Watkins (claiming to be a champion of "free speech") pressed charges.

Even a few hours in a jail cell, especially in the Philippines, would almost certainly lead to Brennan's demise. His nickname "Hot Wheels" refers to the fact that he is wheelchair bound. He is a very small man and was born with osteogenesis imperfecta ("brittle bone disease" such as depicted in the film *Unbreakable*). Knowing he would not survive jail, Brennan fled from the Philippines just in the nick of time.

Brennan is very knowledgeable and has been a staunch fighter of the Q cult. Brennan has had firsthand experience with Watkins and the tumult and controversy surrounding Watkins and his unapologetic grifts. Brennan was never a believer in QAnon, yet his journey shares similarities with those who have left the Q cult, and like the stories that stem from Q, his journey is still unfolding.

With QAnon, the battle is one that never ends and one that can never be won. It is a real-time epic fight between good and evil, with the goal posts constantly moving and the enemies ever increasing. The damage it causes to people can be so life-altering that there is no turning back, but as with Jadeja and Hay as proof, there is hope for recovery and a life out of the darkness—a life that is productive and filled with logic, science, and reason.

Stephanie Kemmerer is a researcher and writer for the podcast Even the Podcast Is Afraid and an occasional contributor for the Southern Oddities podcast, both owned by Ordis Studios (https://www.ordisstudios.com/). She currently resides in Bisbee, Arizona, and is interested in true crime, the paranormal, politics, and conspiracy culture. The podcasts are available on all streaming services. Her Twitter handle is @mcpasteface.

CHAPTER 15

Pizzagate and Beyond
USING SOCIAL RESEARCH TO UNDERSTAND CONSPIRACY LEGENDS

Jeffrey S. Debies-Carl

Vol. 41, No. 6
November/December 2017

It happened less than a year ago. On December 4, 2016, customers were sitting down for a Sunday afternoon meal in the Washington, D.C., pizzeria Comet Ping Pong. Known locally for its quirky atmosphere, live music, and of course its ping pong, on this day the restaurant would make national headlines. Shortly before 3 PM, a man walked in bearing an assault rifle. The man took aim in the direction of one employee, who quickly fled, before discharging his firearm. Law enforcement promptly responded to calls, and officers were able to take the man into custody without further incident. They found two firearms on the suspect and another in his vehicle. Fortunately no one was hurt, but the event has left many people shaken, and not only for the obvious reasons. The accused had apparently not intended to commit a mass shooting, nor had he intended to rob the restaurant. The truth, such as it is, turned out to be quite strange nonetheless.

The accused shooter was twenty-eight-year-old Edgar Maddison Welch, a father of two daughters and resident of Salisbury, North Carolina. After his arrest, he told police that he had made the 350-mile drive up to the capital to investigate claims regarding a conspiracy theory, circulating online, that quickly came to be called "Pizzagate" (Metropolitan Police Department 2016). According to this outlandish set of claims, leaked emails from Hillary Clinton's campaign manager, John Podesta, contained coded signs and messages revealing that Comet Ping Pong was actually a front for an occult, child sex slave ring involving the owner of the restaurant, James Alefantis, Podesta, and Clinton herself. For several days before the presidential election, claims of this sort proliferated across the Internet. Alefantis and his employees began receiving menacing messages via social media, including overt death threats (Kang 2016). The events seemed to reach a climax with Welch's misadventure. He subsequently told police that his intention was to investigate these claims in person and, if he found them to be true, rescue the children held captive there.

This story is admittedly bizarre in many ways, and learning more about the shooter's motivations does not seem to shed much light on it. Likewise, while others have documented the origins and spread of the groundless Pizzagate conspiracy theory (e.g., Kang 2016), neither does this mapping necessarily help us understand how something so ludicrous found traction in a surprisingly wide audience, nor why it would motivate anyone to investigate in person. These events—preposterous as they are—can be understood by applying well-established lessons from social research.

First, there is the peculiar nature of the conspiracy theory itself. Unlike many of the stories one might encounter in everyday life, stories of this sort can be understood as legends. According to folklorists, a legend is a type of story about supposed past events told as though it *might* be real: a "legend is a legend once it entertains debate about belief" (Dégh 2001, 97). Unlike a fable or literature, the events described are presented as possible, even if they are bizarre and not necessarily plausible. For example, one legend theme that used to be told frequently involves encounters with an exotic and dangerous animal in a typically safe and familiar place. People telling the legend usually claim that a friend-of-a-friend, or some other indirect acquaintance, had gone shopping for a carpet at a local department store (Brunvand 1981). He or she put their hand inside a rolled-up rug and felt a sudden, sharp pain. They had been bitten by a snake hiding in the carpet, an exotic species that had apparently been imported, by accident, along with the carpet from some far off, foreign land.

Two characteristics of the story indicate its status as a legend with little or no factual basis. First, no firsthand witnesses can ever be found by researchers. Second, multiple versions of it can be found with varying details. This is because legends constantly change to suit the narrator and the locale in which they are told—the type of shop, the animal, the protagonist—and they can circulate over the years, becoming associated with different people and places (e.g., Radford 2016). Both characteristics apply to the claims about Comet Ping Pong. No witnesses, victims, or perpetrators have come forward, and similar stories have been told about other places at other times. Similar allegations and threats occurred in 2015 in regard to a day care center in Salt Lake City (Peterson 2016), for example. Going further back into history, unfounded panic over alleged occult sexual abuse of children ran rampant during the 1980s and 1990s, and many places and people became the target of groundless accusations (Victor 1993). It appears that the pizzeria was just the latest target of a perennial fear.

Given the ambiguous nature of such stories, people rarely find it easy to determine their credibility. Rather, they must invest a degree of thought and emotional engagement into the narrative while they appraise its merits. One legend, for example, suggests that Martha Washington accidentally invented ice cream when she "left a bowl of cream outside one cold night for a neighborhood kitty and found it frozen solid in the morning" (Ellis 2009, 59). Is this legend true? No, but it sounds like it *could* be, and the central claim is compelling. Consequently, if listeners are engaged with the story, they will frequently seek out further information and participate in intense discussions with others over the legend and its claims. This was certainly the case with the Pizzagate theorizing and, as is the case with conspiracy theories, further information and discourse is likely to be found online and from sources as dubious as the original story. However, the abundance of sympathetic websites and the sheer number of credulous user posts dealing with the topic may appear to be, themselves, evidence that a credible claim has been made. As social psychological research has illustrated, "we determine what is correct by finding out what other people think is correct" (Cialdini 2009, 99).

Participation in legends is not necessarily limited to discussion alone. Sometimes the action they provoke manifests in a form called "legend-tripping" (Hall 1973). Inspired by a legend, a person may travel to the alleged site of the story to investigate its validity directly. In doing so, participants enter into the legend itself, acting out a part of it as one of its characters, and thereby "telling" its narrative through the process of ostension—through their behaviors rather than through words (Dégh and Vázsonyi 1983). This usually takes on a fairly innocuous form, such as when adolescents visit a reputedly haunted graveyard and reenact certain behaviors that legend claims will invoke the spirit, such as calling its name at midnight. However, this sort of legend-tripping is precisely what the shooter did as well, albeit much less innocently. Welch saw himself as the potential hero of the story—a rescuer of children. Instead, he put them at risk, since the only danger present was the danger that he brought with him. As one news headline correctly pointed out, "Fake News Brought Real Guns" (Kang and Goldman 2016). Similarly, legends are not simply stories about events that supposedly occurred in the past. They also serve as "maps for action" (Ellis 2003, 325). As such, they can tell more about the future than the past. Perhaps it should have been expected that the Internet threats against the pizzeria would eventually escalate into something much more serious once sufficient and sustained interest was aroused. The peak in Internet chatter before the election, in hindsight, was a likely warning.

Given the level of absurdity involved in this episode and the lack of anything approaching evidence to corroborate the claims made (LaCapria 2016), it might be reasonably expected that most people would soon realize there was never any truth to the story and move on after its exposure in the media. This, however, is not the nature of the legend process nor was it what happened in this case. Conspiracy theories in particular are notoriously resilient to criticism (Goertzel 2011). Many people remained convinced of Pizzagate and—as is typical with conspiracy theories—public disconfirmation only served to convince diehards of a cover-up in the works. To believers, it seems the media doth protest too much.

As frustrating as this stance may be to those wishing to falsify absurd theories, such a mindset is far from abnormal. Psychological research illustrates how it is difficult for most people to admit they were wrong when they have committed strongly to a belief or course of action. Leon Festinger and colleagues (1956) famously documented how members of a UFO cult doubled down on their belief system after their predicted apocalypse failed to show up on December 21, 1954. The group was saved the trouble (and the cognitive dissonance) of having to admit they were wrong when their prophet conveniently received a last-minute revelation from God via automatic writing. It turned out that the Almighty decided to postpone Armageddon thanks to the cult's faith and devotion. Alex Jones, the extreme right-wing radio show host and conspiracy theorist well-known for promoting claims that the September 11 attacks were hoaxed, inadvertently offered an example of this sort of revisionist postscript shortly after the media storm that followed Edgar Welch's misguided adventure into the pizzeria. In a video posted to the infowars.com website, Jones conceded that the story about a sex ring in the basement was "absurd" without going so far as to disavow it, then promptly suggested that it was a smoke screen used by the media to cover up the "real" revelations found within the Podesta Wikileaks (Jones 2016). By planting and subsequently debunking an absurd story, Jones claimed, the media makes all of the "real" and damning content of the emails seem false by association. This allows him and his devotees to step away from a debunked claim and simultaneously not have to admit they were wrong. All this, despite the fact that the allegedly "real" information in the email is no less absurd and no more substantiated than the sex ring claim (e.g., high level involvement in secret cults, black magic rituals, and so forth). Jones also conveniently overlooked the fact that he himself was one of the primary disseminators of the claims against Comet Ping Pong in the first place. According to his own logic, this must mean that he is actually part of the conspiracy he claims to oppose.

At first blush, the Pizzagate drama seemed so bizarre that it was beyond the bounds of comprehension. It is easy to discount those involved as mentally ill, unintelligent, or perhaps bright but manipulative hucksters. While tempting, doing so would misdiagnose conspiracy theorists, most of whom are mentally healthy individuals (Bost 2015). Moreover, this would result in a missed opportunity to gain a deeper understanding of the situation and others like it. After all, this story is far from unique in its outlandish claims. Whatever the truth may be about the person or persons who initiated the legend, the fact that they took root at all in a wider audience reveals something about them that may help us understand those who entertained the possibility of a Clinton-linked sex slave ring. Legends are generally false in a literal sense, but they also reveal deeper truths about those who tell them, reflecting their "hopes, fears, and anxieties" (Brunvand 1981, 2). Legends about finding a mouse's tail in a soda bottle may not be literally true, but they reveal real concerns about health and safety in industry. People who rank highly in conspiracy ideation also report high levels of support for democratic values and strongly negative attitudes toward authority (Swami et al. 2011). Pizzagate, as a conspiracy legend, reflected these concerns: fears over the trustworthiness of big government, big media, and elites that represent excessive authority and seem to threaten democratic values.

There is a less savory side of the concerns involved as well. A *Slate* article correctly suggested that the very characteristics of Comet Ping Pong that make local leftists love it are what made it a focus for the fears and concerns of the far Right. The place is a haven for artists, punks, gays, and other marginal groups: a tangible emblem of inclusivity, tolerance, and other progressive values that are threatening to the conspiracy-prone alt-Right (Cauterucci and Fischer 2016). Tellingly, the physical signs of these competing values are read differently by those who do not share them. For example, in a "mural of people and faces by an artist who's played the Comet stage, conspiracy theorists see a depiction of a child being strangled. In run-of-the-mill bathroom graffiti, they see secret sexual messages. In the lack of labeling for the gender-neutral bathrooms, haters with a political agenda see 'secret rooms'" (Cauterucci and Fischer 2016). In a previous era, ice cream parlors evoked a similar fear in some Anglo-Americans (Ellis 2009). Distrustful of the foreign, Italian immigrants who frequently owned the parlors, legend had it that young women risked a morally and physically dangerous slippery slope into drugs and forced prostitution if they visited them. The parallels are striking and troubling. A legend such as Pizzagate can only spread if the regressive values it reflects—nativism, racism, and xenophobia—are alive and well and resonate with a sympathetic audience. Strangely, it also indicates that these values may paradoxically be expressed by the same people who support democracy and anti-authoritarianism, odd bedfellows that may find common cause in populism (Panizza 2005).

Legends, just like fake news, can lead to real-world consequences. In addition, these outcomes can themselves reinvigorate the original legend and encourage its further transmission. Discussion of the Pizzagate claims led to action: online threats and an active shooter. These in turn sparked further debate on social media and in the mainstream media. Whether intentionally or not, this continued discussion may encourage further exploits. Hopefully lessons can be learned from all this. While reputations have been damaged, fortunately no one was physically harmed this time. But rumors continued to circulate over online blogs and videos, threatening comments continued to be posted to the Comet Ping Pong Facebook page, and the possibility for further disturbances inspired by dubious legends remains strong. Within days of the shooting, a fifty-seven-year-old woman named Lucy Richards was arrested for texting death threats to another woman who had lost a child in the Sandy Hook School shootings of 2012. According to the Department of Justice, Richards was convinced by conspiracy claims that the shootings were a hoax and, presumably, that the unnamed victim was somehow in on it (Boxley 2016). With an understanding of the social processes at work in these matters, it can be hoped that we will be better-prepared for the next outbreak of conspiracy-inspired legend-tripping.

Jeffrey Debies-Carl, PhD, is associate professor of sociology at the University of New Haven. His research examines the social significance of physical spaces and space-based behaviors and has appeared in various scholarly journals. He is the author of *Punk Rock and the Politics of Place* **(Routledge, 2014). He is not a part of the conspiracy—or so he claims.**

References

Bohn, Kevin, Daniel Allman, and Greg Clary. 2016. Gun-brandishing man sought to investigate fake news story site, police say. CNN (December 5). Available online at http://www.cnn.com/2016/12/04/politics/gunincidentfakenews/index.html.

Bost, Preston R. 2015. Crazy beliefs, sane believers: Toward a cognitive psychology of conspiracy ideation. *Skeptical Inquirer* 39(1). Available online at www.csicop.org/si/show/crazy_beliefs_sane_believers_toward_a_cognitive_psychology_of_conspiracy_id.

Boxley, Mark. 2016. Florida woman calls sandy hook massacre a "hoax," threatens to kill parent of victim, officials say. WFV 9 ABC (December 7). Available online at http://www.wftv.com/news/local/florida-woman-calls-sandy-hook-massacre-a-hoax-threatens-to-kill-parent-of-victim-officials-say/473918395.

Brunvand, Jan Harold. 1981. *The Vanishing Hitchhiker: American Urban Legends and Their Meanings*. New York: Norton.

Cauterucci, Christina, and Jonathan L. Fischer. 2016. Comet is D.C.'s weirdo pizza place. Maybe that's why it's a target. *Slate* (December 6). Available online at http://www.slate.com/blogs/outward/2016/12/06/comet_ping_pong_is_a_haven_for_weirdos_and_now_a_target.html.

Cialdini, Robert B. 2009. *Influence: Science and Practice*. New York: Pearson.

Dégh, Linda. 2001. *Legend and Belief: Dialectics of a Folklore Genre*. Bloomington, IN: Indiana University Press.

Dégh, Linda, and Andrew Vázsonyi. 1983. Does the word "dog" Bite? Ostensive action: A means of legend-telling. *Journal of Folklore Research* 20(1): 5–34.

Ellis, Bill. 2003. *Aliens, Ghosts, and Cults: Legends We Live*. Jackson, Mississippi: University Press of Mississippi.

———. 2009. Whispers in an ice cream parlor: Culinary tourism, contemporary legends, and the urban interzone. *Journal of American Folklore* 122(483): 53–74.

Festinger, Leon, Henry Riecken, and Stanley Schachter. 1956. *When Prophecy Fails: A Social and Psychological Study of a Modern Group that Predicted the Destruction of the World.* New York: Harper and Row.

Goertzel, Ted. 2011. The conspiracy meme. *Skeptical Inquirer* 35(1) (January/February): 28–37. Available online at http://www.csicop.org/si/show/the_conspiracy_meme.

Hall, Gary. 1973. The big tunnel: Legends and legend-telling. *Indiana Folklore* 6(2): 139–73.

Jones, Alex. 2016. Pizzagate is a diversion from the greater crimes in Podesta Wikileaks: Why not cover the hundreds of other dastardly deeds in the emails? *Infowars* (December 5). Available online at http://www.infowars.com/pizzagate-is-a-diversion-from-the-greater-crimes-in-podesta-wikileaks/.

Kang, Cecilia. 2016. Fake news onslaught targets pizzeria as nest of child-trafficking. *New York Times* (November 21). Available online at http://nyti.ms/2f0L9G9.

Kang, Cecelia, and Adam Goldman. 2016. In Washington pizzeria attack, fake news brought real guns. *New York Times* (December 5). Available online at http://nyti.ms/2h8nPmp.

LaCapria, Kim. 2016. Chuck E. Sleaze. *Snopes* (December 4). Available online at http://www.snopes.com/pizzagate-conspiracy.

Metropolitan Police Department. 2016. Arrest made in assault with a dangerous weapon (gun): 5000 block of connecticut avenue, northwest. December 5. Available online at https://mpdc.dc.gov/release/arrest-made-assault-dangerous-weapon-gun-5000-block-connecticut-avenue-northwest.

Panizza, Francisco (ed.). 2005. *Populism and the Mirror of Democracy.* New York: Verso.

Peterson, Eric. 2016. This Salt Lake City day care has become a magnet for conspiracy theories. *Vice* (February 23). Available online at http://www.vice.com/read/the-online-conspiracy-theories-about-a-salt-lake-city-daycare.

Radford, Benjamin. 2016. Mistaken memories of vampires: Pseudohistories of the chupacabra. *Skeptical Inquirer* 40(1) (January/February): 50–54.

Siddiqui, Faiz, and Susan Svrluga. 2016. N.C. man told police he went to D.C. pizzeria with gun to investigate conspiracy theory. *Washington Post* (December 5). Available online at https://www.washingtonpost.com/news/local/wp/2016/12/04/d-c-police-respond-to-report-of-a-man-with-a-gun-at-comet-ping-pong-restaurant/?utm_term=.a04d60bd78da.

Swami, Viren, Rebecca Coles, Stefan Steiger, et al. 2011. Conspiracy ideation in Britain and Austria: Evidence of monological belief system and associations between individual psychological differences and real-world and fictitious conspiracy theories. *British Journal of Psychology* 102(3): 443–63.

Victor, Jeffrey S. 1993. *Satanic Panic: The Creation of a Contemporary Legend.* Chicago: Open Court.

CHAPTER 16

The Scientific Frauds Underlying the False MMR Vaccine–Autism Link

Peter N. Steinmetz

Vol. 44, No. 6
November/December 2020

With the production and distribution of the film *Vaxxed* and its successor *Vaxxed 2*, plus the notorious anti-vaccination/conspiracy video *Plandemic*, it has again become fashionable in some anti-vaccination circles to maintain that vaccines are medically ill-advised, provide little benefit given their risks, and are possibly pushed by a big government–Big Pharma cabal for the primary purpose of optimizing profits.

Of course, there are valid grounds for concerns regarding civil liberties given proposed government-mandated vaccination programs in taxpayer-funded schools. But to bolster the cabal theory, many want to claim that the original report by Dr. Andrew Wakefield of an association between autism and MMR vaccination was correct. They believe the subsequent outright retraction of that paper and its labeling as a fraud (*Lancet* editors 2010) was the work of this cabal to discredit a badly victimized Wakefield.

While Wakefield's theory is almost completely discredited within the biomedical research community, the adherents of the cabal theory simply regard that as proof of the strength of the influence of Big Pharma funding on biomedical research. To help set the record straight, let's review the facts surrounding the frauds in Wakefield's 1998 paper. We will look as close to the primary sources as reasonably possible. That way, anyone can review these sources to make their own determinations regarding these frauds.

On February 28, 1998, Wakefield was the lead author on a paper in the British medical journal *Lancet* titled "Ileal-Lymphoid-Nodular Hyperplasia, Non-Specific Colitis, and Pervasive Developmental Disorder in Children" (Wakefield et al. 1998), which reported an association in twelve children between treatment with the combined measles, mumps, rubella (MMR) vaccine and subsequent development of colitis and autism.

For our purposes, *scientific fraud* (Norwegian National Research Ethics Committees N.d.), or scientific misconduct, will be defined per the U.S. Office of Research Integrity (N.d.) as "fabrication, falsification, or plagiarism in proposing, performing, or reviewing research, or in reporting research results." The ORI further defines "(a) Fabrication is making up data or results and recording or reporting them. (b) Falsification is manipulating research materials, equipment, or processes, or changing or omitting data or results such that the research is not accurately represented in the research record."

Six Fabrications and Falsifications

The primary fabrications and falsifications in the paper occur in five main areas. There is a sixth form of falsification in Wakefield's response (Wakefield 1998) to criticisms of the paper, which will be discussed later. The first three areas of falsification and fabrication concern the reporting of the scientific findings in the article.

FRAUD 1. FINDINGS OF NON-SPECIFIC COLITIS

The paper reported in its Table 1 that eleven of the twelve children examined had "non-specific colitis." This was apparently a phrase used by Wakefield in final revisions to summarize the results of the histopathological examination of the biopsies collected during ileocolonoscopy.

These slides were originally examined by the clinical pathologists at the Royal Free Hospital in London and were determined to be essentially normal (Deer 2010). Given this result, the research team decided to have the slides reexamined by medical school faculty. In this review, specific histological findings were scored on a 0–3 scale by Dr. A.P. Dhillon (Godlee 2011) along with a checkbox at the bottom for other findings, such as "non-specific" or "normal." In eleven of the twelve children, the "non-specific" box was checked for at least one biopsy site.

Evidently the checking of these boxes was then reported as "chronic non-specific colitis" by Wakefield in making final revisions to the paper (Deer 2010). The checkbox on the form filled out by Dhillon, however, may have simply meant that the findings on the slide were of uncertain significance.

When reviewed by two independent specialists in 2011, Geboes (2011) reported that "I see no convincing evidence of 'enterocolitis,' 'colitis,' [or a] 'unique disease process.'" Bjarnason (2011) reported that he and his colleagues "came to an overwhelming and uniform opinion that these reports do not show colitis."

The direction of each of the eleven errors is consistent in tending to overstate the association, and this is unlikely to be due to chance. The errors also included technical medical terminology implying a particular condition is present when it was not in most cases, though Wakefield was a gastroenterologist who knew the meaning of these terms.

It thus appears that Wakefield falsified the results presented in Table 1 of the paper by stating these were examples of non-specific colitis when in fact the totality of the data available at that time indicated something non-specific or of uncertain significance was present.

FRAUD 2. TIMING OF MMR VACCINE ADMINISTRATION AND FIRST BEHAVIORAL SYMPTOMS

The paper's Table 2 lists the "Interval from exposure to first behavioral symptoms." In one case this is listed as "immediately," two cases within twenty-four hours, one case within forty-eight hours, two cases within two weeks, one case within one month, and one case within two months. These reported temporal associations were used in the paper to bolster the case that there was an association between vaccine administration and subsequent development of behavioral problems, such as autism.

Brian Deer reviewed the hospital admission notes (Deer 2011) and the Medical Research Council (MRC) hearing transcripts (MRC Transcripts N.d.) and reported that of the eight of twelve cases that were reported as having first behavioral symptoms within one week, only two could be confirmed in the records. In some cases, such as Child 11, where Table 2 of the paper stated behavioral symptoms developed one week after vaccine administration, the hospital discharge note stated that behavioral symptoms began one month *before* administration of the MMR vaccine.

The direction of all eight of these errors is consistent in tending to indicate a temporal association. It is unlikely that these were simply accidental copying errors. It indicates that the authors falsified the temporal associations between MMR vaccine administration and development of behavioral symptoms.

FRAUD 3. FINDINGS OF REGRESSIVE AUTISM

Table 2 in the paper also lists for all twelve children their "Behavioral Diagnosis." This table lists nine of the twelve children as having autism and one additional child as possibly having autism.

Per reports reviewed by Deer (2011), including the MRC hearing transcripts (MRC Transcripts N.d.), only one child clearly had a diagnosis of regressive autism. Six of the nine listed as having regressive autism did not have this diagnosis, and five of the nine so listed had uncertain behavioral diagnoses.

The lack of underlying documentation of most of the children listed in Table 2 of the Wakefield paper as having regressive autism, when this is one of the main points of the paper, arguably rises to the level of fabrication of these results, insofar as documentation is just missing. Certainly it points to falsification of the data presented in Table 2.

Two additional areas of fraudulent representations within the 1998 Wakefield et al. paper are not in the scientific findings but have to do with other scientific publication issues.

Child	Behavioural diagnosis	Exposure identified by parents or doctor	Interval from exposure to first behavioural symptom	Features associated with exposure	Age at onset of first symptom	
					Behaviour	Bowel
1	Autism	MMR	1 week	Fever/delirium	12 months	Not known
2	Autism	MMR	2 weeks	Self injury	13 months	20 months
3	Autism	MMR	48 h	Rash and fever	14 months	Not known
4	Autism? Disintegrative disorder?	MMR	Measles vaccine at 15 months followed by slowing in development. Dramatic deterioration in behaviour immediately after MMR at 4-5 years	Repetitive behaviour, self injury, loss of self-help	4-5 years	18 months
5	Autism	None—MMR at 16 months	Self-injurious behaviour started at 18 months		4 years	

MMR=measles, mumps, and rubella vaccine.

Neuropsychiatric Diagnosis. *Wakefield, 1998.*

The data presented in Table 2 of the 1998 Wakefield paper (above) lack documentation, which points to possible falsification.

FRAUD 4. ETHICS CONSENT STATEMENT

The paper stated that "Ethical approval and consent investigations were approved by the Ethical Practices Committee of the Royal Free Hospital NHS Trust, and parents gave informed consent." A statement of this type is required for all medical and scientific publications to help prevent abuse of subjects in human subject research studies.

After questions were first raised by Deer (Horton 2004) and others regarding the nature of the investigations and whether they had been approved by the appropriate ethical practices committee, Murch (2004), one of the coauthors, stated that "The protocol for the 1998 *Lancet* paper was submitted on September 16, 1996" and "This protocol formed the basis for all children investigated in the 1998 *Lancet* paper, and all were investigated." Hodgson (2004) stated, "The investigation of these children was properly submitted to and fully discussed by the Ethical Practices Committee at the Royal Free Hampstead in 1996."

This issue was the focus of much investigation in the MRC hearings (MRC Transcripts N.d.), because many of the subjects in the paper were admitted to the hospital for studies prior to December 18, 1996, the date on which that research protocol was approved. On the basis of this and other ethical and practice violations, the General Medical Council struck (or revoked) the medical licenses of both Dr. Andrew Wakefield and Dr. John Walker-Smith. While both initially appealed these findings to the Administrative Court (England and Wales) High Court of Justice Administrative Court, Wakefield dropped out of the appeal. The primary argument in Walker-Smith's defense on appeal was that no such ethics committee approval was required because the investigations were for the clinical benefit of the children and were covered by a prior study approval for his work (Mitting 2012, #91, #93). This defense directly contradicts the statements of both Murch and Hodgson in 2004. Nonetheless, based on other evidence, Mr. Justice Mitting determined on appeal that it was not proven to the requisite criminal standard of proof that Walker-Smith had carried out the investigations without ethics board approval (Mitting 2012 #186, pp. 60–61). Regarding the ethics approval statement in the paper, however, Mr. Justice Mitting found, "This statement was untrue and should not have been included in the paper" (Mitting 2012, #153, p. 47).

In finalizing the paper, there was a discussion of the wording of the ethics consent statement among the authors (Mitting 2012, #153, p. 46). Following this, Wakefield evidently inserted this standard language of an ethics consent approval statement. This was a falsification of the actual record to facilitate publication of the paper.

FRAUD 5. CONFLICT OF INTEREST STATEMENT

In 1998, at the time of the paper's submission, *Lancet*, like most medical journals, required that the authors sign a statement disclosing any actual conflicts of interest and any items that could be perceived as conflicts of interest. Wakefield declared no conflicts of interest with respect to the publication.

Unbeknownst to the editors or readers at the time, however, Andrew Wakefield had filed a patent for virological testing in 1995 (Wakefield 1995). He had been engaged as an expert by lawyer Richard Barr since February 1996 to work on a potential lawsuit against vaccine manufacturers (*Sayer et al. vs. Smithkline et al.* 2007). He was paid in total £435,643 (about $568,700 at current exchange rates) for this work (Deer 2007). Both the editor of *Lancet* (Horton) and a vice dean of the Royal Free and University College School of Medicine (Hodgson) stated in writing that this conflict should have been disclosed. Failing to disclose such an obvious potential monetary conflict of interest in the outcome was a form of falsification of the record to facilitate publication and improve the perceived impact of the findings.

There was a last form of fraud committed by Andrew Wakefield in connection with this paper, but it was not in the paper itself.

FRAUD 6. METHODS OF PATIENT REFERRAL

Immediately after the paper was published, criticisms were raised regarding a possible strong bias in patient selection (Rouse 1998). Nearly all the patients were originally contacted through an anti-vaccine campaign and the solicitors attempting to sue the vaccine manufacturers (Deer 2011).

In the paper itself, this was described as, "We investigated a consecutive series of children with chronic enterocolitis and regressive developmental disorder" and "12 children, consecutively referred to the department of paediatric gastroenterology . . . ." In a subsequent response to this critique, Wakefield (1998) stated, "These children have all been seen expressly on the basis that they were referred through the normal channels (e.g., from general practitioner, child psychiatrist, or community paediatrician) on the merits of their symptoms."

When this was examined in detail during the MRC hearings (MRC Transcripts N.d., #35, p. 47), the committee found that Wakefield's statement in the response was dishonest and irresponsible. The case of referral of Child 12 was examined in detail as the mother testified and revealed that the mother was supplied with a "fact sheet" written by Wakefield prior to being seen. The levels of biasing in the findings for that child as revealed in the MRC transcripts are discussed in detail on the *lbrb* blog (Cary 2012).

This issue was also addressed with respect to Wakefield's coauthor, Dr. Walker-Smith, during the appeal of the MRC findings. In that appeal, Mr. Justice Mitting found that the finding of the MRC panel was not correct with respect to Walker-Smith's coauthorship of the paper (Mitting 2012, #158–159, pp. 62–63). It is important to note that this finding on appeal did not address the MRC finding with respect to the dishonesty of Wakefield's separate response to criticism of the paper.

The significance of the findings in the paper depended on the route of referral. The findings would be stronger if they were found in a consecutive series of children who came to the clinic; they would be weaker if they were found in a set of children chosen to potentially have the significant findings. By claiming that the referrals were through normal channels, when in fact the cases were selected for the findings prior to referral and the parents were prompted with the desired findings, Wakefield falsified this aspect of the scientific record.

OTHER TYPES OF FRAUD

Commentators often simply state that Andrew Wakefield committed fraud in the study that was published in 1998. Other than the scientific fraud discussed above, there are other common meanings of the term. *Fraud* often refers to either criminal fraud or civil fraud, a tort. Wakefield was never tried for either type with respect to the 1998 paper and study.

Criminal fraud has several elements that must be proved to support a conviction (which depend in detail on the jurisdiction in question). These are 1) misrepresentation of a material fact; 2) by someone who knows that the material fact is false; 3) with intent to defraud; 4) to a person or entity who justifiably relies on the misrepresentation; and 5) actual injury or damages result from that reliance on the false representation. In both the United States and the United Kingdom, each of these elements would have to be proven beyond a reasonable doubt (Criminal Fraud N.d.).

In the case of the 1998 paper, it is unclear who the parties would be who were injured or damaged by the scientific frauds in that paper. Assuming such parties existed, it seems it would be difficult to prove beyond a reasonable doubt the third element, that Wakefield engaged in the fraud with the intent to defraud the person injured.

Civil fraud as a tort generally has as requisite elements the intentional misrepresentation or concealment of an important fact upon which the victim is meant to rely, and in fact does rely, to the harm of the victim. People who invested money in Wakefield's business proposal or the attorneys who paid him a large amount of money as a consultant for their lawsuits might have some claim for a monetary injury. Because the standard of evidence in a civil case is simply the preponderance of the evidence, the review of the scientific frauds above suggests such a lawsuit may have succeeded; however, none was ever brought.

Conclusion

The scientific frauds in Wakefield's 1998 paper are clear from the readily available records, and it is clear why this paper was eventually retracted when the full record became available. Whether these would rise to the level of a civil or criminal fraud is unknown, as these scientific issues were never adjudicated in a court of law. While there are good reasons to consider the safety and efficacy of vaccines and for patients to be fully informed before being vaccinated, the alleged link between the MMR vaccine and autism is not one of them.

Peter N. Steinmetz is a research neurologist and chief scientist at the Neurtex Brain Research Institute. He received his MD and PhD from the Johns Hopkins University School of Medicine. He has a long-standing interest in skeptically examining claims of the magical and other fringe-science phenomena. He lives in Phoenix, Arizona.

References

Bjarnason, I. 2011. Commentary: We came to an overwhelming and uniform opinion that these reports do not show colitis. *BMJ* 343: d6979. Available online at https://www.bmj.com/content/343/bmj.d6979.

Cary, M. 2012. Transcripts from the GMC hearings. *lbrb* (February 2). Available online at https://leftbrainrightbrain.co.uk/2012/02/02/transcripts-from-the-gmc-hearings.

Criminal Fraud. N.d. Everything you need to know about fraud crimes and fraud law. Find Law. Available online at https://criminal.findlaw.com/criminal-charges/fraud.html.

Deer, B. 2007. Revealed: Undisclosed payments to Andrew Wakefield at the heart of vaccine alarm. Available online at http://briandeer.com/wakefield/legal-aid.htm.

———. 2010. Wakefield's 'autistic enterocolitis' under the microscope. *BMJ* 340: c1127. Available online at https://www.bmj.com/content/340/bmj.c1127.

———. 2011. How the case against the MMR vaccine was fixed. *BMJ* 342: c5347. Available online at https://www.bmj.com/content/342/bmj.c5347.

Geboes, K. 2011. Commentary: I see no convincing evidence of "enterocolitis," "colitis," or a "unique disease process." *BMJ* 343: d6985. Available online at https://www.bmj.com/content/343/bmj.d6985.

Godlee, F. 2011. Institutional research misconduct. *BMJ* 343: d7284 (data supplement). Available online at https://www.bmj.com/content/343/bmj.d7284.

Hodgson, H. 2004. A statement by the Royal Free and University College Medical School and the Royal Free Hampstead NHS Trust. *Lancet* 363(9411): 824. Available online at https://www.thelancet.com/journals/lancet/article/PIIS0140673604157115/fulltext.

Horton, R. 2004. A statement by the editors of *The Lancet*. *Lancet* 363(9411): 820–821. Available online at https://www.thelancet.com/journals/lancet/article/PIIS0140-673604)15699-7/fulltext.

Lancet editors. 2010. Retraction—Ileal-lymphoid-nodular hyperplasia, non-specific colitis, and pervasive developmental disorder in children. *Lancet* 375(9713): 445. Available online at https://www.thelancet.com/journals/lancet/article/PIIS0140-6736(10)60175-4/fulltext.

Mitting, J. 2012. *Walker-Smith v. General Medical Council.* Available online at http://www.bailii.org/ew/cases/EWHC/Admin/2012/503.html.

MRC Transcripts 2007–2010. N.d. Casewatch. Available online at http://steinmetz.org/peter/Medical/wakersTranscripts.zip.

Murch, S. 2004. A statement by Dr. Simon Murch. *Lancet* 363(9411): 821–822. Available online at https://www.thelancet.com/journals/lancet/article/PIIS0140-6736(04)15708-5/fulltext.

Norwegian National Research Ethics Committees. N.d. Fraud and plagiarism. Available online at https://www.etikkom.no/en/library/topics/integrity-and-collegiality/fraud-and-plagiarism/.

Rouse, A. 1998. Correspondence. *Lancet* 351(9112): 1356. Available online at https://www.thelancet.com/journals/lancet/article/PIIS0140-6736(05)79082-6/fulltext.

Sayer et al. v. Smithkline et al. 2007. MMR and MR Vaccine Litigation Sayers and others v. Smithkline Beecham plc and others. All ER (D) 30 (Jun).

U.S. Office of Research Integrity. N.d. Definition of research misconduct. Available online at https://ori.hhs.gov/definition-misconduct.

Wakefield, A. 1995. Diagnosing Crohn's disease or ulcerative colitis by detection of measles virus. (UK patent application 2 300 259 A). UK Patent Office.

———. 1998. Autism, inflammatory bowel disease, and MMR vaccine. *Lancet* 351(9112): 1356. Available online at https://www.thelancet.com/journals/lancet/article/PIIS0140-6736(05)79083-8/fulltext.

Wakefield, A. et al. 1998. Ileal-lymphoid-nodular hyperplasia, non-specific colitis, and pervasive developmental disorder in children. *Lancet* 351: 637–41. Available online at https://www.thelancet.com/journals/lancet/article/PIIS0140-6736(97)11096-0/fulltext.

CHAPTER 17

Wildlife Apocalypse

HOW MYTHS AND SUPERSTITIONS ARE DRIVING ANIMAL EXTINCTIONS

Bob Ladendorf and Brett Ladendorf

Vol. 42, No. 4
July/August 2018

Sporting AK-47 assault rifles and axes, the group of men stalk a black rhino through the African bush. They soon bring it down with powerful volleys. While still alive, the rhino peers at the men as they approach. The poachers quickly use the axe to sever its horn from its head, not caring that they are inflicting great pain as they hit a nerve and leave the rhino dehorned and its head a pulpy mess. It dies, leaving its own family behind because of human greed. "'Rhino have a particularly plaintive cry,' (conservationist Ian) Player[1] wrote (in *The White Rhino Saga*), 'which once heard is never forgotten. The screams of agony from rhino that have had their horns chopped off while still alive should reach into the hearts of all of us'" (Rademeyer 2017).

The poachers sell that horn to a middleman, who may be working for yet another smuggler, a criminal syndicate, or even terrorists. Government border agents and officials are bribed as the horn makes its way to countries such as China and Vietnam, where the horn is used in Traditional Chinese Medicine (TCM) to treat various ailments, none of them proven scientifically to work.[2]

While most rhino horns are ground into powder and used as medicine to supposedly cure cancer, impotence, or, as an illegal wildlife trade monitor says, "you name it," people in Asia have begun wearing beads or bangles made from rhino horns thought to cure ailments as well as for status symbols. Some horns are fashioned into ceremonial cups (Kolata 2018).

Why is the illegal supply and demand for rhino horns so pervasive? Rhino horn, after all, is mainly composed of keratin, the same substance in human hair and fingernails. But it's as valuable as gold or heroin. A kilogram, for instance, can sell for $60,000 (Kolata 2018).

The killing of rhinos is just the tip of the iceberg in the ever-increasing destruction of wildlife for dubious reasons. Not only rhinos are facing extinction but also African elephants; certain species of lions, tigers, and wolves; Grauer's gorillas; and even giraffes. All this is done primarily at the hands of humans despite courageous efforts by conservation groups, governments, and individuals to stop the attacks. Some wildlife, such as rhinos and wolves, among many others, faced extinction when trade in animal parts was legal, but they now face that possibility again with illegal trading and other extinction pressures.

"Leading international wildlife crises involve illegal poaching of rhinos, elephants, and sharks for their body parts, to be sold on the Asian black market for exorbitant prices and used for medicinal purposes or art," stated Cristina Eisenberg, chief scientist at Earthwatch Institute in Boston and author of *The Carnivore Way: Coexisting with and Conserving North America's Predators*.

The myth underlying this illegal bone trade runs very deep. Proponents tout rhino horn, shark fin (cartilage), and elephant tusk medicinal uses, as tonics, blood-purifiers, or aphrodisiacs. But ultimately, it's about money—these illegal products are primarily seen as status symbols in Asia. While the purported medicinal use of these items has not been proven by science, the profound negative consequence of poaching has been thoroughly documented and is decimating populations of rhinos, elephants, and sharks, leaving them at or near extinction (Eisenberg 2018).

As of 2016, there were only 29,500 rhinos left in the world, 70 percent of them in South Africa. There are five species of rhinos—most of them endangered—with two subspecies going extinct in 2011 (Gwin 2012). Just a century ago, there were an estimated one million rhinos in Africa (Ellis 2005).

Some 30,000 elephants are poached yearly for their ivory (Showing That Every Elephant . . . 2017). *The Ivory Game* documentary warns that African elephants may become extinct in fifteen years. Biologists estimate that total loss of large mammals in Africa went up to 60 percent between 1970 and 2013 (Paterniti 2017). In the "Scientists' Warning to Humanity: Second Notice" last year, signed by more than 15,000 scientists in 184 countries, a highlight of the document was a 29 percent reduction in the numbers of mammals, reptiles, amphibians, birds, and fish since the publication of the first notice in 1992 (Houtman 2017). The global black market in live animals and parts is the fourth largest in the world, with an estimated $20 billion in profits (Tackling Wildlife Trafficking 2017).

"Traders in ivory actually want extinction of elephants, and that is probably the biggest danger," warns Craig Millar, head of security for the Big Life Foundation/Kenya, in *The Ivory Game*. "The less elephants there are, the more the price rises. The more the price rises, the more people want to kill them. And this is an ever ongoing circle that is just going to end up bringing about exactly what they want—extinction." The same could be said about rhinos, lions, gorillas, and many other animal species.

Myths and Superstitions

While the trade in rhino horn is banned under the Convention on International Trade in Endangered Species of Fauna and Flora (CITES), the black market fueled by demand particularly from China and Vietnam is lucrative and primarily recent. In 2005, according to the organization Save the Rhino International, about sixty rhinos were killed for their horns or as trophies in Africa. Since then, more than 7,000 have been killed, with 1,346 in 2015 alone (Poaching in numbers 2017). In South Africa alone, poaching increased 9,000 percent from thirteen in 2007 to 1,215 in 2014 (Juskalian 2017; Save the Rhino International 2018).

Connecting a real animal with a mythical one is a task undertaken by marine biologist Richard Ellis, author of *Tiger Bone and Rhino Horn: The Destruction of Wildlife for Traditional Chinese Medicine*. He is a research associate at the American Museum of Natural History. "The use of rhino horn . . . can be traced to the unicorn, another animal with a horn growing from a totally unsuspected place" (Ellis 2005). He also wrote this for the European Association of Zoos and Aquaria's rhino campaign in 2005:

> It is not clear that rhino horn serves any medicinal purpose whatsoever, but it is a testimony to the power of tradition that millions of people believe that it does. Of course, if people want to believe in prayer, acupuncture, or voodoo as a cure for what ails them, there is no reason why they shouldn't, but if animals are being killed to provide nostrums that have been shown to be useless, then there is very good reason to curtail the use of rhino horn. . . . It is heartbreaking to realize that the world's rhinos are being eliminated from the face of the earth in the name of medications that probably don't work. (Save the Rhino International 2017)

While the scientifically unproven medicinal uses of rhino horn have driven the eastern Asian black market, there are additional extinction drivers, including the superstitious beliefs in the efficacy of rhino horn for hangover cures and as aphrodisiacs. While the media reports were actually wrong about Asians using rhino horn as a sexual stimulant, the attention paid to that error ironically sparked interest in using it for that equally scientifically unproven purpose! Elizabeth Kolbert pointed out in *The Sixth Extinction* that rhino horn in recent years is "even more sought-after as a high-end party 'drug'; at clubs in southeast Asia, powdered horn is snorted like cocaine" (Kolbert 2015).

An even more sensational claim is that rhino horn cures cancer, fueling even more demand. There's no scientific basis for that claim. The cause was likely a rumor started in Vietnam a decade ago that rhino horn had cured cancer in a near-death South Vietnamese Communist Party official. The rumor spread rapidly, and the price of rhino horn surged (Rademeyer 2017). This myth prompted poachers to increase their efforts at killing rhinos in Africa, some even using helicopters to track them down (Watts 2011).

Of course, there may be a placebo effect for some users of rhino horn. "Belief in a treatment, especially one that is wildly expensive and hard to get, can have a powerful effect on how a patient feels," stated Mary Hardy, medical director of Simms/Mann UCLA Center for Integrative Oncology and "a traditional medicine expert," according to *National Geographic* magazine (Gwin 2012).

While TCM does include a lot of vegetable- and herbal-based medicines, as well as non-endangered animal parts, the use of critically endangered animal parts that it promotes for scientifically unproven treatments and cures has been a major factor in the decline and extinction of animal species. Numerous articles in science publications, including this magazine, confirm that these purported remedies have no basis in fact. The late Robert Carroll wrote in his *Skeptic's Dictionary* that "Magical thinking is clearly the basis for some of these concoctions, e.g. deer penis to enhance male virility. Many of the medicinals lead to the suffering and unnecessary maiming and killing of many animals." As examples, Carroll relates how thousands of bears are kept in cages throughout Asia so their bile can be tapped and sold to cure various ailments. "Other animals are treated with equal disdain: sharks for their fins, rhinos for their horns, and tigers and tortoises for various body parts" (Carroll 2018).

As TCM continues the pressure on the illegal use of rhino horn, other connected factors help to reduce the numbers of these animals, as well as other wildlife. Some 73–100 million sharks are killed yearly, primarily for their fins for shark fin soup in Vietnam and China (Masson 2014; Defenders of Wildlife 2018) There's no scientific evidence that the soup treats any medical condition, including cancer. It's primarily a luxury item in Chinese culture, although consumption of the soup has been reduced in recent years with the introduction of an imitation shark fin soup (Shark fin soup 2018).

The vaquita, the smallest marine mammal that lives exclusively in the upper Gulf of California in Mexico, is almost extinct because they get caught in gillnets used to catch Mexican shrimp. Because of a high demand in China for its dried swim bladders "for their supposed medicinal properties," the endangered totoaba fish is caught in the illegal gillnetting in the Gulf. A campaign urging consumers to boycott Mexican shrimp and asking the Mexican government to ban all gillnetting to save the vaquitas and totoaba has not been successful. "The Mexican government is putting shrimp industry profits over saving this tiny porpoise from its freefall into extinction," says Alejandra Goyenechea, senior international counsel for the Defenders of Wildlife organization (Boycott Mexican Shrimp to Save Vaquitas! 2017). In 2018, the Elephant Action League's Sea Shepherd ship continued its battle with fishermen and the illegal nets; one of their anti-poaching camera drones was shot down there in late December 2017 (Tillman 2017).

Another mammal under assault for its dubious medicinal qualities is the pangolin, who rolls up in a ball for defense with scales on the outside. While not currently endangered, the pangolin may be the most illegally trafficked animal in the world, with some estimates as high as 2.7 million yearly. The pangolin scales are sold for as much as $750 a kilogram. "Most . . . end up in China and Vietnam," reports *The Economist*. "In these countries pangolins' meat is a treat and their scales are used in folk medicine, even though the scales are made of keratin . . . and thus have no medicinal value" (A problem of scale 2018).

Some providers and consumers of sharks and other endangered species in East Asian countries may argue that the animals are killed for calories and protein, in addition to dubious medicinal practices, and continue to be needed to help feed growing populations. As far as they are concerned, animal species may be low-hanging fruit, whether endangered or not. They also may question whether those in the West who are critical of their eating habits should deal with their own issues of overfishing in the Gulf of Mexico, the waters of the Pacific Northwest, and Chesapeake Bay. Then there is the religious argument that humans have "dominion" over the animals, as mentioned in Genesis 1.

The Price of Poaching

With retail prices per kilo in the tens of thousands of dollars, the $20 billion black market hosts brazen players trying to make a buck. The mastermind sellers in animal parts, with methods for extraction, distribution, and financing, are likely to operate their networks similar to that of international drug cartels or arms dealers. Along the supply chain exist financial incentives for personnel with wealth accumulating to those who can control most of the network. From poachers to wholesalers to dealers to art merchants to buyers at the retail level, the profit margins drive incentives. Enforcers of the parts trade accumulate wealth but so do those at the retail end who can distribute to mass markets, whether in the form of "medicine" or in the form of "art." For example, poachers will receive as little as $7 per kilo of ivory for an African elephant tusk. In the documentary *The Ivory Game*, an arrested Tanzanian poacher received from a dealer such a sum—a couple hundred dollars—for two tusks weighing fifteen kilos each. The dealer then parlayed his purchase into $3,000 per kilo in China.

On the streets across the world, there's significant variance in the economic value of the tusks, or rhino horns, driving the incentive for wholesalers to move more product. In one instance, the ivory tusks were found in a Chinese retail shop that was selling a painted tusk for $330,000, or $22,000 per kilo. If that same tusk had been extracted by the Tanzanian poacher, that would be more than 3,000 times the price paid at the source.

With the recent banning of the ivory trade in China, prices for the legal selling of tusks dropped, but it's too early to determine that impact on the black market. However, documented evidence of the illegal trade, such as that high-lighted in *The Ivory Game*, is shining a light on the amounts involved along the supply chain. As a tusk, or a rhino horn, travels from the animal carcass on the plains of sub-Saharan Africa to the medicine cabinet of an East Asian retiree, the price increase has been phenomenal in recent years.

Other Animal Extinction Pressures

According to a 2013 survey by TRAFFIC, an organization that monitors illegal wildlife trade for the World Wildlife Fund and the International Union for Conservation of Nature (IUCN)—known for its Red List of Endangered Species—rhino horn also is a status symbol for the rich in countries such as China and Vietnam. "The motivation for consumers buying rhino horn (are) the emotional benefits rather than medicinal, as it reaffirms their social status among their peers. Image and status (are) important to these consumers," and "they tend to be highly educated and successful people who have a powerful social network and no affinity to wildlife. Rhino horns are sometimes bought for the sole purpose of being gifted to others; to family members, business colleagues or people in positions of authority" (Save the Rhino International 2017).

Even war is bad for wildlife, as shown by researchers Rob Pringle and Joshua Daskin in their recent *Nature* article. They conclude that wars do wildlife more harm than good, exposing animals to bombs and landmines and increasing the demand for ivory and bushmeat that are used to finance and feed armies (Conflict's other casualties 2018; Kaplan 2018).

The emotional impact on chimpanzees and gorillas was well illustrated in the documentary *Virunga*, which showed the heroic struggles of Virunga National Park caretakers and military rangers to protect the animals from the intrusions of armies as well as poachers. Seeing the fear in the animal faces as they clung to the caretakers as bombs exploded nearby shows the difficulties faced by both wildlife and humans.

An additional pressure on wildlife and their ecosystems is the proposed completion of the U.S.-Mexico border wall by President Trump. According to studies, some 700 vertebrate species, such as jaguars, Mexican gray wolves, ocelots, mountain lions, and black bears, rely on the borderland habitat—and more than 180 of the borderland species are already listed as endangered or threatened. A wall also would keep those animals from natural crossings—wildlife corridors (State of the border 2018). While U.S. laws could help protect endangered species, Congress passed a law in 2005 giving the Department of Homeland Security authority to waive all laws when constructing a wall. The agency already has used its authority to waive forty laws, such as the Clean Air Act and the Endangered Species Act, in constructing 650 miles of barrier in past years (Schlyer 2018).[3] There have been border protests against the new wall and pending lawsuits by environmental and animal rights organizations (Against the Wall 2017).

One area of dispute as to whether it contributes to the decline in large animals is trophy hunting. Hunters who paid a lot of money for permits to shoot and import elephant and other wildlife trophies argue that the money helps in the conservation of animals in Africa, while animal rights groups say the trophy hunting "causes immense suffering and fuels the demand for wild animal products" (Pearce 2017b). For a big game hunt, for instance, a hunter might pay up to $200,000 for a rifle and $80,000 for a fourteen-day single elephant hunt (Paterniti 2017). A portion of the fee is paid to community members, such as the San in Namibia, and a portion for a conservation fund. An African trophy hunt for a leopard may bring in as much as $55,000, while a lion fetches up to $76,000. While some people still hunt to eat, sport hunters are in it for the thrill and to show off their "trophies," although some face severe criticism as did the American who killed the well-known Cecil the Lion (Paterniti 2017).

President Trump planned to partially reverse an Obama-era ban last November by allowing hunters to import trophies from Zambia and Zimbabwe, then he reversed himself and postponed the decision after an outcry from citizens and lawmakers. California Rep. Ed Royce, a Republican Congressman, pointed out that the political turmoil in Zimbabwe could spell doom for wildlife. "Elephants and other big game in Africa are blood currency for terrorist organizations, and they are being killed at an alarming rate," he said (Pearce 2017a). In that country, points out Vanda Felbab-Brown, a senior fellow at the Brookings Institution and author of *The Extinction Market*, authorities seize the hunting preserves and keep the profits; they don't reinvest in conservation. She said the trophy hunting business "becomes very commercialized and the profits are captured by elites. You can also end up with trophy hunting serving as a cover for trafficking" (Nuwer 2017).

Climate change provides additional pressures on wildlife, such as polar bears coping with the shrinking of Arctic ice. There are many other effects. "As the seasonal cycles in temperature and rainfall shift," writes climate scientist Prof. Michael E. Mann, "altering by different amounts the timing of the hatching of insects and the arrival of birds, entire food webs are in danger of disruption. Plants and animals possess a certain amount of behavioral elasticity, but the more rapid the changes, the more likely this intrinsic adaptive capacity will be exceeded, and the more likely that we humans will be responsible for one of the most devastating extinction events in Earth's history" (Mann and Toles 2016).

These additional pressures—added to the demand for certain wildlife, such as rhinos and elephants, based on myths and superstitions—may indeed produce a wildlife apocalypse.

Live Wild Pet Trade

While China and Vietnam have been the main drivers for the extinction of rhinos and elephants, the United States and Europe have surprisingly major black markets for the trade in wild, exotic pets. Birds and snakes from overseas are stuffed into soda bottles for transit to the Western countries. Tragically, 90 percent of these animals die in transit (Wild Matters 2017). Many of the same black marketers in wild animal parts, such as rhino horn, also spark the trade in live animals (Conniff 2017).

"Many of these people who were doing the traditional medicine trade are now branching out because the high-end pet trade in China has grown immensely," commented Brian Horne, a herpetologist for the Wildlife Conservation Society. Critically endangered adult ploughshare tortoises that live only in Madagascar cost $100,000 each, which now draws in criminal elements. For example, thieves broke into a captive breeding facility in Thailand—set up by conservationists to rebuild populations of endangered species—and stole six ploughshare tortoises. The trade in exotic pets, according to conservation biologist David S. Wilcove, has "the potential to drive species to extinction even when they have suitable habitat, and to do so without anyone being aware of it" (Conniff 2017).

How Smartphones Decimated Grauer's Gorillas

Just when anyone interested in preserving species on the verge of extinction feels comfortable that many efforts are being made to fight back through the work of governments, nongovernmental organizations (NGOs), and concerned individuals, the disquieting news is the human demand for cell phones is the cause of at least one mammal's near extinction. Grauer's gorillas in the Congo have suffered a 77 percent decline in the past two decades because of the consumer electronics explosion. How?

One of the key components of a cell phone is the mineral coltan, and 80 percent of it is found in mines in the Congo. Those mines that destroy the land to unearth coltan and other minerals often use young children. These are "artisanal" operations, meaning that the mining requires not machinery but laborers digging craters into stream beds by hand. Amnesty International reports that as many as 40,000 children may be mining for coltan in the Congo.

"To feed these people, wildlife is hunted from the surrounding forests," said Tara Stoinski, president and chief scientist of the Dian Fossey Gorilla Fund International. "This includes gorillas, chimpanzees, elephants, and many other species." Trade in bushmeat is illegal, but the Congo is a war-torn region that makes such laws unenforceable (Posada 2017).

Fighting Back

While the outlook is dire for many species, including giraffes in Africa that have seen their numbers decline nearly 40 percent from 1985–2015 to less than 100,000 now, the good news is that many governments, NGOs, conservation organizations, and individuals are banding together to save as many species as possible. As of January 1, 2018, China has banned all trade in ivory, which follows the lead of the United States in 2016 (Giraffes newly classified 2017). Hong Kong also announced in late January that it would ban all ivory trade by 2021. Just this past July, 7.2 tons of new elephant tusks were found under frozen fish in Hong Kong and confiscated. Only ivory acquired before 1970 is legal there (May 2018).

In 2017, Operation Thunderbird, a sixty-nation global seizure of illegal wildlife and floral trade, identified 900 suspects, with 1,300 seizures worth $5.1 million (*Wild Matters: Tackling Wildlife Trafficking* 2017). More than 1,000 rangers have given up their lives from 2004–2014, primarily in Africa, protecting wildlife from poachers (Chancellor 2014).[4] A conservation organization, The Nature Conservancy, partnered with the Northern Rangelands Trust to reduce poaching in Northern Kenya (Oluchina 2014). However, the fight against poachers in Africa received a setback when famous American conservation investigator Esmond Bradley Martin, seventy-five, was stabbed to death at his home in a possible murder that may have been disguised as a robbery of the long-time activist who uncovered illegal global trafficking of ivory and rhino horn (Dixon 2018).

In January 2018, Ivory Coast officials said they broke up an international ivory-smuggling network, the second such bust on the continent that month. They arrested six people and confiscated more than half a ton each of ivory and pangolin scales, as well as leopard parts. The network hid ivory parts in hollowed-out logs that were resealed and shipped to Asian countries. The suspects had made calls to tax-haven countries, leading officials to suspect money laundering. In another bust in Gabon, officials said they also broke up a smuggling network that had ties to a cell of Boko Haram, the Islamic militant group responsible for numerous murders and kidnappings in northern Nigeria and bordering countries (Searcey 2018).

Undercover NGO investigators and journalists have been instrumental in identifying companies, merchants, and corrupt businessmen involved in the illegal wildlife trafficking trade, as shown in the documentaries *The Ivory Game* and *Virunga*. There are many organizations working to save wildlife, from long-standing ones such as The Sierra Club and Defenders of Wildlife to newer ones such as the Elephant Action League, Wildleaks, and United for Wildlife, which was created by the Duke (Prince William) and Duchess (Catherine) of Cambridge and Prince Harry. Others, such as Earthwatch, engage citizen scientists in worldwide expeditions to provide data for scientific studies on wildlife, climate change, and other matters.

There's even a new tactic in wildlife conservation: horn and tusk forensics. Like the genetic fingerprinting methods in the criminal justice system, scientists are making efforts to match the DNA of a rhino or elephant with its horn or tusk in possession of a poacher. A scientific database called Rhodis (modeled after the FBI's Codis system) has been established with some 20,000 samples taken from rhinos by Dr. Cindy Harper, a veterinarian at the University of Pretoria, and her colleagues (Kolata 2018).

These efforts may be too little, too late for some species, but they give hope to others. Not only is there some success in reducing poaching, but there is also increasing awareness in the public about the wildlife trafficking issues. The false beliefs that have driven poaching and decimation of various species need to be corrected, and the "profits captured by elites," as termed by Felbab-Brown, need to be stopped (Nuwer 2017). Even John Hume, the controversial rhino rancher behind the rhino-ranching movement to legalize the rhino horn trade in South Africa and the subject of the controversial documentary *Trophy*, thinks rhino horn medicinal uses are bunk. It doesn't matter to him that rhino horn is snake oil when it comes to treating serious maladies. "I'm not ashamed that the rhino horn I make available to the world could possibly be ingested by somebody who's got cancer and he dies anyway. It's not going to help them" (Christy 2016).

It's hard not to feel sad for the brutality inflicted on animals for purposes of human beliefs in myths and superstitions, for status and appetites, and for plain old greed. A lasting image of the horrible legacy of inhumane treatments of animals can be seen in The Reliquary, a U.S. government warehouse outside Denver that holds 1.3 million products made from animals, many of them threatened or endangered species. Many were donated, but most were seized upon entry. Just 10 percent of global trade in banned wildlife is intercepted. In the repository, you'll see an African elephant footstool, tiger teeth and claws fashioned into jewelry, a hat made of black bear skin, Tibetan antelope shawls, and a rhinoceros snout and horns on a wooden platter (Spinski 2017).

The fight continues to save endangered animals, and we can only hope that all humans realize the necessity for animal biodiversity and the need for scientific evidence in the use of medicines. "Too many animals, from sea horses to rhinoceroses, are endangered by the demands of traditional Chinese medicine," says author Richard Ellis.[5] "Of course, TCM is not the only factor in the endangerment of these animals, but it plays an enormous part. If present trends continue, tigers and rhinos will become extinct in the wild, perhaps in our lifetime and almost certainly in the lifetime of our children's children" (Ellis 2005).

From the savannahs of Africa to the ports of North America, the black market trade in animal parts is lucrative for top smugglers. Demand is driven for many reasons, of which belief in false medicines can perhaps have the best chance of being reduced through educational outreach and policies guided by progressive studies of human behavior. Government programs and public-private agency partnerships can and have demonstrated success in nudging consumer behavior in a direction that can produce positive outcomes for the self and the community. It can start with something as little as a contest for an anti-littering slogan along Texan highways to change human behavior. It can be a program to frame better choices for consumers who desire certain attributes from parts of animals. Though affecting the behavior of those who demand parts for status or for value may prove the hardest work, moving humans toward awareness through education and science may have the most profound effect on a mass scale.

For Carl Sagan, it would be "far better to grasp the Universe as it really is than to persist in delusion, however satisfying and reassuring" (Sagan 1997). Though Sagan focused on the possibility of life beyond Earth, he knew that the greatest dangers to our own well-being and to that of our environment came from within ourselves. For our planet, the reduction and loss of species from these delusions of grandeur is tragic. It also would be a tragedy if we weren't able to fight off the AK-47s and machetes with better knowledge on why people reject science in favor of the dark.

Bob Ladendorf is a freelance writer, former chief operating officer at the Center for Inquiry Los Angeles, and coauthor of an article on "The Mad Gasser of Mattoon" for *Skeptical Inquirer*. He also recently reviewed the documentary *The Pathological Optimist* and Michael E. Mann's book on climate change for this magazine.

Brett Ladendorf has worked in the financial markets for more than twelve years and currently has a financial services consulting practice for alternative

investment managers and financial technology firms. He holds a BA in economics from the University of Wisconsin-Madison and an MBA in finance and accounting from the University of Chicago Booth School of Business. His coauthor is his father.

Notes

1. Ian Player is credited with saving South Africa's rhinos from extinction in the 1960s.

2. A 2015 article in *Skeptical Inquirer* by Harriet Hall, for instance, casts doubt on TCM versus science-based medicine: "Traditional Chinese Medicine (TCM) Didn't Win a Nobel Prize, Scientific Medicine Did."

3. The value and extent of wildlife corridors in North America is explained by Cristina Eisenberg in her book *The Carnivore Way: Coexisting with and Conserving North America's Predators.*

4. For a more detailed account of how rangers face dangers from poachers, see Robyn Dixon's "Elephant Men," *Los Angeles Times*, December 22, 2017. Another article on the rhino horn legal trade controversy is Robyn Dixon's "It's Cruelty beyond Words," *Los Angeles Times*, August 2, 2017.

5. In Ellis's *Tiger Bone & Rhino Horn: The Destruction of Wildlife for Traditional Chinese Medicine*, chapter 3 ("Chinese Medicine, Western Medicine") discusses TCM in detail, while chapter 4 ("Horn of Plenty") details the history of the "unicorn" and its connection to the supernatural and the reality of real animal horns.

References

A problem of scale. 2018. *The Economist* (February 3).

Against the wall. 2017. *Audubon* (Winter).

Boycott Mexican shrimp to save vaquitas! 2017. *Defenders* (Summer).

Carroll, Robert. 2018. *The Skeptic's Dictionary* (online). Available online at http://skepdic.com/tcm.html.

Chancellor, David. 2014. Save the animals. *Time* (August 18).

Christy, Bryan. 2016. Deadly trade. *National Geographic* (October).

Conflict's other casualties. 2018. *The Economist* (January 13).

Conniff, Richard. 2017. Loved to death. *Scientific American* (October).

Defenders of Wildlife. 2018. Email to Bob Ladendorf (February 9).

Dixon, Robyn. 2017a. It's cruelty beyond words. *Los Angeles Times* (August 2).

———. 2017b. Elephant men. *Los Angeles Times* (December 22).

———. 2018. Ivory activist fatally stabbed. *Los Angeles Times* (February 6)

Eisenberg, Cristina. 2014. *The Carnivore Way: Coexisting with and Conserving North America's Predators*. Washington: Island Press.

————. 2018. Letter to Bob Ladendorf (January 29).

Ellis, Richard. 2005. *Tiger Bone & Rhino Horn: The Destruction of Wildlife for Traditional Chinese Medicine.* Washington, D.C.: Island Press.

Giraffes newly classified as vulnerable. 2017. *Population Connection* (March).

Gwin, Peter. 2012. Rhino wars. *National Geographic* (March).

Houtman, Nick. 2017. Concerned scientists—world scientists' warning to humanity: Second notice. *Oxford University Press* blog. Available online at https://blog.oup.com/2017/12/concerned-scientists-world-scientists-warning-humanity-second-notice/; accessed February 14, 2018.

Juskalian, Russ. 2017. Last chance to be. *Discover* (November).

Kaplan, Karen. 2018. Wild animals highly sensitive to effects of war, researchers find. *Los Angeles Times* (January 18).

Kolata, Gina. 2018. A poaching scene is a crime scene. *New York Times* (January 9).

Kolbert, Elizabeth. 2015. *The Sixth Extinction.* New York: Picador.

Mann, Michael E., and Tom Toles. 2016. *The Madhouse Effect: How Climate Change Denial Is Threatening Our Planet, Destroying Our Politics, and Driving Us Crazy.* New York: Columbia University Press.

Masson, Jeffrey Moussaieff. 2014. *Beasts: What Animals Can Teach Us about the Origins of Good and Evil.* New York: Bloomsbury USA

May, Tiffany. 2018. Hong Kong moves to ban all ivory sales, closing a loophole. *New York Times* (January 31). Available online at https://www.nytimes.com/2018/01/31/world/asia/hong-kong-elephant-ivory.html.

Nuwer, Rachel. 2017. Elephants as trophies? An endless debate. *New York Times* (December 5).

Oluchina, Charles. 2014. The price of poaching. *Nature Conservancy* (August/September).

Paterniti, Michael. 2017. Should we kill animals to save them? *National Geographic* (October).

Pearce, Matt. 2017a. Trump postpones plan to allow elephant trophy imports. *Los Angeles Times* (November 19).

————. 2017b. U.S. to allow elephant trophy hunting. *Los Angeles Times* (November 17).

Poaching in numbers: 2006–2017. 2017. *The Horn 2017.* Available online at https://issuu.com/savetherhinointernational/docs/sr3260_thehorn17-book-web.

Posada, Brenda. 2017. *Call of the wild. Zoo View* (Summer). (A quarterly magazine of the Greater Los Angeles Zoo Association.)

Rademeyer, Julian. 2017. *Killing for Profit: Exposing the Illegal Rhino Horn Trade.* South Africa: Zebra Press.

Sagan, Carl. 1997. *The Demon-Haunted World.* New York: Ballantine Books.

Save the Rhino International. 2017. Poaching for rhino horn. Available online at https://www.savetherhino.org/rhino_info/threats_to_rhino/poaching_for_rhino_horn.

————. 2018. Poaching statistics. Available online at https://www.savetherhino.org/rhino_info/poaching_statistics; accessed February 2, 2018.

Schlyer, Krista. 2018. Walled off: A 'big beautiful wall' would devastate wildlife populations on both sides. *Defenders* 93(2) (Winter).

Searcey, Dionne. 2018. Ivory Coast arrests six in crackdown on smuggling. *New York Times* (January 26).

Shark fin soup. 2018. Wikipedia. Available online at https://en.wikipedia.org/wiki/Shark _fin_soup#Imitation_shark_fin_soup; accessed February 15, 2018.

Showing that every elephant—and every voice—counts. 2017. *National Wildlife* (October–November).

Spinski, Tristan. 2017. The reliquary. *New York Times* (July 11).

State of the border. 2018. *Defenders* 93(2) (Winter).

Tackling wildlife trafficking. 2017. *Defenders* (Winter).

Tillman, Laura. 2017. Anti-poaching drone shot down in Mexico. *Los Angeles Times* (December 27).

Watts, Jonathan. 2011. "Cure for cancer" rumour killed off Vietnam's rhinos. *The Guardian* (November 25). Available online at https://www.theguardian.com/ environment/2011/nov/25/cure-cancer-rhino-horn-vietnam.

Wild matters: Tackling wildlife trafficking. 2017. *Defenders* (Summer).

CHAPTER 18

Quackery at WHO

A CHINESE AFFAIR

Cees N. M. Rencken and Thomas P. C. Dorlo

Vol. 43, No. 5
September/October 2019

> *Traditional Chinese medicine is a "gem" of the country's scientific heritage.*
>
> —President of China Xi Jinping

Since its founding in 1948, maintenance of the *International Classification of Diseases* (ICD) report is among the many tasks allocated to the World Health Organization (WHO). The ICD was initially started by the U.S. International Statistical Institute as the International List of Causes of Death. More than 100 countries worldwide use the ICD currently for morbidity and mortality statistics. But the ICD also serves an important financial role: the ICD is being used as a basis for reimbursement policies and allocation of national funding for almost 70 percent of global healthcare costs.

On June 18, 2018, the WHO published a draft of the eleventh version, ICD-11. New additions to the ICD-11 included "gaming disorder," a reclassification of gender dysphoria, and the chance to register diagnostic terminologies and syndromes used in so-called traditional medicine (TM). In January 2019, representatives of WHO member states further prepared the embedding of the proposed ICD-11, which was presented to the World Health Assembly in May 2019. If everything goes as planned, the new ICD-11 will be put into use on January 1, 2022. Curiously, according to member state representatives, no discussion was expected in the Assembly on this highly remarkable extension of the ICD. And indeed their prediction came true: the proposal was accepted unanimously and without any debate.

Traditional Chinese Medicine in the ICD-11

The WHO has dedicated a full chapter in ICD-11 to traditional Chinese medicine (TCM) diagnoses and syndromes. This means that from 2022 onward, official reporting can be performed for diagnoses such as the *bladder meridian syndrome*, which is supposedly characterized by severe headache, neck- and lower-back pain, excessive tearing, a stuffy nose, and a numb little toe. Another example: *triple energized meridian dysfunction*, a syndrome that it claims is characterized by deafness, tinnitus, swelling and obstruction of the throat, and reduced use of the ring finger (Hong-Zhou et al. 2013).

Given the pseudoscientific character of TCM and the impossibility of integrating this Chinese taxonomy and its concepts into modern medicine, we had expected a substantial turmoil following the announcement of ICD-11 by the WHO (WHO 2018). However, it attracted, with a few exceptions, very little attention and commotion, even in the medical community (Gorski 2018; *Scientific American* Editors 2019; WHO 1984; WHO 1993; Maassen 2018).

What Preceded ICD-11?

A retrospective look at the history of the WHO illustrates how TM and in particular TCM gained a foothold within the WHO. The Alma Ata conference in 1978, with its slogan "Health for all in the year 2000," stated that TM had a place in primary care. Between 1984 and 2007, the WHO made several attempts to standardize the various nomenclatures used within TCM as practiced in China, Japan, and Korea with the aim of promoting its acceptance. The last official WHO-endorsed standard on this topic dates from 2007 and included the description of fourteen meridians, 361 classical acupuncture points, eight extra meridians, forty-eight extra points, and fourteen acupuncture lines on the skull (WHO 2007).

The WHO Traditional Medicine Strategy 2002–2005 policy plan was published in May 2002. The document praised countries such as China, North Korea, and South Korea because they managed to have TM fully integrated into their health system. It presented the native flora as an undiscovered source of new medicines, of which the supposed benefits should be protected as intellectual property. The resentment against "Western" medicine can be felt on every page. Dr. Xiaorui Zhang, a former barefoot doctor[1] who later studied medicine in the United States, was one of the main persons responsible for this policy plan, which became "the first global strategy on traditional and alternative medicine." When we confronted then Dutch Minister of Health Hans Hoogervorst with this scandalous WHO endorsement of TCM, he plainly replied that every member state is free to ignore the recommendations by the WHO, which has no supranational powers.

The report *Acupuncture: Review and Analysis of Reports on Controlled Clinical Trials*, also under WHO auspices, appeared in 2003. It asserted that the efficacy of acupuncture in acute dysentery, hay fever, leucopenia, anovulation, and rheumatoid arthritis, among other ailments, had been proven. This uncritical review was performed without any kind of peer review by non-acupuncturists. In 2014, this scandalous report was tacitly removed from the WHO website, but it was never formally retracted. At the end of November 2004, Zhang distributed a similar draft report on homeopathy. It was offered for review to a few homeopaths, but its contents were leaked. This overview, *Homeopathy: Review and Analysis of Reports on Controlled Clinical Trials*, was as prejudiced as the report on acupuncture: various indications were mentioned for which efficacy of homeopathy was said to have been demonstrated, e.g., tropical diarrhea in children, hay fever, incipient flu, fibromyalgia, and intestinal paralysis after abdominal surgery. The report also proposed unscientific and mythological explanations for the mechanism of action of submolecular diluted substances (diluted to the point where not even a single molecule of the alleged effective ingredient should remain). It never came to a final version of the report, possibly also as a result of criticism originating from Dutch and Flemish quackery fighters (Renckens 2005). WHO TM coordinator Zhang wrote a weak reply in the Dutch daily newspaper *NRC Handelsblad* in answer to our criticism (Zhang 2005), but nevertheless the WHO homeopathy report was aborted. We reported this affair in the *Skeptical Inquirer* (Renckens et al. 2005).

In 2014, the WHO Traditional Medicine Strategy 2014–2023 appeared with content comparable to that of its predecessor from 2002: no reliance on evidence; instead, an emphasis on "real-life studies," reimbursements by health insurers, and commercialization of TCM (Dorlo et al. 2015).

TCM as Part of the ICD

Already in 2010, the WHO had indicated that it was aiming to merge TCM-diagnoses into the ICD (WHO 2010). This planned integration of the International Classification of Traditional Medicine (ICTM) into the "family of other WHO-classifications" would eventually "enable unification of the conventional and the traditional medicine classifications" and "will facilitate enhanced acceptance" of TM, as stated in the WHO Background Document on ICTM (WHO 2010). On March 3, 2011, the WHO published the list of experts who should further give shape to the ICTM. Among a variety of Asian TCM-enthusiasts there was also the Dutch ethnopharmacologist Peter de Smet. The planned ICTM was going to be based on the WHO International Standard Terminologies on Traditional Medicine in the Western Pacific from 2007, which contains around 3,000 items. What the exact details will be of the TM-modules that will be featured in the ICD-11 is still unknown.

Outside the WHO domain, the Chinese government is pursuing a similar agenda—TCM as an export product—by attempts to have an International Organization for Standardization (ISO) certification for, for example, TCM herbs and sterilization procedures for acupuncture needles (Dorlo and Timmerman 2009). These attempts fit the overall goal of the Chinese government to enlarge China's herbal exports and to gain recognition for Chinese herbs, given that the Chinese herbs will never be able to receive a formal medical product registration for European or U.S. markets due to the "strict" requirements to demonstrate efficacy and safety. At this moment, ISO standards provide a false aura of reliability to thirty-three TCM products and "activities" of planting, from the sowing of ginseng seeds to an infrared moxibustion device, and another forty-three standards are in the making (International Organization for Standardization 2019). In a direct meeting between then WHO Director-General Margaret Chan and President Xi Jinping, the latter said straightforwardly that he counted on a good collaboration between China and WHO and that he expected the WHO would help with "promoting TCM and Chinese herbs to foreign countries." The Chinese government lobbied Chan repeatedly while attempting to increase TCM's acceptancy and suitability for export. This culminated, among other things, in the publication of purely commercial paid advertising supplements in *Nature* in 2011 and *Science* in 2014, in which the pseudoscientific articles received an approval in a preface written by then WHO Director-General Margaret Chan. (See David Gorski, "*Science* Sells Out: Advertising Traditional Chinese Medicine in Three Supplements," *Skeptical Inquirer*, May/June 2015.) In 2017, the value of the growing Chinese export of medicinal herbs had peaked to $295 million (Cyranoski 2018).

Promoting Quackery

There are no public debates in the World Health Assembly around WHO policy plans, such as the ICD-11, but these plans are prepared by WHO administrators, and their acceptance is the result of consensus. The Dutch representatives at WHO do not hesitate to point out that policy recommendations by WHO are not binding for any of its member states and the TM modules in the ICD-11 will therefore probably remain unused in the Netherlands and many other Western countries. Nevertheless, these kinds of consensus approvals and the apparent cynicism among WHO member state representatives grants international and highly regarded status to quacks and makes their practices potentially more viable for monetary reimbursement. This is especially so as the ICD-11 is used as a blueprint for this in many countries worldwide. Countries with less functional medical regulatory authorities might embrace TCM on the basis of the explicit WHO approval. TCM is already spreading widely across Africa (*The Economist* Editors 2018). If patients with HIV, tuberculosis, or malaria seek treatment at the increasing number of TCM clinics, they are risking their lives. For HIV, it has been shown that the complementary use of TM—in this case traditional medicine of African origin—has a negative effect on the success of conventional antiretroviral treatment, even if both approaches are combined (Moshabela et al. 2017).

Instead of cheering the use of TM, the WHO should strive to ensure that truly effective medical care, unhindered by TM mythology, becomes available and accessible worldwide, a situation that can be achieved by a fair and global distribution of wealth and economic growth. Critical WHO officials and member state representatives who favor evidence-based medicine should operate with this state of mind and should be instructed by their governments to distance their countries—for instance with minority reports disagreeing with reprehensible WHO initiatives.

Our Published Paper and a Rebuttal

An article similar to this one was published by us in Dutch in the Dutch medical journal *Medisch Contact* on February 28, 2019 (Renckens and Dorlo 2019). Peter de Smet, the aforementioned expert member of the ICTM Project Advisory Group (WHO 2011), reacted and defended the inclusion of TM in the ICD by arguing that TM diagnoses and interventions are still widely used worldwide and that this implementation could yield useful epidemiological information. The exact details of the TM inclusion in the ICD-11 are still shrouded with mystery, but the fact that even interventions would be part of the ICD-11 was a surprise to us. Apparently, the aforementioned WHO International Standard Terminologies on Traditional Medicine in the Western Pacific (2007) will be used for this purpose. In this document, twenty-one therapeutic principles and 347 treatment methods are mentioned. Two examples of these treatments: the *eight methods*, a collective term for diaphoresis, emesis, purgation, mediation, warming, clearing, tonification, and resolution; or: *disperse wind and discharge heat*, a therapeutic method to treat externally contracted wind with interior heat by using exterior-releasing medicinals and heat-clearing medicinals in combination. In addition, de Smet insisted that the efficacy of TM interventions for certain TM diagnoses could be investigated. However, to his regret, creation of the list of TM interventions had been delayed.

Our publication also caught the attention of Henk van Gerven, a member of the Dutch parliament on behalf of the Socialist Party. He asked the Minister of Health various written questions about the WHO policy and general state of affairs. The Minister stated that, within WHO, the Netherlands did not provide an opinion on TM in the ICD and left that to member state Romania, which responded on behalf of all the EU member states. The Western countries chose to comply with the wishes of the countries where TM is still being used. He also pointed out that entering the TM data for national health statistics is optional and that the Netherlands will further ignore this option. The Minister also stated that China's push for expansion of the ICD had been driven, at least partially, by commercial considerations, i.e., export of TCM products.

Conclusion

As expected, the ICD-11 was adopted at the 72nd World Health Assembly, and it will be implemented as planned in 2022. Although WHO spokesman Tarik Jasarevic stated that the inclusion of TM in the ICD is not an endorsement of its scientific validity, proponents of complementary and alternative medicine (CAM) will of course misuse the inclusion as such (Hunt 2019). A WHO News Release on this subject stated that the "ICD-11 is now fit for many uses, including clinical recording, primary care, patient safety, antimicrobial resistance, resource allocation, reimbursement, casemix, in addition to mortality and morbidity statistics." The enormous goodwill for TM and CAM within the WHO is evidenced by the warm welcome given quacky organizations such as EUROCAM (representing, among others, various homeopathy associations), the World Naturopathic Federation, and the World Chiropractic Association (World Naturopathic Federation 2019). All these organizations were invited by the WHO to attend the World Health Assembly, where it released the WHO Global Report on Traditional and Complementary Medicine 2019. This report calls the fact that 88 percent of all member states have formally developed policies, laws, regulations, programs, and offices for TM and CAM "a unique milestone."

That political and economic considerations play a more important role within WHO than medical evidence-based science remains difficult to comprehend and accept. This unfortunate fact should lead to discussion and reflection on the role and position that member states have within the WHO. That prescientific mythological concepts now have gained a serious position in the WHO morbidity and mortality classification and statistics can be regarded as a direct failure of the political consensus-strategy.

Cees N. M. Renckens, PhD, board member of the Dutch Society against Quackery (www.kwakzalverij.nl).

Thomas P. C. Dorlo, PhD, is a senior scientist and clinical pharmacology expert at the Netherlands Cancer Institute and member of the board of the Dutch Society against Quackery.

Note

1.   Barefoot doctors are farmers who received minimal basic medical and paramedical training and worked in rural villages in China. Their purpose was to bring health care to rural areas where urban-trained doctors would not settle.

References

Cyranoski, D. 2018. Why Chinese medicine is heading for clinics around the world. *Nature* 561(7724): 448–50.

Dorlo, T. P. C., W. Betz, and C. N. M. Renckens. 2015. WHO's strategy on traditional and complementary medicine. A disgraceful contempt for evidence-based medicine. *Skeptical Inquirer* 39(3): 42–45.

Dorlo, T. P. C. and H. Timmerman. 2009. China vaart wel bij kruidenexport. *Medisch Contact* 64(46): 1900–3.

Gorski, D. 2018. The World Health Organization: Embracing traditional Chinese medicine pseudoscience in ICD-11 (blog entry). *Respectful Insolence* (October 4). Available online at https://respectfulinsolence.com/2018/10/04/the-world-health-organization -embracing-traditional-chinese-medicine-pseudoscience-in-icd-11/.

Hong-Zhou, W., F. Zhao-Qin, and C. Pan-Ji. 2013. *Introduction to Diagnosis in Traditional Chinese Medicine.* Hackensack: World Century Publishing Corporation.

Hunt, K. 2019. Chinese medicine gains WHO acceptance but it has many critics. CNN. Available online at https://edition.cnn.com/2019/05/24/health/traditional-chinese -medicine-who-controversy-intl/index.html.

International Organization for Standardization. 2019. ISO/TC 249 Traditional Chinese Medicine. Available online at https://www.iso.org/committee/598435.html.

Maassen, H. 2018. WHO erkent traditionele Chinese geneeskunde. *Medisch Contact* 72(40):6.

Moshabela M., D. Bukenya, G. Darong, et al. 2017. Traditional healers, faith healers, and medical practitioners: The contribution of medical pluralism to bottlenecks along the cascade of care for HIV/AIDS in Eastern and Southern Africa. *Sexually Transmitted Infections* 93(Suppl 3): e052974. doi:10.1136/sextrans-2016-052974.

Renckens, C. N. M. 2005. Hoed u voor de homeopaten bij de WHO. *NRC Handelsblad* (March 31).

Renckens, C. N. M., and T. P. C. Dorlo. 2019. Hoe China de WHO misbruikt voor zijn eigen agenda. *Medisch Contact* 73(9): 23–25.

Renckens, C. N. M., T. Schoepen, and W. Betz. 2005. Beware of quacks at the WHO. *Skeptical Inquirer* 29(5): 12–14.

Scientific American Editors. 2019. The World Health Organization gives the nod to traditional Chinese medicine. Bad idea. *Scientific American* 320(4): 6.

The Economist Editors. 2018. La Clinique Chinoise. Chinese medicine is on the rise in Africa. *The Economist* (November 8).

WHO. 1984. Standard Acupuncture Nomenclature. Geneva: WHO.

———. 1993. Standard Acupuncture Nomenclature Second Edition. Manilla: WHO.

———. 2007. International Standard Terminologies on Traditional Medicine in the Western Pacific. Geneva: WHO.

———. 2010. Background Document on ICTM. Traditional Medicine in Health Information Systems: Integrating Traditional Medicine into the WHO Family of International Classifications. WHO /HSI/CTS 2010/03. Available online at https://sites.google.com/site/whoictm/home/ICTMProjectPlan.pdf.

———. 2011. ICTM Expert List of Participants. Available online at https://docs.google.com/viewer?a=v&pid=sites&srcid=ZGVmYXVsdGRvbWFpbnx3aG9pY3RtfGd4Oj Q2MWZjMDExZmYwNmNmZWY.

———. 2018. Press release: WHO Releases New International Classification of Diseases. Available online at http://www.who.int/news-room/detail/18-06-2018-who-releases-new-international-classification-of-diseases-(icd-11).

World Naturopathic Federation. On Facebook at https://www.facebook.com/1444688405800272/posts/2422802201322216?s=736930180&sfns=mo.

Zhang X. 2005. WHO propageert homeopathie niet. *NRC Handelsblad* (April 8).

CHAPTER 19

Everything Means Something in Viking

Brian Regal

Vol. 43, No. 6
November/December 2019

With the twenty-first century resurgence in racist bigotry in the Anglo-American and European world, racists have appropriated nineteenth-century Viking pseudohistory to bolster their belief in white supremacy.

The growing racist, white supremacist, alt-right movement (not to mention a plethora of recent television documentaries about ancient America) has taken on an incongruous aspect. They have embraced ancient Nordic, Viking culture. As strange as this sounds, it is nothing new. The fondness for Vikings has a long pedigree. In the early nineteenth century, some in the United States began to believe that rather than Christopher Columbus, Vikings—Leif Eriksen in particular—had discovered America. There was little evidence to support this notion, and what supporters held up as such was dubious at best. The belief that Eriksen discovered America was largely confined to the realms of America's wealthy, Protestant, "Native Born," big-city Caucasian elites. Theorists such as Harvard chemist-turned-amateur-historian Eben Norton Horsford published books and articles, lobbied politicians, surveyed sites, and put up statues declaring Leif Eriksen the "true" discoverer of America. Underneath the insistence upon the Viking theory of American Discovery was a deep-seated fear and hatred of immigrants—especially the Irish and Italian Catholic immigrants then pouring in—who were seen by the Viking fans as degenerate outsiders ruining the country. That underlying bigotry is also behind the current resurgence in Viking infatuation by those who believe that everything means something in Viking.

There are two Viking stories here. Genuine archaeological evidence of Norse explorers who reached North America centuries before Columbus was discovered in the 1960s. The Norse reached L'Anse aux Meadows in Newfoundland and established settlements there. There is no evidence, however, that they traveled much farther south than that. The other Viking theory, established long before L'Anse aux Meadows was discovered, is problematic. The original idea of the Norse discovery of America had specific individuals, most famously Leif Eriksen, having come to the new world and exploring and setting up encampments and even cities throughout New England and as far south as Florida. These claims have no evidence for them aside from a few Old Norse texts that are more literary and romantic than factual and a few supposed artifacts with even less veracity. These are the Vikings some have used to inspire dark, racist, anti-immigrant feelings.

Norse raiders—often erroneously referred to as Vikings—did wreak havoc upon the British Isles, Northern Europe, and even as far south as the Middle East during the tenth and eleventh centuries. They were widely and rightly feared. By the 1300s, however, these raids had ended, and the Norse had settled down across the region and become part of several separate, indigenous populations. So many people from Scandinavia had settled in northern France, for example, that the region came to be known as Normandy. Slowly, farming and commerce replaced conflict. The fearsome raiders who appeared as wild-eyed defilers all but vanished from the Western imagination. There are no mentions of the Norse, for example, in the Arthurian legends, and Beowulf was unknown in the United Kingdom until the nineteenth century.

Until the latter twentieth century, the only "evidence" that Norse people came to the New World prior to Columbus was a collection of late medieval manuscripts collectively known as the Vinland Sagas. This name comes from the belief that in his explorations of the lands west of Greenland, Leif Eriksen (who may or may not have been a historical personage) called the places he found Vinland. It was not until the latter part of the 1600s, however, that a few scholars from Iceland and Norway began to suggest that Vinland may have been North America. The texts tell a rousing story of adventure, warfare, love, murder, friendship, valor, and cowardice just as one would expect of a text meant to inspire, enthrall, and entertain an audience.

Violent warrior culture was only a part of Viking life. Most of these people—who had no homogeneous ethnicity or culture—were traders and farmers. In the Vinland Sagas, Leif Eriksen and his homicidal sister Freydis take center stage, but it is his friend, the farmer Thorfinn Karlsefni, who holds the story and the settlements together. Thorfinn and his wife, Gudrid, have a child named Snorri (if this is historically accurate, then Snorri Karlsefni—rather than Virginia Dare—was the first European born in the Americas). The modern-day alt-right Viking poseurs have little interest in farmers. They prefer the supposed manly mannishness of the warriors (though their choice of khaki pants, polo shirts, and tiki torches does have a tendency to undermine that image).

The nineteenth-century idea of the Vikings merged smoothly with a growing interest in the mythology of the Aryans, where it was adopted by Germanic nationalists and anti-Semites. For some, the image of the Viking represented the

Nordic or Aryan ideal of racial superiority. In America, Norse theory drew several proponents who were anti-Catholic and anti-immigrant bigots and used the idea of Viking discovery to bash Columbus in particular and the Catholic Church in general. In a fascinating twist on the current fantasy of Vikings as pagan sorcerers, in his *The Goths in New England* (1843) George P. Marsh argued that they were more than the stereotyped fur-covered ravagers. The Norse, or Goths as he called them, were not "the savage destructors and devastators that popular error has made them." They were bringing Protestantism to sweep away "the spiritual and intellectual tyranny of Rome." He argued the Puritans who came to New England were children of the Goths, "the noblest branch of the Caucasian race."

The great proponent of nineteenth-century Viking fandom was Eben Norton Horsford (1818–1893). Having become convinced that Leif Eriksen had discovered America, laid down a personal settlement, and helped found a city, Horsford tirelessly searched the region around his home in Cambridge, Massachusetts, and claimed to have found these very places. He and his daughter Nellie tirelessly promoted this idea. Though popular with the general public, few in the professional world of history and archaeology accepted his arguments. The Norse theory of American discovery quickly became part of a wider anti-immigrant, anti-Catholic movement.

A friend of Horsford, Thomas Gold Appleton (1812–1884), was a Boston lawyer-turned-Aryan aficionado. He said that Eriksen and the others were Aryans and as such "the greatest race in the history of Mankind." In musings on life he published as *A Sheaf of Papers* (1875), Appleton randomly evaluated the different ethnic groups coming to America. "The Latin races," he says, "are now being weighed in the balance and found wanting." Erik the Red, on the other hand, was the "flowering of the Alpine [read *Aryan*] Rose." He felt that "not without meaning, at the head of that swarm [of dark-skinned people] God has placed the Anglo-Saxon race."

Another virulent anti-Catholic Viking proponent was Maria Shipley (1843–1900). She argued that the Norse discovery of America concerned more than just whose feet first stepped upon the New World. She saw, as did several pro-Norse authors, that the very survival of the United States was at stake. To allow Columbus to continue to be credited with the European discovery of America would allow the Catholic Church inroads to power and influence in the United States that would lead ultimately to the country's downfall at the hands of a repressive religion and church hierarchy. Indeed, her book is less a reasoned presentation of facts and evidence supporting the Norse theory than a long rant against the evils of Catholicism (and eventually Protestant Christianity as well). She placed Nordic culture at the heart of American governmental and political development. She referred to the Vikings as a kind of equivalent to the "Pilgrim Fathers."

A far cry from today's misogynist Viking fans, Shipley viewed the Norse as a kind of fairy tale pagan, occult society with a nature-loving philosophy, female equality, and sexual license. Shipley combined her passion for Vikings with a growing admiration for "pagan" religion and an ardent feminism coupled with a consuming distaste for the Catholic Church and what she perceived as a Catholic conspiracy to take over America. Her book *The Icelandic Discoverers of America* (1888) was another long rant against all things Columbus and Catholic. She saw Columbus as little more than a cowardly slaver, whereas the Vikings were virile, manly men. Columbus, she argued, was a bit of an effeminate wuss. Current alt-right fans echo this sentiment in their deep-seated fears of a supposed growing effeminacy of modern man, emphasized by their odd obsession with milk and soy.[1]

By the early twentieth century, the Anglo-American Viking Theory obsession faded, only to return by the early twenty-first century. The interest in the Vikings became a part of the growing alt-right, white supremacist, and Neo-Nazi resurgence in America, England, and Western Europe. White supremacists have tried their awkward best to argue that Norse, like medieval, history shows a bygone world of Caucasian homogeneity where people of color were totally absent. They believe that this romantic white pseudohistory is one where no influence of other cultures has intruded. They argue that Norse culture is the fount of all racial superiority in the world. It is part of a wider phenomenon of pseudohistory in which amateur theorists have tried to revise the understanding of Medieval Europe and Civil War–era America. Like the Viking theorists of the nineteenth century, the twenty-first century enthusiasts use their belief to bolster their fear of immigrants, Jews, and Muslims—rather than Irish and Italians—while attempting to prop up their own masculinity. The alt-right firebrand Richard Spencer once said that women prefer alt-right men because of their virility. Women, Spencer said, prefer strong men "like cowboys or Vikings." YouTuber Mike Peinovich, while discussing the supposed conquest of Europe by immigrant Muslims, said, "It's time to become Vikings and wipe these people out."

Pseudohistory, like pseudoscience, is an intellectual endeavor that purports to follow the rules of academic historical research but does not. For pseudohistorians, a thin veneer of superficial scholarship covers the deeper rot at its heart. Academic historians pursue knowledge about the past by tracking down and digesting primary source documents, writings, books, and artifacts. They follow this material, while contextualizing it, so as to learn about the past. They do it without preconceived ideas or answers. They let the facts push them one way or another and arrive at some insight regardless of the political, cultural, or religious consequences. Pseudohistorians dispense with all this and begin with a "truth" they want confirmed. In the case of Viking theory, they believe Viking culture was as lily white as the snows they believe they hailed from. All efforts then go into proving this preconceived idea regardless of the facts and evidence. This approach is at the heart of the Neo-Viking movement.

To live their Viking fantasies, some alt-right aficionados like to dress up in odd combinations of modern helmets and body armor with their idea of period garb in a form of the worst cosplay ever. Despite their increasing use of violence, the overall effect of their costumes is more amusing than intimidating. Little shields with ersatz runic inscriptions proliferate. Their admiration owes more to nineteenth-century Germanic theater and twentieth-century cinema than any historical or archaeological reality. Scholars argue that the modern imagery these people have adopted is more Victorian than Viking.

The mash-up of Viking regalia and political violence and hate speech is not new. The Nazis had a barely contained arousal at Viking culture. They embraced the runic inscriptions as their own and styled much of their symbolism on what they considered their forebears. There was a Nazi army unit called The Viking Division. They too saw in the fantasy of Viking culture the mirage of their own antecedents of racial superiority. This helps account for the confluence of racist, Neo-Nazi, and Neo-Confederate imagery often seen at their gatherings and protests today (the preference for bogus history and science also means beliefs such as flat-earth, anti-vaccination, and UFOlogy have begun to bleed into them).

While most of the Viking aficionados are mere sound and fury, some of them have gone to violent extremes. Anti-racist protests have been met by violent alt-right counter protesters who quickly resort to physical group violence. Individuals have also taken up the Viking sword. In May 2017, Jeremy Joseph Christian, who stabbed and murdered people on the Portland, Oregon, light rail, posted "Hail Victory, Hail Vinland" online before his rampage. When the gangs of white supremacists marched in Charlottesville the following August, some carried cardboard shields with Norse runic symbols on them. One of them, James A. Fields Jr., intentionally drove his car into anti-Nazi protesters, killing Heather Heyer. He was later convicted and sent to prison. He will likely feel at home there. White supremacist prison organizations have used Viking symbolism and embraced the historical religion of the Vikings, Odinism, perverting it in the process. A more recent organization to come out of the same white supremacist/prison nexus is the Vinland Folk Resistance.

In an interesting turn that few of the modern Viking theory fans recognize, nineteenth-century Viking researchers rejected the pagan Odinism of tenth- and eleventh-century Norse people and claimed they were really Christians. Today's Viking enthusiasts embrace the pagan religious aspects of their heroes and try to package it as some sort of eternal Caucasian spirituality not ruined by Jews and other "mud people" they consider inferior. This has resulted in an unlikely backlash.

The racist embrace of Norse religions such as Odinism is as fantastical and misunderstood by its appropriators as any other hijacked aspect of Norse or medieval history. Asatru is a modern revival of the Old Norse religion. In one last ironic twist to this story, the modern followers of Asatru—which began in Iceland in the 1970s and has since grown around the world—have denounced the racist embrace. They argue Odinism/Asatru is a peaceful religion that does not recognize racial hatred. Indeed, they argue that the Viking era was multi-ethnic and had no concept of race as it is interpreted today. They have actively fought to distance themselves from what they consider the violent bastardization of their religion, and they consider themselves anti-racist Vikings.

The embrace of Viking history by the alt-right is a disturbing example of the attempt to subvert history to some evil end. It fits in with the "Irish Slave" myth, the delusion that the Democratic Party is the genuine party of racism and fascism, the suggestion that African slaves fought voluntarily for the Confederacy, and other forms of historical memetic nonsense. The idea that Leif Eriksen discovered America was, and still is, a desperate claim that America, as well as the United Kingdom and Western Europe in general, is a whites-only homeland settled right from the start by Nordic supermen. Native Americans are discounted as little more than brown-skinned, savage heathens who only got in the way of proud white expansion, and non-white, non-Christians are seen as savage defilers of all that is good and beautiful who must be stopped at all costs. While we do know now that the earliest Europeans to come to America (of which we are aware) were Norse, the particular Leif Eriksen story of Vinland has no supporting evidence. It is a wish-fulfillment fantasy meant to satisfy a troubling emotional need for everything to mean something in Viking.

If the genuine Vikings of the tenth and eleventh centuries could see what is going on now in their names, they would likely be puzzled. But Thorfinn and Gudrid are together in eternity. They hear not the call of the alt-right to a white paradise that never existed.

Brian Regal teaches the history of science at Kean University, New Jersey. His latest book is *Searching for Sasquatch: Crackpots, Eggheads and Cryptozoology* (Palgrave, 2011).

Note

1. In one of the odder aspects of current white supremacist Viking fandom, some alt-right types go on about how cow's milk is a symbol of white supremacy. It is genuine; it is what the Vikings drank. Soy, on the other hand, is for weaklings and the effeminate. Milk enhances male virility while soy destroys it. Also, veganism is for sissy boys while meat-eating is for real warriors.

Part III

SCIENCE AND SKEPTICISM

CHAPTER 20

The Scientist's Skepticism

Mario Bunge

Vol. 44, No. 6
November/December 2020

Those of us who question the beliefs in ghosts, reincarnation, telepathy, clair-voyance, psychokinesis, dowsing, astral influences, magic, witchcraft, UFO-abductions, graphology, psychic surgery, homeopathy, psychoanalysis, and the like, call ourselves "skeptics." By so doing, we wish to indicate that we adopt Descartes's famous methodical doubt. This is just initial distrust of extraordinary perceptions, thoughts, and reports. It is not that skeptics close their minds to strange events but that, before admitting that such events are real, they want to have them checked with new experiences or reasonings. Skeptics do not accept naively the first things they perceive or think; they are not gullible. Nor are they neophobic. They are just critical; they want to see evidence before believing.

Two Kinds of Skepticism

Methodical doubt is the nucleus of *methodological skepticism.* This kind of skepticism must be distinguished from *systematic skepticism*, which denies the possibility of any knowledge and therefore entails that truth is inaccessible, and the search for it vain. The skeptics of both varieties criticize naiveté and dogmatism, but whereas methodological skepticism urges us to investigate, systematic skepticism blocks research and thereby leads to the same result as dogmatism, namely, stagnation or worse.

The craftsman and the technologist, the manager and the organizer, as well as the scientist and the authentic philosopher, behave as methodological skeptics even if they have never heard about this approach—and even if they behave naively or dogmatically after work hours. In fact, in their professional work they are not gullible, nor do they disbelieve everything, but they mistrust any important idea that has not been put to the test and demand the control of data as well as the test of conjectures. They look for new truths instead of remaining content with a handful of dogmas, but they also hold certain beliefs.

For example, the electrician makes some measurements and tests the installation before delivering it; the pharmacologist tests the new drug before advising to proceed to its manufacture in industrial quantities; the manager orders marketing research before launching a new product; the editor asks for advice of referees before sending a new work to the printer; teachers test their students' progress before evaluating them; mathematicians attempt to prove or disprove their theorems; physicists, chemists, biologists, and psychologists design and redesign experiments by means of which they test their hypotheses; the sociologist, serious economist, and the political scientist study random samples of the populations they are interested in before announcing generalizations about them—and so on. In all these cases, people search for truth or efficiency and, far from admitting uncritically hypotheses, data, techniques, and plans, they bother to check them.

On the other hand, the theologians and school philosophers, the neoclassical economist and messianic politicians, as well as the pseudoscientists and counterculture gurus, indulge in the luxury of repeating dogmas that either are untestable or have failed rigorous tests. The rest, those of us who make a living working with our hands, producing or diffusing knowledge, organizing or managing organizations, are supposed to practice methodological doubt.

Methodological skepticism is a methodological, practical, and moral stance. Indeed, those who adopt it believe that it is foolish, imprudent, and morally wrong to announce, practice, or preach important ideas or practices that have not been put to the test or, worse, that have been shown in a conclusive manner to be utterly false, inefficient, or harmful. (Note the restriction to *important* beliefs; by definition, trivialities are harmless even when false.)

Because we trust research and action based on research findings, we are not systematic skeptics. We disbelieve falsity and suspend judgment concerning whatever has not been checked, but we believe, at least temporarily, whatever passed the requisite tests. At the same time, we are willing to give up whatever beliefs prove to be groundless. In sum, methodological skeptics are constructive. . . .

The Scientist's Skepticism

It is impossible to evaluate an idea in and by itself, independently of some system of ideas that is taken as a basis or standard. When examining any idea, we do so in the light of further ideas that we do not question at the moment; absolute doubt would be as irrational as absolute belief. Hence systematic or radical skepticism is logically untenable. By the same token, every methodological skeptic has some creed or other, however provisional it may be.

For example, we evaluate a mathematical theorem in light of its premises and the laws of logic—and in turn the latter are evaluated by their fertility and consistency with mathematics. We judge a physical theory by its logical consistency and its mathematical tidiness as well as according to its harmony with other physical theories and its correspondence with the relevant empirical data. We evaluate a chemical theory according to the physical theories it takes for granted and according to whether or not it jibes with other chemical theories as well as with the relevant experimental data. We proceed in a similar manner with the remaining sciences. In particular, we demand that psychology does not violate any biological laws and that the social sciences respect psychology and harmonize with one another. (The fact that mainstream economics and politology do not care for other social sciences is precisely a point against them.)

In other words, the scientist's skepticism is methodological and partial, not systematic and total. Serious researchers are neither gullible nor nihilistic; they do not embrace beliefs uncritically, but do admit, at least until new notice, a host of data and theories. Their skepticism is constructive, not just critical.

Moreover, in every case the methodological skeptic presupposes—albeit seldom explicitly—that scientific theories and methods satisfy certain philosophical requirements. These are *(a) materialism*: everything in the universe is concrete or material, though not necessarily corporeal, and everything behaves lawfully; *(b) realism*: the world exists independently of those who study it, and moreover it can be known at least partially and gradually; *(c) rationalism*: our ideas ought to be internally consistent and they should cohere with one another; *(d) empiricism*: every idea about real things should be empirically testable; and *(e) systemism*: the data and hypotheses of science are not stray but constitute a system (for details, see Bunge 1983a, 1983b).

No doubt, few scientists realize that these five principles are indeed presupposed in scientific research. However, it does not take much to show that *(a)* if any of the preceding principles were relinquished, scientific research would miscarry, and *(b)* the main difference between science and pseudoscience is not so much that the former is true and the latter false, but that pseudoscience does not abide by those principles—as a consequence of which it seldom delivers truth and it never corrects itself.

Not all skeptics share these philosophical principles. Most of them believe that only scientific method applied to data-gathering is required to conduct scientific research. However, it is possible to apply the scientific method to a nonscientific investigation, such as trying to measure the speed of ghosts or the intensity of the action of mind on matter. To yield knowledge, the scientific method must be accompanied by a scientific worldview: materialist, realist, rationalist, empiricist, and systemic. This is the core of the skeptic's credo.

Conclusion

Methodological skeptics are not gullible, but they do not question everything at once either. They believe whatever has been demonstrated or has been shown to have strong empirical support. They disbelieve whatever clashes with logic or with the bulk of scientific knowledge and its underlying philosophical hypotheses. Theirs is a qualified, not indiscriminate skepticism.

The methodological skeptics uphold many principles and, above all, they trust humans to advance even further in the knowledge of reality. Their faith is critical, not blind; it is the explorer's faith, not the believer's. They do not believe anything in the absence of evidence, but they are willing to explore bold new ideas if they find reasons to suspect that they have a chance (see, for example, Bunge 1983c). They are open-minded but not blank-minded; they are quick to filter out intellectual rubbish.

For example, methodological skeptics find no reason to engage in experimental investigations before denying that pure thought may energize machines, that surgery can be performed by sheer mental power, that magical incantations or solutions of one part in ten raised to the 100th power have therapeutic power, that perpetual-motion machines can be built, or that there are solve-all theories. All such beliefs can be demolished by wielding some well-tested scientific or philosophical principles. This strategy is certainly cheaper than naive empiricism.

To conclude: Pseudoscience and pseudotechnology are the modern versions of magical thinking. They must be subjected to critical scrutiny not only to clean up culture but also to prevent quacks from cleaning out our pockets. And to criticize them it is not enough to show that they lack empirical support—for, after all, one might believe that such support could be forthcoming. We must also show that those beliefs in the arcane or the paranormal clash with either well-established scientific theories or fertile philosophical principles. For this reason, the criticism of magical thinking should be a common endeavor of scientists, technologists, philosophers, and educators. Given the massive commercial exploitation of junk culture, as well as the current decline in the teaching of science and technology in numerous countries, unless we work harder on debunking pseudoscience and pseudotechnology, we shall be in for a sharp decline of modern civilization (Bunge 1989).

Mario Bunge, who died in February 2020 at the age of 100, was both a physicist and a philosopher. He spent many decades in the Foundations and Philosophy of Science Unit at McGill University in Montreal. This article was excerpted by permission from "A Skeptic's Beliefs and Disbeliefs," the lead article of a special issue of *New Ideas in Psychology* devoted to "Mario Bunge on Nonscientific Psychology and Pseudoscience: A Debate" (vol. 9, no. 2,

1991). That journal's editor at the time, Pierre Moessinger, wrote in a leadoff editorial, "Mario Bunge is one of the few great philosophers of science of our times. More precisely he is both a philosopher and a scientist. . . . What is so extraordinary about Bunge is the breadth of his intellectual interests and his ability to coordinate and synthesize problems." Bunge's paper was originally presented at a CSICOP conference on "Magical Thinking and Its Prevalence in the World Today" in Mexico City in 1989. Bunge was a longtime CSICOP/ CSI fellow and frequent contributor to *Skeptical Inquirer*.

References

Bunge, M. 1983a. *Exploring the World*. Dordrecht, The Netherlands: Reidel

————. 1983b. *Understanding the World*. Dordrecht, The Netherlands: Reidel.

————. 1983c. Speculation: Wild and sound. *New Ideas in Psychology* 1: 3–6.

————. 1989. The popular perception of science in North America. *Transactions of the Royal Society of Canada*, ser. V, 4: 269–280.

CHAPTER 21

The Selfish Gene Revisited

Richard Dawkins

Vol. 41, No. 2
March/April 2017

Scientists, unlike politicians, can take pleasure in being wrong. A politician who changes his mind is accused of "flip-flopping." Tony Blair boasted that he had "not got a reverse gear." Scientists on the whole prefer to see their ideas vindicated, but an occasional reversal gains respect, especially when graciously acknowledged. I have never heard of a scientist being maligned as a flip-flopper.

In some ways I would quite like to find ways to recant the central message of *The Selfish Gene*. So many exciting things are fast happening in the world of genomics, it would seem almost inevitable—even tantalizing—that a book with the word "gene" in the title would, forty years on, need drastic revision if not outright discarding. This might indeed be so, were it not that "gene" in this book is used in a special sense, tailored to evolution rather than embryology. My definition is the population geneticists' definition adopted by George C. Williams, one of the acknowledged heroes of the book, now lost to us along with John Maynard Smith and Bill Hamilton: "A gene is defined as any portion of chromosomal material that potentially lasts for enough generations to serve as a unit of natural selection." I pushed it to a somewhat facetious conclusion: "To be strict, this book should be called . . . *The slightly selfish big bit of chromosome and the even more selfish little bit of chromosome*." As opposed to the embryologist's concern with how genes affect phenotypes, we have here the neo-Darwinist's concern with changes in frequencies of entities in populations. Those entities are genes in the Williams sense (Williams later called that sense the "codex"). Genes can be counted and their frequency is the measure of their success. One of the central messages of this book is that the individual organism doesn't have this property. An organism has a frequency of one, and therefore cannot "serve as a unit of natural selection" (not in the same sense of *replicator* anyway). If the organism is a unit of natural selection, it is in the quite different sense of gene "vehicle." The measure of its success is the frequency of its genes in future generations, and the quantity it strives to maximize is what Hamilton defined as "inclusive fitness."

A gene achieves its numerical success in the population by virtue of its (phenotypic) effects on individual bodies. A successful gene is represented in many bodies over a long period of time. It helps those bodies to survive long enough to reproduce in the environment. But the environment means not just the external environment of the body—trees, water, predators, etc.—but also the internal environment, and especially the other genes with which the selfish gene shares a succession of bodies through the population and down the generations. It follows that natural selection favours genes that flourish in the company of other genes in the breeding population. Genes are indeed "selfish" in the sense promoted in this book. They are also *cooperative* with other genes with which they share, not just the present particular body, but bodies in general, generated by the species' gene pool. A sexually reproducing population is a cartel of mutually compatible, cooperating genes: cooperating today because they have flourished by cooperating through many generations of similar bodies in the ancestral past. The important point to understand (it is much misunderstood) is that the cooperativeness is favoured, not because a group of genes is naturally selected as a whole, but because individual genes are separately selected against the background of the other genes likely to be met in a body, and this means the other genes in the species' gene pool. The pool, that is, from which every individual of a sexually reproducing species draws its genes as a sample. The genes of the species (but not other species) are continually meeting each other—and cooperating with each other—in a succession of bodies.

We still don't really understand what drove the origin of sexual reproduction. But a consequence of sexual reproduction was the invention of *the species* as the habitat of cooperating cartels of mutually compatible genes. As explained in the chapter called "The Long Reach of the Gene," the key to the cooperation is that, in every generation, all the genes in a body share the same "bottlenecked" exit route to the future: the sperms or eggs in which they aspire to sail into the next generation. *The Cooperative Gene* would have been an equally appropriate title for this book, and the book itself would not have changed at all. I suspect that a whole lot of mistaken criticisms could have been avoided.

Another good title would have been *The Immortal Gene*. As well as being more poetic than "selfish," "immortal" captures a key part of the book's argument. The high fidelity of DNA copying—mutations are rare—is essential to evolution by natural selection. High fidelity means that genes, in the form of exact informational copies, can survive for millions of years. Successful ones, that is. Unsuccessful ones, by definition, don't. The difference wouldn't be significant if the potential lifespan of a piece of genetic information was short anyway. To look at it another way, every living individual has been built, during its embryonic development, by genes which can trace their ancestry through a very large number of generations, in a very large number of individuals. Living animals have inherited the genes that helped huge numbers of ancestors to survive. That is why living animals have what it takes to survive—and reproduce. The details of what it takes vary from species to species—predator or prey, parasite or host, adapted to water or land, underground or forest canopy—but the general rule remains.

A central point of the book is the one developed by my friend the great Bill Hamilton, whose death I still mourn. Animals are expected to look after not only their own children but other genetic relatives. The simple way to express it, and the one that I favour, is Hamilton's Rule: a gene for altruism will spread if the cost to the altruist, C, is less than the value, B, to the beneficiary, devalued by the coefficient of relatedness, r, between them. r is a proportion between 0 and 1. It has the value 1 for identical twins; 0.5 for offspring and full siblings; 0.25 for grandchildren, half-siblings, and nieces; 0.125 for first cousins. But when is it zero? What is the meaning of zero on this scale? This is harder to say, but it is important and it was not fully spelled out in the first edition of *The Selfish Gene*. Zero does *not* mean that the two individuals share no genes in common. All humans share more than 99 percent of our genes, more than 90 percent with a mouse, and three-quarters of our genes with a fish. These high percentages have confused many people into misunderstanding kin selection, including some distinguished scientists. But those figures are not what is meant by r. Where r is 0.5 for my brother (say), it is zero for a *random member of the background population with whom I might be competing*. For purposes of theorizing about the evolution of altruism, r between first cousins is 0.125 only when compared to the reference background population (r = 0), which is the rest of the population to whom altruism potentially might have been shown: competitors for food and space, fellow travellers through time in the environment of the species. The 0.5 (0.125, etc.) refers to the *additional* relatedness *over and above* the background population, whose relatedness approaches zero.

Genes in the Williams sense are things you can count as the generations go by, and it doesn't matter what their molecular nature is; it doesn't matter, for instance, that they are split up into a series of "exons" (expressed) separated by mostly inert "introns" (ignored by the translation machinery). Molecular genomics is a fascinating subject but it doesn't heavily impinge on the "gene's eye view" of evolution, which is the central theme of the book. To put the point another way, *The Selfish Gene* is quite likely a valid account of life on other planets even if the genes on those other planets have no connection with DNA. Nevertheless, there are ways in which the details of modern molecular genetics, the detailed study of DNA, can be gathered into the gene's eye fold and it turns out that they vindicate my view of life rather than casting doubt on it. I'll come on to this after what may seem like a radical change of subject, beginning with a specific question, which obviously stands for any number of similar questions.

How closely related are you to Queen Elizabeth II? As it happens, I know I'm her fifteenth cousin twice removed. Our common ancestor is Richard Plantagenet, 3rd Duke of York (1412–1460). One of Richard's sons was King Edward IV, from whom Queen Elizabeth is descended. Another son was George, Duke of Clarence, from whom I am descended (allegedly drowned in a butt of Malmsey wine). You may not know it but you are very probably closer to the Queen than fifteenth cousin and so am I and so is the postman. There are so many different ways of being somebody's distant cousin, and we are all related to each other in many of those ways. I know that I am my wife's twelfth cousin twice removed (common ancestor George Hastings, 1st Earl of Hunting-don, 1488–1544). But it is highly probable that I am a closer cousin to her in various unknown ways (various pathways through our respective ancestries), and it is absolutely certain I am also her more distant cousin in many more ways. We all are. You and the Queen might simultaneously be ninth cousins six times removed, and twentieth cousins four times removed, and thirtieth cousins eight times removed. All of us, regardless of where in the world we live, are not only cousins of each other. We are cousins in hundreds of different ways. This is just another way of saying we are all members of the background population among whom r, the coefficient of relatedness, approaches zero. I could calculate r between me and the Queen using the one pathway for which records exist, but it would, as the definition requires, be so close to zero as to make no difference.

The reason for all that bewildering multiplicity of cousinship is sex. We have two parents, four grandparents, eight great-grandparents, and so on, up to astronomical numbers. If you go on multiplying by two back to the time of William the Conqueror, the number of your ancestors (and mine, and the Queen's and the postman's) would be at least a billion, which is more than the world population at the time. That calculation alone proves that, wherever you come from, we share many of our ancestors (ultimately all if you go sufficiently far back) and are cousins of each other many times over.

All that complexity disappears if you look at cousinship from the gene's point of view (the point of view advocated, in different ways, throughout this book) as opposed to the individual organism's point of view (as has been con-ventional among biologists). Stop asking: What kind of cousin am I to my wife (the postman, the Queen)? Instead, ask the question from the point of view of a single gene, say my gene for blue eyes: What relation is my blue eye gene to the postman's blue eye gene? Polymorphisms such as the ABO blood groups go way back in history, and are shared by other apes and even monkeys. The A gene in a human sees the equivalent gene in a chimp as a closer cousin than the B gene in a human. As for the SRY gene on the Y chromosome, which determines maleness, my SRY gene "looks upon" the SRY gene of a kangaroo as its kissing cousin.

Or we can look at relatedness from a mitochondrion's point of view. Mitochondria are tiny bodies teeming in all our cells, vital to our survival. They reproduce asexually and retain the remnants of their own genomes (they are remotely descended from free-living bacteria). By the Williams definition, a mitochondrial genome can be thought of as a single "gene." We get our mitochondria from our mothers only. So if we were now to ask how close is the cousinship of your mitochondria to the Queen's mitochondria there is a single answer. We may not know what that answer is, but we do know that her mitochondria and yours are cousins in only one way, not hundreds of ways as is the case from the point of view of the body as a whole. Trace your ancestry back through the generations, but always only through the maternal line and you follow a single narrow (mitochondrial) thread, as opposed to the ever branching thread of "whole organism pedigrees." Do the same for the Queen, following her narrow maternal thread back through the generations. Sooner or later the two threads will meet and now, by simply counting generations along the two threads, you can easily calculate your mitochondrial cousinship to the Queen.

What you can do for mitochondria, you can in principle do for any particular gene, and this illustrates the difference between a gene's point of view and an organism's point of view. From a whole organism point of view you have two parents, four grandparents, eight great-grandparents, etc. But, like a mitochondrion, each gene has only one parent, one grandparent, one great-grandparent, etc. I have one gene for blue eyes and the Queen has two. In principle we could trace the generations back and discover the cousinship between my blue eye gene and each of the Queen's two. The common ancestor of our two genes is called the "coalescence point." Coalescence analysis has become a flourishing branch of genetics and very fascinating it is. Can you see how congenial it is to the "gene's eye view" that this whole book espouses? We are not talking about altruism anymore. The gene's eye view is flexing its muscles in other domains, in this case looking back at ancestry.

You can even investigate the coalescence point between two alleles in one individual body. Prince Charles has blue eyes and we can assume that he has a pair of blue eye alleles opposite each other on Chromosome 15. How closely related to each other are Prince Charles' two blue eye genes, one from his father, one from his mother? In this case we know one possible answer, only because royal pedigrees are documented in ways that most of our pedigrees are not. Queen Victoria had blue eyes and Prince Charles is descended from Victoria in two ways: via King Edward VII on his mother's side; and via Princess Alice of Hesse on his father's side. There's a 50 percent probability that one of Victoria's blue eye genes peeled off two copies of itself, one of which went to her son,

Edward VII, and the other to her daughter Princess Alice. Further copies of these two sibling genes could easily have passed down the generations to Queen Elizabeth II on one side and Prince Philip on the other, whence they were reunited in Prince Charles. This would mean the "coalescence" point of Charles's two genes was Victoria. We do not—cannot—know whether this actually is true for Charles' blue eye genes. But statistically it has to be true that many of his pairs of genes coalesce back in Victoria. And the same kind of reasoning applies to pairs of your genes, and pairs of my genes. Even though we may not have Prince Charles' well-documented pedigree to consult, any pair of your genes could, in principle, look back at their common ancestor, the coalescence point at which they were "peeled off" from a single ancestral gene.

Now, here's something interesting. Although I can't establish the exact coalescence point of any particular allelic pair of my genes, geneticists can in principle take all the pairs of genes from any one individual and, by considering all possible pathways back through the past (actually not all possible pathways because there are too many, but a statistical sample of them), derive a pattern of coalescences across the whole genome. Heng Li and Richard Durbin of the Sanger Institute in Cambridge realized a remarkable thing: the pattern of coalescences among pairs of genes in the genome of a single individual gives us enough information to reconstruct demographic details about datable moments in the prehistory of an entire species.

In our discussion of coalescence between pairs of genes, one from the father and one from the mother, the word *gene* means something a bit more fluid than the normal usage of molecular biologists. Indeed, you could say that the coalescence geneticists have reverted to something a bit like my "slightly selfish big bit of chromosome and even more selfish little bit of chromosome." Coalescence analysis is studying chunks of DNA which might be larger or even smaller than a molecular biologist's understanding of a single gene but which can still be seen as cousins of each other, having been "peeled off" from a shared ancestor some definite number of generations ago.

When a gene (in that sense) "peels off" two copies of itself and gives one to each of two offspring, the descendants of those two copies may, over time, accumulate differences due to mutation. These may be "under the radar" in the sense that they don't show up as phenotypic differences. The mutated differences between them are proportional to the time that has elapsed since the split, a fact that biologists make good use of, over much greater time spans, in the so-called "molecular clock." Moreover, the pairs of genes whose cousinship we are calculating needn't have the same phenotypic effects as each other. I have one blue eye gene from my father paired with one brown eye gene from my mother. Although these genes are different, even they must have a coalescence somewhere in the past: the moment when a particular gene in a shared ancestor of my two parents peeled off one copy for one child and another copy to its sibling. That coalescence (unlike the two copies of Victoria's blue eye gene) was a long time ago, and the pair of genes has had a long time to accumulate differences, not least the difference in the eye colours that they mediate.

Now, I said that the coalescence pattern within one individual's genome can be used to reconstruct details of demographic prehistory. Any individual's genome can do this. As it happens, I am one of the people in the world who has had his complete genome sequenced. This was for a television programme called *Sex, Death and the Meaning of Life* which I presented on Channel Four in 2012. Yan Wong, my co-author of *The Ancestor's Tale*, from whom I learned everything I know about coalescence theory and much else besides, seized upon this and did the necessary Li/ Durbin style calculations using my genome, and my genome alone, to make inferences about human history. He found a large number of coalescences around 60,000 years ago. This suggests that the breeding population in which my ancestors were embedded was small 60,000 years ago. There were few people around, so the chance of a pair of modern genes coalescing in the same ancestor back at that time was high. There were fewer coalescences 300,000 years ago, suggesting that the effective population size was larger. These figures can be plotted as a graph of effective population size against time. Here's the pattern he found, and it is the same pattern as the originators of the technique would expect to find from any European genome.

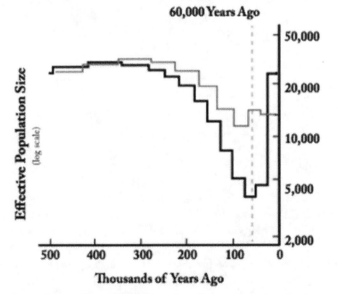

A comparison of breeding population over time suggests a large number of coalescences around 60,000 years ago in the author's genome. *Courtesy of Y. Wong, 2006.*

The black line shows the estimates of effective population size at various times in history based upon my genome (coalescences between genes from my father and my mother). It shows that the effective population size among my ancestral population plummeted around 60,000 years ago. The grey line shows the equivalent pattern derived from the genome of a Nigerian man. It also shows a drop in population around the same time, but a less dramatic one. Perhaps whatever calamity caused the drop was less severe in Africa than in Eurasia.

Incidentally Yan was my undergraduate pupil in New College, Oxford, before I started learning more from him than he learns from me. He then became a graduate student of Alan Grafen, whom I had also tutored as an undergraduate, who subsequently became my graduate student and whom I have described as being now my intellectual mentor. So Yan is both my student and my grandstudent—a neat memetic analogue to the point I was making earlier about how we are related in multiple ways—although the direction of cultural inheritance is more complicated than this simple formulation implies.

To summarize, the gene's eye view of life, the central theme of this book, illuminates not just the evolution of altruism and selfishness, as expounded in previous editions. It also illuminates the deep past, in ways of which I had no inkling when I first wrote *The Selfish Gene* and which are expounded more fully in relevant passages (largely written by Yan, my co-author) of the second edition (2016) of *The Ancestor's Tale*. So powerful is the gene's eye view, the genome of a single individual is sufficient to make quantitatively detailed inferences about historical demography. What else might it be capable of? As foreshadowed by the Nigerian comparison, future analyses of individuals from different parts of the world could give a geographic dimension to these demographic signals from the past.

Might the gene's eye view penetrate the remote past in yet other ways? Several of my books have developed an idea which I called "The Genetic Book of the Dead." The gene pool of a species is a mutually supportive cartel of genes that have survived in particular environments of the past, both distant and recent. This makes it a kind of negative imprint of those environments. A sufficiently knowledgeable geneticist should be able to read out, from the genome of an animal, the environments in which its ancestors survived. In principle, the DNA in a mole *Talpa europaea* should be eloquent of an underground world, a world of damp, subterranean darkness, smelling of worms, leaf decay, and beetle larvae. The DNA of a dromedary, *Camelus dromedarius*, if we but knew how to read it, would spell out a coded description of ancient ancestral deserts, dust storms, dunes, and thirst. The DNA of *Tursiops truncatus*, the common bottle-nose dolphin, spells, in a language that we may one day decipher, "open sea, pursue fish fast, avoid killer whales." But the same dolphin DNA also contains paragraphs about earlier worlds in which the genes survived: on land when the ancestors escaped the attentions of tyrannosaurs and allosaurs long enough to breed. Then, before that, parts of the DNA surely spell out descriptions of even older feats of survival, back in the sea, when the ancestors were fish, pursued by sharks and even eurypterids (giant sea scorpions). Active research on "The Genetic Book of the Dead" lies in the future. Will it colour the epilogue of the fiftieth anniversary edition of *The Selfish Gene?*

Richard Dawkins, FRS, is a renowned evolutionary biologist and emeritus professor of the Public Understanding of Science at Oxford. He is a senior editor and columnist for *Free Inquiry* and author of many books on science and atheism.

CHAPTER 22

Authority and Skepticism

Daniel C. Dennett

Vol. 40, No. 5
September/October 2016

When I was a child, the conversations around the dinner table in our house were especially vigorous and impassioned (as I soon learned when I discovered that I had to adjust my manner when I was a guest in other homes). When points of factual disagreement arose, this triggered the only grounds for being excused from the table in the middle of dinner: to look up the answer in the *World Book Encyclopedia*.

I remember vividly many occasions of triumphantly returning to the table, bulky blue volume in hand, to quote my vindication, and just as vividly the other occasions when I conceded defeat. I was wrong, and it says so right here in the *World Book*. The question of whether the *World Book* itself might be wrong seldom arose but was a recognized possibility.

Where would grownups turn today to settle similar disagreements? One of the unwelcome side effects of the mostly wonderful democratization of knowledge that has been ushered in by the age of the Internet is that we are losing consensus on what to consult when settling a bet. Sources of information that are *mutually recognized* as reliable—not perfect but reliable—are very useful assets for a society. If in the past we have often been overly submissive in the face of the epistemic authorities, today we risk swinging too far in the other direction and becoming knee-jerk, all-purpose skeptics. Skepticism across the board often sours into cynicism, and while a scattering of tolerated cynics in a society is probably a sign of health, when cynicism explodes into a pandemic, it can sap the enthusiasm of a people and threaten the security and coordination that permits a free society to operate. What people tend to forget is that all-purpose skepticism is too easy, a shtick that disables trust and makes resolute action on shared information more difficult.

We skeptics have a very important role to play of policing the epistemic environment, rooting out falsehoods and myths, and exposing charlatans and propagandists, but we cannot effectively do that job without endorsing and illustrating the *contrast* between these inferior and toxic products and methods and the (I use the word advisedly) *authority* of investigations done right. Does that mean that a good skeptic has to be some kind of *authoritarian*? Some kind of *elitist*? Yes. The good kind. Meritocracy has its place, and best practices are (usually) rightly called best practices. We must not be cowed by the chorus of oh-so modern (and postmodern) believers in epistemological democracy who decry the category of *expert*, replacing the distinction between amateur and professional with a lazy relativism that refuses to take sides.

How should we defend our acceptance of authorities when faced with the disapproving murmurs—sometimes rising to a roar—of *vox populi*? Carefully. We need to walk the tightrope between *appropriate* impatience with self-congratulatory ignorance on the one hand and on the other open-mindedness so tolerant that nonsense is granted "respect" that is mere lip service in any case. Like grade inflation, politically correct respect for all points of view threatens the quality control in thinking that modern society depends on, to put it bluntly. Not all points of view are equally "valid." One effective tactic is pointing out that these hyper-egalitarians know better than to trust their surgeries to amateurs or even novice professionals and wax indignant when they discover that somebody has *made a mistake* designing their car or advising them on how to make out their income tax returns. They settle for nothing less than expertise when it comes to the arrangements that affect their health, security, and comfort. They should have the consistency to honor expertise in other arenas as well. Moreover, when they point to the lapses and foibles in science, they invariably cite the policing campaigns of science itself, the most systematically self-critical institution the world has ever known. That should be enough (though of course it seldom is).

Daniel C. Dennett is university professor, professor of philosophy, and director of the Center for Cognitive Studies at Tufts University. Among his books is *Breaking the Spell: Religion as a Natural Phenomenon*; he is a fellow of the Committee for Skeptical Inquiry.

CHAPTER 23

Why Skepticism?

Steven Novella

Vol. 40, No. 5
September/October 2016

Twenty years ago, I became actively involved in the skeptical movement when I and several others founded a humble local skeptical group. We were inspired by CSICOP (now CSI) and *Skeptical Inquirer* to add what we could to efforts to make the world a more skeptical place.

Over the past two decades, the skeptical landscape has changed quite a bit, but one constant has been the endless question: What is skepticism? What exactly do we do and why? As the movement has grown and diversified, the question has become only more complex.

What Is the Mission of the Skeptical Movement?

I have come to understand that scientific skepticism is a weird beast that is often difficult to understand, especially from the outside. We are not exactly scientists or journalists or lobbyists or educators, and yet we are all of those things to some extent.

I think the best way to explain scientific skepticism is that it is expertise in everything that can go wrong with science and belief, and it includes execution, communication, education, and regulation. It combines knowledge of science, philosophy, and critical thinking with special expertise in flawed reasoning and deception.

To understand this better, here is a list of what scientific skeptics promote and do.

RESPECT FOR KNOWLEDGE AND TRUTH

Skeptics value reality and what is true. We therefore endeavor to be as reality-based as possible in our beliefs and opinions. This means subjecting all claims to a valid process of evaluation.

METHODOLOGICAL NATURALISM

Skeptics believe that the world is knowable because it follows certain rules or laws of nature. The only legitimate methods for knowing anything empirical about the universe follows this naturalistic assumption. In other words, within the realm of the empirical you don't get to invoke magic or the supernatural.

PROMOTION OF SCIENCE

Science is the only set of methods for investigating and understanding the natural world. Science is therefore a powerful tool and one of the best developments of human civilization. We therefore endeavor to promote the role of science in our society, public understanding of the findings and methods of science, and high-quality science education. This includes protecting the integrity of science and education from ideological intrusion or antiscientific attacks. This also includes promoting high-quality science, which requires examining the process, culture, and institutions of science for flaws, biases, weaknesses, conflicts of interest, and fraud.

PROMOTION OF REASON AND CRITICAL THINKING

Science works hand-in-hand with logic and philosophy, and therefore skeptics also promote understanding of these fields and the promotion of critical thinking skills.

SCIENCE VS. PSEUDOSCIENCE

Skeptics seek to identify and elucidate the borders between legitimate science and pseudoscience, to expose pseudoscience for what it is, and to promote knowledge of how to tell the difference.

IDEOLOGICAL FREEDOM/FREE INQUIRY

Science and reason can flourish only in a secular society in which no ideology (religious or otherwise) is imposed upon individuals or the process of science or free inquiry.

NEUROPSYCHOLOGICAL HUMILITY

Being a functional skeptic requires knowledge of all the various ways in which we deceive ourselves, the limits and flaws in human perception and memory, the inherent biases and fallacies in cognition, and the methods that can help mitigate all these flaws and biases.

CONSUMER PROTECTION

Skeptics endeavor to protect themselves and others from fraud and deception by exposing fraud and educating the public and policy-makers to recognize deceptive or misleading claims or practices.

ADDRESSING SPECIFIC CLAIMS

Skeptics combine all of the above to address specific claims that are flawed, biased, or pseudoscientific and to engage in the public discussion of these claims.

CULTURAL MEMORY

Skeptics as a whole act as the cultural memory for pseudosciences and scams of the past. Such beliefs tend to repeat themselves, and remembering the past can be very useful in quickly putting such beliefs into their proper perspective.

SCIENCE JOURNALISM

Many skeptics spend a large portion of their time doing straight science communication and journalism, which is important because science is so central to our mission. This is also an important skill to explore and develop because it is so rarely done well. Correcting and criticizing bad science news reporting, especially in the Internet age, has become a large part of what skeptics do.

What Topics Do We Cover?

Traditional skepticism addresses a very broad range of topics: all of alternative medicine, parapsychology, cryptozoology, conspiracy theories, scams, postmodernism, self-help, education, science and the media, neuroscience and self-deception, fringe science, and a long list of topics that have political, religious, or social implications: genetically modified foods, organic farming, free energy and other energy issues, climate change, creationism, miracle claims, faith-healing, prophesy, channeling—the list is massive.

There has been frequent discussion about which topics skeptics "should" cover. My approach has always been that everyone, of course, should feel free to cover whatever topics suit their interests, motivations, and talents. There are no right or wrong topics to cover.

There are, however, many considerations worth discussing. Skepticism is a method of applying science and critical thinking to all areas. It is worth thinking about how those methods relate to any particular topic of interest.

Here are some of the factors I consider when deciding what topics to address as part of my skeptical activism.

TEACHABLE MOMENT

One very important criterion is this: Would addressing a claim or topic provide a useful teachable moment? Since one (if not the) primary goal of skepticism is education, this is a crucial criterion, and in fact it is often sufficient reason to address a topic.

This is the primary reason I have never addressed issues such as ghosts, Bigfoot, astrology, or the Bermuda Triangle (classic skeptical topics all). I honestly don't care at all about ghosts, and I agree that this has extremely low priority as an issue. However, ghost hunters engage in a variety of pseudoscientific activities and defend their claims with numerous logical fallacies.

There are many generic lessons about science and critical thinking that can be learned by examining any pseudoscience, and often the most obvious ones are the best examples.

I have also found that by examining the full spectrum of pseudoscience, I have been able to see recurring patterns that enable me to understand pseudoscience much more thoroughly and then apply those lessons to more important areas such as medicine.

INTEREST

Related to the teachable moment criterion is public interest. The whole point is to engage the public, and one technique for doing so is to go to where the people already are. The public is interested in ghosts, cryptids, and UFOs, and in fact they often learn pathological science from popular treatments of these topics.

If we leave these popular subjects to the charlatans, they will happily spread scientific illiteracy unopposed. This is, however, a great opportunity to teach the public about how science actually operates, mechanisms of self-deception, how to tell if a claim is valid, and how to detect pseudoscience.

Addressing pseudoscience and the paranormal is a way to popularize science, such as writing about the physics of *Star Trek* or the philosophy of *The Simpsons*. Ghosts and UFOs are the hook; the payoff is scientific literacy and the ability to think a bit more critically.

IMPACT

The relative impact or importance of an issue is definitely important, and nothing I write here should be interpreted as dismissing or minimizing that point. In fact, as the skeptical movement has matured over the past few decades I have noticed a definite shift to issues of greater social importance.

My primary issue is alternative medicine, the abject infiltration of fraud and pseudoscience into the institutions of healthcare. This results in the wasting of billions of dollars and diverting of research funds, and it causes direct harm to the health of individuals.

Other important issues we tackle regularly are vaccine refusal, global climate change, genetically modified foods, our energy infrastructure, future technology, teaching creationism and other pseudoscience in science classes, issues surrounding mental illness, the self-help industry, scams, racial or gender pseudoscience, and other issues that have a direct impact on people's lives and our civilization. We also may consider how much of an effect we can have. Some issues are more amenable to scientific information than others.

EXPERTISE

The world needs all kinds of experts, and scientific skepticism is a legitimate area of expertise. It involves a deep knowledge of pseudoscience, the philosophy of science, mechanisms of deception, neuropsychological humility, scams, logic, and other aspects of critical thinking. This includes knowledge of the history of pseudoscience.

Within skepticism, individuals also tend to focus their writing and speaking on their area of scientific expertise. So, skeptical doctors focus on medicine, astronomers on astronomy, biologists on issues such as evolution and creation, physicists on free energy, and so on.

If we have a bias, it is toward the areas of expertise that also tend to attract people to the skeptical movement itself, but this is hard to avoid. It is also not simple to correct, and straying outside of our areas of expertise is not a good solution. At the very least, it takes a lot more work to address an issue about which I am not already fairly expert.

FILLING A NEED

Very relevant to the question of what targets skeptics choose is who else, if anyone, is already addressing those problems. For example, reviewing evidence and establishing a standard of care for a particular issue within mainstream medicine is very important, but there are already professional societies that do that. All physicians and scientists should be skeptical, but in areas where mainstream scientists are already doing just fine at addressing misperceptions, physicians who have an expertise in skepticism are not needed and have nothing particular to add.

We tend to focus our efforts where there is the most need, meaning where there is a current lack of attention. Fringe ideas tend not to get attention from scientists, who don't want to waste their time. Whether or not this is a reasonable position is debatable, but meanwhile skeptics are happy to fill the void. As a skeptical neurologist, for example, I am not going to spend my time delving into and engaging in debate over the possible mechanisms of Parkinson's disease. There are scientists who are doing that. But I will engage with those claiming that near-death experiences are evidence for an afterlife because most scientists don't bother to do that.

JOURNALISTIC INTEGRITY

This last criterion is a bit of a personal choice. Some journalists and outlets unapologetically advocate for a political ideology. Everyone knows the *Huffington Post* is a liberal news source, for example. Some journalists, however, try to be as politically neutral as possible so that they will be viewed as a fair arbiter of factual information and analysis.

Similarly, some skeptics combine their skeptical activism with ideological activism. I have no problem with this, and most are upfront about it. Some skeptics, however, choose to be political or ideologically neutral in their activism, except for a defense of science and reason. I think this can be helpful.

While I certainly do have political opinions, I try to keep them separate from questions of science and evidence. If, for example, I am discussing global warming, I want to focus on the science and not be dismissed as liberal. Or if I am writing about GMOs, I do not want to be dismissed as conservative or libertarian. That can still happen, sometimes simultaneously, because people make unwarranted self-serving assumptions, but it helps when it is untrue. My opinions on these and similar topics are informed by the science, not my politics. This becomes a harder sell when you are also advocating for a political position. I also think it is helpful to have a movement that is based upon evidence and logic and is agnostic toward ideological positions or values, which are tangential.

Finally, it is more challenging to be neutral and unbiased when dealing with an issue about which you have a passionate ideological belief. You can make a reasonable argument for steering clear of such issues when trying to communicate objective science, or at least proceeding especially carefully. Otherwise you risk damaging your reputation as a science communicator.

What about Religion?

More often than not, the question of what skeptics do and what topics we address comes to the topic of religion. While I believe I have addressed all the relevant issues above, this is a common enough question that it is worth special mention.

No skeptical activist I know treats religious claims differently from any other type of claims. Any claim to empirical truth or scientific knowledge, whether based ultimately in religious ideology or social or political ideology, is fair game. The criteria I outlined above apply. Philosophical arguments are also fair game, as the tools of logic and critical thinking apply.

However, it is important to recognize that faith statements are simply different from scientific or philosophical statements. This does not mean they are exempt from critical thinking; it just means you need to be aware of the context and address them properly. This distinction, however, is often misinterpreted as avoiding religious claims, which is patently not true.

When a believer states that they believe something to be true based upon their own personal faith, there are a number of valid approaches. It is important to point out that the principles of freedom of religion and separation of church and state require that their personal faith not be imposed upon others. They don't have a right to make other people follow their faith, to deprive their children of the basic necessities of life, or expect government to legislate their faith.

It is also useful to point out that beliefs based purely on faith are not subject to scientific analysis, and therefore they do not belong in the arena of science. They therefore cannot mix faith and science. Either the science stands on its own or it doesn't. You cannot legitimately use faith to rescue bad science from refutation or to render it immune to falsification.

This applies to faith-based beliefs that are not overtly religious. It is not uncommon for believers in alien visitation or extrasensory perception (ESP) to retreat to faith-based claims when the evidence does not support their position. They are effectively refuted by simply stating that they have left the arena of science and therefore have ceded this territory. If they wish to have a religion of ESP, then so be it, but they cannot simultaneously claim to be backed by science.

In other words, skeptics can address faith claims epistemologically without making the same epistemological error as the believer in order to falsely claim that empirical methods can disprove beliefs that are not empirically based.

Conclusion

Being part of the skeptical movement for most of my adult life has been extremely fulfilling and tremendously educational. After two decades, I still find it among the most rewarding work that I do. There is also a continued great need for the expertise and work of activist skeptics. We are engaged in work that will never be completed. Our goals are generational, to slowly move our species in the direction of science and reason.

Count me in until entropy prevails over my temporary biological processes.

Steven Novella, MD, is an assistant professor of neurology at Yale University School of Medicine. He is the host of the *Skeptics' Guide to the Universe* podcast, author of the *NeuroLogica* blog, executive editor of the *Science-Based Medicine* blog, and president of the New England Skeptical Society.

CHAPTER 24

Why Parapsychological Claims Cannot Be True

Arthur S. Reber and James E. Alcock

Vol. 43, No. 4
July/August 2019

The July/August 2018 issue of American Psychologist *contained an article titled "The Experimental Evidence for Parapsychological Phenomena: A Review" by Etzel Cardeña. Cardeña is known for research on hypnosis and consciousness, parapsychology, and, interestingly, for his work in theater as an actor and director. The paper prompted us to examine and critique the science behind parapsychology (Reber and Alcock, 2020). This article is a summary of our arguments.*

The *American Psychologist* is the flagship publication of the American Psychological Association (APA), the largest and most influential professional organization in our field, and it is sent to its nearly 120,000 members. An article being published within it is equivalent to granting the imprimatur of the APA. Interestingly, this wasn't the first time the APA had entered this controversial domain of psychology; in 2011, another of its respected journals, the *Journal of Personality and Social Psychology*, published a paper by Daryl Bem of Cornell University that purported to show evidence of precognition. Bem's paper lit a small firestorm largely because Bem was well-known for research in fields outside of parapsychology. It was applauded enthusiastically by psi researchers and, of course, was immediately subjected to efforts at replication by other labs (which almost uniformly failed) and well-honed criticisms, including one by one of us (Alcock 2011).

Cardeña's paper was, to the eye untutored in the world of the paranormal, an impressive effort. It reviewed the data for psi, focusing mainly on meta-analyses of published papers that showed small or marginal effects and, importantly, acknowledged the fact that there is no coherent theory for psi. Cardeña, in an effort to find a causal mechanism through which to understand the paranormal, brought in quantum mechanics (QM) and, to a lesser extent, relativity theory and the recently proposed notion of a "block universe" model in which past, present, and future all enjoy a simultaneous coexistence. The effort failed—mainly from some unfortunate misunderstandings of QM, relativity theory, and the fact that the block universe notion is little more than creative speculation.

While the paper bothered us on several levels, our primary concern was that it was symptomatic of a larger, more important issue that was being missed. It is not a matter of reviewing the existing database, scratching at the marginal and highly suspect findings of meta-analyses for something that passes the "< .05" cutoff point. It is not a matter of rummaging around in arcane domains of theoretical physics for plausible models. It is more basic than that: parapsychology's claims cannot be true. The entire field is bankrupt—and has been from the beginning. Each and every claim made by psi researchers violates fundamental principles of science and, hence, can have no ontological status.

We did not examine the data for psi, to the consternation of the parapsychologist who was one of the reviewers. Our reason was simple: the data are irrelevant. We used a classic rhetorical device, *adynaton*, a form of hyperbole so extreme that it is, in effect, impossible. Ours was "pigs cannot fly"—hence data that show they can are the result of flawed methodology, weak controls, inappropriate data analysis, or fraud. Examining the data may be useful if the goal is to challenge the veracity of the findings but has no role in the kinds of criticism we were mounting. We focused not on Cardeña specifically but on parapsychology broadly. We identified four fundamental principles of science that psi effects, were they true, would violate: *causality*, *time's arrow*, *thermodynamics*, and the *inverse square law*.

Causality

Effects have causes. Bridging principles identify the causal links for observed effects. The appropriate response to circumstances that lack such a mechanism is skepticism or an existential agnosticism—and, historically, this has been the case. Newton's notion of gravity as "action at a distance" was considered suspect until rescued by Einstein's relativity theory; mystics' claims to control autonomic functions were thought to be scams until the discovery of biofeedback; Wegener's theory of continental drift was viewed skeptically until mid-ocean ridges and sea-floor spreading were discovered.

Within the study of psi, there are no causal mechanisms, and none have been hypothesized. Worse, there is virtually no discussion over whether the claimed effects have singular or multiple causal mechanisms or why the purported findings lack coherency. If psychokinesis affects the roll of dice in a psi lab, why not at craps tables? If telepathy exists, why are our brains not constantly abuzz with the thoughts of those around us? To maintain that the future puts in appearances—but only in psychology labs at Lund or Cornell—is to strain credulity to the snapping point. There are no patterns here. As we noted in our paper, "It is as if actors from a dozen different plays have appeared on the same stage in a discordant farrago."

Time's Arrow

Within parapsychology time is turned upon itself, most glaringly in precognition. Psi researchers have tried to explain this through QM, in particular the "entanglement" effect. It won't work. It's true that the spin of two particles separated in space are *entangled* (the state of one is simultaneously aligned with the other), but there is no *reversal* of time, simply concurrent effects. In the so-called "twin paradox" the twins age at different rates, but neither gets younger. As we argued, "The notion that the strangeness of the quantum world harbors an explanation of the strangeness of parapsychology is a false equivalency." QM is a physical theory but not in the ordinary, Newtonian sense that we confront in daily life. As Nobelist Richard Feynman has quipped on many an occasion, "It is safe to say that nobody understands quantum mechanics"—outside of the mathematics. And as quantum theorist Jonathan Dowling noted in his 2013 book *Schrödinger's Killer App*, there is nothing in QM that entails paranormal effects. If anything, it rules them out (Dowling 2013).

Thermodynamics

Again, take precognition. If the future affected the present, it would violate the thermodynamic principle that energy cannot be created or destroyed in an isolated system. The act of choosing a card from a fixed array, a common procedure used in psi research, involves neurological processes that use measurable biomechanical energy. The choice is presumed to be caused by a future that, having no existential reality, lacks energy. And it won't work to argue that by virtue of time reversal the system isn't closed. If that were true, all systems would be subject to this "borrowing" of energy from the future, leading to the incoherent conclusion that the First Law no longer applies anywhere.

Inverse Square Law

In telepathy, the distance between the two linked persons is never reported to be a factor, a claim that violates the principle that signal strength falls off with the square of the distance traveled. Psi researchers again import the "entanglement" effect as a possible explanation for this, but it won't work. In QM, there is no transmission of energy between the separated particles; it is only that they are "entangled."

In short, parapsychology cannot be true unless the rest of science isn't. Moreover, if psi effects were real, they would have already fatally disrupted the rest of the body of science. If one's wishes and hopes were having a psychokinetic impact on the world—including computers and lab equipment—scientists' findings would be routinely biased by their hopes and beliefs. Results would differ from lab to lab whenever scientists had different aims. The upshot would be empirical chaos, not the (reasonably) ordered coherent picture developed over the past several centuries.

At the close of our paper, we wondered why parapsychology still exists as a field of study. Why are some scientists still focused on the impossible? For 150 years, we've witnessed a cycle. Evidence for psi is announced with fanfare then later falls into disregard. A new theory is proposed then abandoned. A novel methodology is introduced but, when findings are not replicated, is discarded. Each time there's a resurgence of interest when another apparently successful result is reported. Lather, rinse, repeat.

This enterprise has involved literally thousands of papers, hundreds of conferences, dozens of review volumes, and *nothing has been learned*. Parapsychology is precisely where it was in the 1880s. Why, we wondered, are researchers still running experiments, using ever-more sophisticated statistical techniques, reaching out to ever-broader realms of science, expanding their analyses into studies of consciousness and mind? This pattern of persistent belief in the anomalous may be the most psychologically interesting phenomenon associated with the study of psi. One of us (Alcock 1985) has argued it is likely linked with a vague sense that science, hard-nosed and physicalist, lacks that mysterianist element found in religious or spiritual realms. The lure of the "para"-normal emerges, it seems, from the belief that there is more to our existence than can be accounted for in terms of flesh, blood, atoms, and molecules. A century and a half of parapsychological research has failed to yield evidence to support that belief.

However, for us, the alluring mysteries are the ones that emerge from the straightforward study of the astonishingly complex, inviting world of normal science in all its mechanistic glory.

Note

Our paper was vetted by two experts in quantum mechanics, one of our choosing and the other chosen by the journal editor. Both assured us that our comments about physics in general and quantum mechanics in particular were correct.

Arthur S. Reber is a professor of psychology at Brooklyn College and at the Graduate Center of the City University of New York.

James E. Alcock is professor of psychology at York University, Toronto, Canada. He is author of *Parapsychology: Science or Magic?* and coeditor of *Psi Wars: Getting to Grips with the Paranormal.* He is a member of the Committee for Skeptical Inquiry's executive council and of the *Skeptical Inquirer* editorial board.

References

Alcock, James E. 1985. Parapsychology as a "spiritual" science. In P. Kurtz (ed.), *A Skeptic's Handbook of Parapsychology.* Amherst, NY: Prometheus Books, 537–565.

———. 2011. Back from the future: Parapsychology and the Bem affair. *Skeptical Inquirer* 35: 31–39.

Bem, Daryl J. 2011. Feeling the future: Experimental evidence for anomalous retroactive influences on cognition and affect. *Journal of Personality and Social Psychology* 100: 407–425.

Cardeña, Etzel 2018. The experimental evidence for parapsychological phenomena: A review. *American Psychologist* 73: 663–677.

Dowling, Jonathan P. 2013. *Schrödinger's Killer App: Race to Build the World's First Quantum Computer.* Boca Raton, FL: Taylor & Francis.

Reber, Arthur S., and James E. Alcock. 2020. Searching for the impossible: Parapsychology's elusive quest. *American Psychologist* 75(3): 391–399. https://doi.org/10.1037/amp0000486.

CHAPTER 25

Behe, Bias, and Bears (Oh My!)

Nathan H. Lents

Vol. 45, No. 2
March/April 2021

After a series of stinging defeats in the U.S. courts, one branch of biblical creationism mutated into a neocreationist movement known as intelligent design (ID) in the late 1980s. The selective pressure that led to this evolution was a single line in the 1987 *Edwards* decision by Supreme Court Justice William Brennan that emphasized differing views of human origins might be permissible in public schools if they were driven by "secular intent." This led some creationists to strip the religious language and present their ideas as purely scientific. No serious person was fooled by this game of rhetorical dress-up, least of all a George W. Bush–appointed federal judge who ruled that ID is not scientific and cannot be divorced from its religious motivations (Jones 2005). (For a history of the modern ID movement, see Barbara Forrest and Paul R. Gross's 2007 book, *Creationism's Trojan Horse: The Wedge of Intelligent Design*.)

Among the most obvious signs of the pseudoscientific nature of ID is the almost complete absence of relevantly qualified experts. Its leaders are mostly theologians, philosophers, and historians, along with some physicians, engineers, and mathematicians, but there are shockingly few scientists—and even fewer biologists—and almost none have research experience in evolutionary or molecular biology. This gives them no pause in attempting to challenge the entire scientific establishment, not through original research but through books and articles published on Evolution News, the most influential ID-supporting website. Although Evolution News presents itself as a science website, it frequently covers social issues such as abortion, religious freedom, gay marriage, euthanasia, and their favorite bogeyman: Marxism.

Another sign of the pseudoscientific nature of ID proponents is their confident assertions regarding the imminent ascendance of ID. Readers are assured that scientists are confused and desperate because the end is nigh for ideas such as the big bang and the whole of evolutionary theory. Claims such as, "Given a 'primal blueprint' that preceded the first life, the solution can only be intelligent design," are offered as scientific certainties even though scientists who actually study the origins of life see no such blueprint. Similarly, the past ten years have seen unprecedented discoveries that help illuminate our evolutionary past, yet, as Evolution News reports it, the discoveries of the 2010s "actually weakened the evidence for human evolution." Not a single active paleoanthropologist would agree.

The Author

One of the very few legitimately credentialed molecular biologists who subscribes to ID is Lehigh University biochemist Michael Behe. While most ID material is unworthy of a response, Behe is a clear and engaging writer with particular command of the molecular details of life. He is neither a rube nor a neophyte and writes with a polished veneer of scientific authority. In his 1996 book *Darwin's Black Box*, Behe asserts that many biological structures harbor the property of *irreducible complexity*, being so intricately complex with multiple interconnected parts that they could not possibly have evolved piecemeal, because the overall structure has no function until all parts are in place. To be sure, the challenge that irreducible complexity posed to the field of molecular evolution was rhetorical, not scientific, but it did draw attention to these important questions and inspired scientists to more fully flesh out the theoretical framework for how some complex molecular structures evolve and to better explain this framework to the public. For that reason, *Darwin's Black Box* could be charitably seen as a contribution to science.

Every single one of Behe's examples of irreducibly complex molecules and structures has since been thoroughly debunked by scientists. Russell Doolittle taught us how the blood clotting cascade likely evolved, with no reliance on incredibly unlikely steps (Doolittle et al. 2008). The bacterial flagellum evolved from the type III secretion machinery (Miller 2004). The vertebrate eye, and its stunningly similar counterpart in cephalopods, evolved in discrete steps of *increasing*, not irreducible, complexity (Lamb et al. 2008). Yet Behe continues to trumpet these examples and declares them "unchallenged."

The Thesis

Recently, I reviewed Behe's latest book, *Darwin Devolves*, for *Science*, the top scientific research journal in the United States (Lents et al. 2019). Behe argues that unguided random mutations serve primarily to damage genes and that doing so is occasionally good for the organism, leading to adaptation through natural selection of these damaging-but-advantageous mutations. Thus, Behe accepts that microevolution through random mutation can diversify organisms into species and genera—and perhaps families—but that something more is needed for large-scale evolutionary transitions.

Behe accepts the true age of the earth and admits that the evidence for the common ancestry of all life, including humans, is overwhelming. He sees a gap in evolutionary theory, however, in explaining the emergence of new kinds of organisms and concludes that it must require the intervention of the designer in some way. He stops short of providing evidence for the intervention and instead presents ID as the default position left standing after he shows how the unguided forces of evolution are insufficient for accounting for the origins and diversity of life.

But Behe does not show that. *Darwin Devolves* is mostly dedicated to explaining, often in great detail, how some high-profile examples of evolutionary research actually favor *his* view, rather than the interpretation of the scientists who did the work. As I will show, his discussion of every single example is misleading, sometimes egregiously so, insofar as he exaggerates the evidence that supports his view and ignores or dismisses the evidence that doesn't. But even more damning is his near-complete omission of any discussion of the molecular and evolutionary forces that are responsible for the very phenomena he focuses on.

The Omissions

If Behe seeks to support his claim that standard evolutionary forces are insufficient to generate adaptive innovations, one would think he would dedicate quite a bit of time to discussing those forces and why scientists are wrong about them. Instead, he takes the opposite approach and either summarily dismisses them or ignores them altogether. For example, one molecular mechanism that has driven otherwise incredibly improbable evolutionary events is horizontal gene transfer, when genetic material moves from one species to another, usually through a virus (Keeling and Palmer 2008). This phenomenon is most famous for the origin of mitochondria and chloroplasts, foundational events in the evolution of complex cells, but it has been shown to occur in more mundane instances as well. For example, deer ticks gained new defenses against bacteria through genes that came from the bacteria themselves (Chou et al. 2015). While certainly not a common occurrence, horizontal gene transfer can, in one momentous instant, have profound effects in the evolutionary potential of a lineage. Despite its importance in the very concepts that Behe tackles, he does not mention this concept even once in *Darwin Devolves*.

Also unmentioned by Behe is exaptation, the co-opting of a structure, be it a molecule or anatomy, for a new function. For example, two of the three bones of the mammalian middle ear were co-opted from jaw bones in our reptilian ancestors. Wings, feathers, and swim bladders are other well-worn anatomical examples, but exaptation is even more prolific at the molecular scale. With the subtlest of tweaks, enzymes can catalyze different reactions, genes can be expressed in different tissues, and proteins can find new binding partners. Though Behe does not bother to address this, the molecular possibilities of exaptation are endless, particularly when gene duplication is involved. In the age of genomics, the evidence for molecular exaptation is abundant.

Speaking of gene duplication, Behe barely mentions this phenomenon either, despite it being a common first step in evolutionary innovation and diversification. When a gene has an essential function, it is tightly constrained by natural selection, and almost any mutation would cause harm and be eliminated. When a gene has been duplicated, however, the original copy can retain essential functions while the randomness of mutation and recombination fiddle around with the new one. In a powerful 2012 article, completely unmentioned by Behe, Dan Andersson and colleagues showed that gene duplication—through random mutations—can lead bacterial cells to evolve the ability to synthesize an amino acid that they were previously unable to (Näsvall et al. 2012). What Behe claims evolution can't do, scientists have already shown it can.

Behe fails to discuss genetic recombination as well. It is beyond frustrating that Behe claims that evolution has no means to generate adaptive innovations when not only do such mechanisms exist but there is a rich literature of research on them that he simply ignores. This is part of a larger pattern by ID proponents, and Behe in particular, of ignoring the very evolutionary mechanisms that they claim do not exist. Furthermore, Darwin Devolves, like Behe's previous two books, frequently discusses evolution and natural selection as though they are synonymous or as though natural selection is the sole engine of evolution. But natural selection is but one evolutionary force, albeit the most famous one. Mutation, genetic drift, sexual selection, recombination, horizontal gene transfer, frequency-dependent selection, and neutral theory all contribute to the toolkit of evolution. When Behe claims that natural selection alone cannot account for the rich molecular biodiversity we observe, he's absolutely correct. Rather than looking to the designer, he should look to some of the discoveries we've made in the past 160 years (Darwin 1859).

The Misinterpretations

To support his view, Behe focuses his attention on several of the most high-profile examples of evolution research and attempts to reinterpret them. For example, he spends nearly an entire chapter explaining how the long-term *E. coli* evolution experiment (LTEE) shows how mutation and natural selection serve only to "break or blunt genes." Richard Lenski, the architect of the LTEE for thirty years and 70,000 bacterial generations, has responded to Behe at length in a series of blog posts (Lenski 2019). It is hard to overstate just how badly Behe misinterprets the LTEE, including the fact that the controlled environment is highly artificial *on purpose*. There is unlimited food, static temperature, and high oxygen, and there are no competitors, pathogens, or immune system to challenge them. It is essentially an endless race at breakneck speed with strong competitive advantage for streamlining, efficiency, and growth acceleration. The LTEE was designed to detect the molecular mechanisms of evolution and has succeeded in spectacular detail, earning Lenski membership in the National Academy of Sciences (Good et al. 2017).

Behe is correct that the bacteria have adapted to this peculiar environment by ditching basically anything that slows them down, especially genes that aren't useful in this setting. He fails to mention that plenty of gene products show enhanced function as well. Further, some of the bacteria even developed the ability to eat citrate, which is included in the growth broth to assist the absorption of iron, not as a food source (Blount et al. 2012). Behe begrudgingly admits this but dismisses it as a "sideshow" and declines to explain to his readers that the bacteria achieved this feat by reactivating a nonfunctional gene through an elaborate genomic rearrangement. Yes, this was a very improbable event, but that is exactly the point. Behe's whole thesis is that the odds of getting new functions from unguided mutations are so low that they just don't happen. But the LTEE proves they do.

Behe also takes aim at Darwin's finches. He describes the remarkable work that Rosemary and Peter Grant have done on these famous birds, especially in regard to their genomes before and after a drought that briefly but significantly reduced their numbers (Grant and Grant 1993). But Behe uses the opportunity to scoff at the diversification of the finches in the first place, which took place when ancestors from the mainland found themselves stranded on the island two million years ago; their descendants have since adapted to the local habitats. He is unimpressed with their diversification into as many as eighteen species across five genera, adapting to a wide variety of foods, including mature leaves, a resource that no other bird species is known to subside on (Grant 1981).

Bizarrely, Behe compares the diversification of the finches on the Galapagos to the adaptive radiation of animals during the Cambrian explosion, more than

500 million years ago, complaining that the finches didn't evolve "at least one crummy new phylum, class, or order." Besides occurring over a much longer period of time and in smaller, simpler, and faster reproducing animals, the Cambrian explosion was a global diversification event that occurred when the animal kingdom was just getting started. Dry land hadn't even been fully colonized by life yet, and new niches were opening up all over the planet. In contrast, most of the niches that Behe is surprised that the finches didn't adapt to were already occupied when they got there. That Behe fails to grasp this reveals a poor understanding of not just ecology but evolutionary theory itself—the idea he has spent the past three decades critiquing.

The Misunderstandings

The opening section of *Darwin Devolves* on the evolution of polar bears may be the most erroneous example in the book. Scientists recently sequenced the full genomes of polar bears and their closest relatives, brown bears, and were able to approximate the common ancestral genome using pandas as a reference (Liu et al. 2014). Because polar bears are a relatively young species—just several hundred thousand years old—the genetic differences between them and their close cousins are relatively modest and easy to scrutinize. The scientists generated lists of polar bear genes that have undergone recent evolution. Not surprisingly, polar bears have undergone much more change than brown bears, who pretty much live in the same way that their ancestors did. As Behe sees it, the engine of polar bear evolution was the accumulation of mutations that break or diminish their genes.

The gene Behe discusses most is *APOB*, which codes for apolipoprotein B, the main protein responsible for clearing cholesterol from our blood following a high-fat meal. Behe declares this "the most strongly selected mutation" (it's actually the second-most, and he later acknowledged this error). *APOB* has several polar bear–specific differences and has undergone strong positive selection, implying that it is important to their survival and success. This makes sense because polar bears live mostly on seal meat, a diet that is very high in saturated fat and cholesterol, much higher than their brown bear cousins who eat leaves, berries, and sometimes fish. Yet, they do not suffer much from the heart disease that we know can often result from a high cholesterol diet. Presumably, polar bear *APOB* must be highly optimized for handling the high cholesterol burden.

Somehow, Behe draws the opposite conclusion. This is puzzling because he notes that humans and mice with mutations that diminish the function of *APOB* are more prone to suffer heart disease. But polar bears do *not* get cholesterol-driven heart disease, so the logical inference to draw from these data is that polar bear *APOB* is *enhanced*, not diminished, by the mutations.

This is where it becomes crystal clear that Behe is blinded by his bias. Although there is no reason to conclude that polar bear *APOB* has been diminished based on the data alone, because that is what Behe was *hoping* to see, he did. In fact, he even seems to believe that the authors of the paper agree, claiming that "they determined that the mutations were very likely to be damaging." Having read the paper from title to references, I can confidently say that the authors determined no such thing. In fact, sentences such as "adaptive changes in *APOB* . . . contribut[e] to the effective clearance of cholesterol from the blood" clearly imply that they believe that polar bear *APOB* is better at its job, not worse (Liu et al. 2014).

Let's look at these mutations. First, they are not evenly distributed throughout the protein. More than half the polar bear–specific changes are in a small region called the *N-terminal domain*, which comprises just 22 percent of the protein. As the authors state, "This domain encodes the surface region and contains the majority of functional domains for lipid transport." Proteins can be damaged by mutations virtually anywhere. Enhancing mutations, however, must usually be placed precisely in the domain that performs the enhanced function. Thus, the location of the *APOB* mutations provides more support that enhancement, not diminishment, has occurred.

The evidence on which Behe bases his claim that polar bear *APOB* has been damaged comes from his reading of the authors' presentation of predictions made by a computer program called PolyPhen-2. This program can predict whether a given mutation diminishes the function of a protein based on how closely it corresponds to human genetic variation that is known to cause pathology. The PolyPhen tool was designed to identify disease-causing mutations in genomic data collected from human patients, a very different purpose than Behe's. Trained with large data sets of human genomic variation, the program predicts whether a given mutation alters the structure of a protein and, if so, labels these possibly or probably damaging. Mutations that likely cause little or no structural change in the protein, PolyPhen calls "benign."

On the other hand, the program has no way to predict enhancing mutations. That would require that it somehow knows all the functions of every protein and how those functions are carried out. There is no database of *enhancing* genetic variation in humans or any other species. Behe and others at Evolution News seem not to know this. They repeatedly insist that the program does not list any "constructive mutations" in polar bears, which is not even a possible output of the program! Importantly, what PolyPhen calls damaging (simply because the structure has been altered) could actually be a new or enhanced function, a point not lost on the study authors. In an interview with *New Scientist*, one of them repeated their conclusion that polar bear *APOB* likely makes cholesterol clearance "more efficient" (Coghlan 2014).

Behe's reading of the polar bear research seems to begin and end with the PolyPhen predictions. This is problematic for several reasons. First, whereas damaging mutations might *occur* more commonly than enhancing ones, the vast majority of them don't *persist*. When studying speciation, the observed mutations have already been scrutinized by natural selection. We therefore expect an enrichment of those rare mutations that PolyPhen calls damaging but are actually constructive, as seems likely to have happened here. In fact, it's possible that *none* of the most strongly selected mutations are damaging. Second, other kinds of empirical data are more reliable than PolyPhen predictions, given that this is not the task the tool was designed to perform. Everything else we know about *APOB* points to the conclusion that the polar bear version must be well optimized.

My colleagues at Peaceful Science and I decided to look even more closely at the polar bear mutations in *APOB*. In a series of four discussion forums with nearly 100 contributions, eight molecular biologists pored over the data from the polar bear genome, as well as DNA sequencing data available in the NCBI database, and we scrutinized the PolyPhen-2 predictions ourselves, the kind of due diligence that Behe would have been wise to do (Swamidass 2021). We found that the picture is even murkier than it already appears. For one thing, if you substitute the human *APOB* protein sequence for the ancestor bear sequence as the reference in PolyPhen, only three of the nine polar bear variants are predicted to be possibly or probably damaging. (Using the human sequence as a reference is particularly useful because data from human disease-causing gene variants is how PolyPhen is trained.)

Even worse, we found that *none* of the nine mutations under scrutiny are actually fixed in the polar bear population. Yes, they are common, but most polar bears do not have all these individual variants. While this fact does not help us understand whether the mutations are damaging or not, it does weaken any claim that *APOB* was crucial in polar bear evolution. And the final nail in the coffin for this reasoning is that polar bears are resistant to atherosclerosis *despite* very high levels of circulating cholesterol. It may be that the polar bear adaptation to a high-fat diet has little to do with *APOB* whatsoever. Although we are left with an unclear picture, there is simply no hard evidence that *APOB* is damaged, let alone that the damage was somehow adaptive. Rather than carefully and objectively examining all available evidence, Behe simply pounced on a chart with some computer predictions that he didn't fully understand.

The Misrepresentation

There is a very unfortunate coda to the polar bear story. After Arthur Hunt and I wrote a blog post rebutting Behe's claims, he responded with an angry rant on Evolution News that was aimed, oddly, more at Jerry Coyne for sharing our post than at us for writing it (Behe 2019). He offered a mere two sentences of defense, posting a chart with "the relevant information" that supported "every actual undistorted claim" that he made. However, instead of presenting the actual chart from the article, he made a new chart, including only the PolyPhen predictions that support *his* version of polar bear evolution and removing those that argue against it.

It's hard to overstate just how misleading this version of the chart is. In the original polar bear article, the complete list of PolyPhen predictions is in Table S7, which includes forty-eight separate mutations and the results of two Poly-Phen predictions based on different genetic datasets (Liu et al. 2014). In one of them, 25 percent of the mutations are predicted to be possibly or probably damaging. Behe chose to leave those data out, preferring instead to show results

Circled are parts that Behe omitted

Gene	Protein position	Ancestral AA	Polar bear AA	HDivPred	HDivProb	HVarPred	HVarProb
ABCC6	655 Q		H	probably damaging	0.96	possibly damaging	0.547
AIM1	530 T		S	benign	0.03	benign	0.018
AIM1	632 I		S	benign	0.00	benign	0.003
AIM1	821 N		K	possibly damaging	0.65	benign	0.122
APOB	716 N		K	possibly damaging	0.80	benign	0.265
APOB	749 D		E	possibly damaging	0.94	possibly damaging	0.807
APOB	2623 D		N	probably damaging		probably damaging	0.989
APOB	3920 T		P	possibly damaging	0.48	benign	0.088
APOB	4418 L		H	probably damaging	0.99	probably damaging	0.915
ARID5B	457 N		K	benign	0.12	benign	0.02
ARID5B	775 H		Q	benign	0.00	benign	0.002
ARID5B	787 S		R	benign	0.41	benign	0.133
ARID5B	875 H		Q	probably damaging	0.99	possibly damaging	0.731
COL5A3	149 R		S	probably damaging	0.99	probably damaging	0.993
COL5A3	694 K		N	probably damaging	0.99	probably damaging	0.991
COL5A3	963 P		T	benign	0.26	benign	0.249
COL5A3	1117 D		E	possibly damaging	0.95	possibly damaging	0.804
CUL7	508 D		N	possibly damaging	0.73	benign	0.159
CUL7	709 D		E	benign	0.00	benign	0.007
CUL7	1477 N		K	probably damaging		probably damaging	0.999
EHD3	269 K		N	benign	0.25	benign	0.197
IPO4	362 R		W	probably damaging	0.97	benign	0.339
LAMC3	791 D		E	probably damaging	0.99	probably damaging	0.971
LYST	1046 D		Y	possibly damaging	0.94	benign	0.272
LYST	1084 A		S	benign	0.04	benign	0.015
LYST	1140 F		L	benign		benign	0
LYST	1770 L		F	benign	0.00	benign	0.007
LYST	2672 N		K	benign	0.03	benign	0.021

This table lists the mutations that have been most strongly selected during polar bear evolution (we are displaying only the top half for readability). Circled are lines and columns Behe removed to create a chart he presented to defend his conclusions, which makes it seriously misleading. *Liu, et al. 2014.*

from the other column, which predicted that 52 percent of the mutations might be damaging. Because this still doesn't look like an "overwhelming tendency" of natural selection to damage genes, Behe took things further and removed *all* the mutations that were not predicted to be damaging—nearly half the data—to create a chart that gives the desired impression. (See accompanying figure.)

When I called out this sleight of hand on Twitter, Evolution News responded in characteristic "google swarm" fashion, publishing a whole series of articles attempting to rescue Behe's claims about polar bear evolution while insulting me and my colleagues. They even dedicated one whole article to the "fake scandal" about the doctored chart, offering two defenses. First, they claim he was merely "saving space," an odd concern for a web-only publication. Second, they claim that he was only providing the "relevant information," that is, the data that shows that many of the mutations are damaging. This, too, strains credulity. When one's entire argument is that the overwhelming trend of unguided evolution is toward the breaking or blunting of genes, one cannot deliberately obscure all contrary evidence and not expect others to cry foul. In this case, an honest and transparent presentation of the data indeed paints a very different picture from the one Behe wanted his readers to see. In the quest to be taken seriously, Evolution News is its own worst enemy.

Conclusion

Darwin Devolves is a case study in how proponents of intelligent design grasp at any evidence that they can interpret in a favorable way, while simultaneously ignoring massive amounts of opposing evidence. Although this strategy is standard fare for creationism, the age of genomics has brought nearly limitless information to our fingertips, making it easier than ever for those poorly trained in genomics to mine the data for nuggets of evidence for their claims. Of course, the wealth of data also provides the full context for debunking weak claims, as we have done with Behe's assertions about polar bears. The careful work of analyzing and interpreting huge genomic data sets takes a great deal of time, training, and repetition. The few scientists who work within the ID framework would be well served to take the time and do the training. But let them be warned: it will be very difficult to remain within the ID framework if they do.

Nathan H. Lents is professor of biology at John Jay College, City University of New York, where his laboratory studies the evolution of the human genome with a special focus on the genetics of human uniqueness. He is also the author of *Not So Different: Finding Human Nature in Animals* and *Human Errors: A Panorama of Our Glitches from Pointless Bones to Broken Genes*.

References

Behe, Michael J. 2019. Coyne and polar bears: Why you should never rely on incompetent reviewers. *Evolution News* (February 14). Available online at https://evolutionnews.org/2019/02/coyne-and-polar-bears-why-you-should-never-rely-on-incompetent-reviewers/.

Blount, Zachary D., Jeffrey E. Barrick, Carla J. Davidson, et al. 2012. Genomic analysis of a key innovation in an experimental Escherichia coli population. *Nature* 489(7417): 513–518.

Chou, Seemay, Matthew D. Daugherty, S. Brook Peterson, et al. 2015. Transferred interbacterial antagonism genes augment eukaryotic innate immune function. *Nature* 518(7537): 98–101.

Coghlan, Andy. 2014. Polar bears evolved to eat junk food. *New Scientist* 2969. Available online at https://www.newscientist.com/article/dn25535-zoologger-polar-bears-evolved-to-eat-junk-food/.

Darwin, Charles. 1859. *On the Origin of Species by Means of Natural Selection*. London: Murray.

Doolittle, Russell F., Yong Jiang, and Justin Nand. 2008. Genomic evidence for a simpler clotting scheme in jawless vertebrates. *Journal of Molecular Evolution* 66(2): 185–196.

Forrest, Barbara, and Paul R. Gross. 2007. *Creationism's Trojan Horse: The Wedge of Intelligent Design*. Oxford University Press.

Good, Benjamin H., Michael J. McDonald, Jeffrey E. Barrick, et al. 2017. The dynamics of molecular evolution over 60,000 generations. *Nature* 551(7678): 45–50.

Grant, Peter R. 1981. Speciation and the adaptive radiation of Darwin's finches: The complex diversity of Darwin's finches may provide a key to the mystery of how intraspecific variation is transformed into interspecific variation. *American Scientist* 69(6): 653–663.

Grant, Barbara Rosemary, and Peter Raymond Grant. 1993. Evolution of Darwin's finches caused by a rare climatic event. *Proceedings of the Royal Society of London. Series B: Biological Sciences* 251(1331): 111–117.

Jones III, John E. 2005. *Tammy Kitzmiller, et al. v Dover Area School District, et al.* Case No. 04cv2688-JEJ. December 20.

Keeling, Patrick J., and Jeffrey D. Palmer. 2008. Horizontal gene transfer in eukaryotic evolution. *Nature Reviews Genetics* 9(8): 605–618.

Lamb, Trevor D., Edward N. Pugh, and Shaun P. Collin. 2008. The origin of the vertebrate eye. *Evolution: Education and Outreach* 1(4): 415–426.

Lenski, Richard E. 2019. Does Behe's 'first rule' really show that evolutionary biology has a big problem? *Telliamed Revisted* (February 15). Available online at https://telliamedrevisited.wordpress.com/2019/02/15/does-behes-first-rule-really-show-that-evolutionary-biology-has-a-big-problem/.

Lents, Nathan H., S. Joshua Swamidass, and Richard E. Lenski. 2019. The end of evolution? *Science* 363(6427): 590.

Liu, Shiping, Eline D. Lorenzen, Matteo Fumagalli, et al. 2014. Population genomics reveal recent speciation and rapid evolutionary adaptation in polar bears. *Cell* 157(4): 785–794.

Miller, Kenneth R. 2004. The flagellum unspun—the collapse of 'irreducible complexity.' In *Philosophy of Biology: An Anthology*, 439–449.

Näsvall, Joakim, Lei Sun, John R. Roth, et al. 2012. Real-time evolution of new genes by innovation, amplification, and divergence. *Science* 338(6105): 384–387.

Swamidass, S. Joshua. 2021. Behe meets the Peaceful Science Forum. *Panda's Thumb*. Available online at https://pandasthumb.org/archives/2021/03/behe-meets-peaceful .html.

CHAPTER 26

Skepticism and the Persuasive Power of Conversion Stories

Scott O. Lilienfeld

Vol. 43, No. 3
May/June 2019

Those of us in the skeptical community have our work cut out for us. In the process of disseminating scientific thinking, we often challenge unsubstantiated beliefs that are held with considerable conviction. Every one of us who has tried to persuade committed believers in astrology or homeopathy that they are mistaken knows just how challenging—and in some cases, how futile—this endeavor can be. We skeptics rarely win popularity contests.

So how can we effectively persuade believers in dubious claims to change their minds, or at least to give our contrary ideas a fair hearing? Traditionally, much of the science communication literature has operated implicitly from the "information deficit model." From this perspective, which is premised on Sir Francis Bacon's principle that knowledge is power, the primary driver of pseudoscience is inadequate scientific literacy. If we could only find a means of better educating the general public about science, this model assumes, most unsupported ideas would lose their stranglehold over the populace. Nevertheless, recent data suggest that the information deficit model, although probably containing a kernel of truth, does not tell the full story. For example, research by Dan Kahan of Yale University and his colleagues reveals that among political conservatives (but not liberals), higher levels of scientific literacy are associated with *greater* skepticism of global warming and its damaging impacts (Kahan et al. 2012). Although such findings are open to multiple interpretations, they raise the possibility that imparting scientific knowledge might in some cases backfire, perhaps by affording individuals who hold unwarranted beliefs the intellectual ammunition to rebut scientific arguments. If corrective science education has its limits, are there better alternatives to debunking erroneous beliefs?

A recent study conducted by Benjamin Lyons and colleagues, published in the peer-reviewed journal *Public Understanding of Science* (Lyons et al. 2019), offers a potential answer. The authors aimed to determine whether presenting participants with "conversion stories"—descriptions of individuals who once held an unsubstantiated belief but who later changed their minds—might be an effective, albeit underutilized, persuasion technique.

To test this hypothesis, they showed participants video clips from a 2013 conference talk by Mark Lynas, a British author and environmental activist who was initially a staunch opponent of genetically modified organisms (GMOs) but later became persuaded that GMOs were safe. Participants, who were 727 American adults recruited for an online survey, were randomly assigned to view three different clips of Lynas: (1) one in which he simply advocated for GMOs, (2) one in which he advocated for GMOs and noted that he had initially been opposed to GMOs, and (3) one in which he advocated for GMOs and explained what had prompted him to change his mind. Both conversion conditions (2 and 3) led to more positive attitudes toward GMOs than did condition 1, with no significant differences between conditions 2 and 3.

It is not known how conversion experiences work as persuasive tactics. Lyons and colleagues found that their two conversion conditions appeared to operate by boosting the strength of the speaker's arguments rather than by enhancing his personal credibility. Other data suggest that an effective means of refuting scientific misconceptions is to supplant false beliefs with an equally or more compelling explanatory narrative (Lewandowsky et al. 2012). For example, if we want to rebut the erroneous belief that vaccines cause autism (now called autism spectrum disorder), presenting individuals with compelling data that autism risk is substantially inherited may help, as may presenting them with narratives describing how and why even intelligent people can fall prey to the erroneous belief.

The use of conversion stories for persuasive purposes is hardly new. Social psychologists who have studied persuasion have long underscored the utility of inoculating people against false claims by first presenting them with information that seemingly supports these claims and then refuting it (Pratkanis 2007). Moreover, legal scholar Cass Sunstein and his colleagues have written of the persuasive power of "surprising validators": people we would not typically expect to support a position but who end up doing so (Glaeser and Sunstein 2014). A fervent believer in astrology who later becomes convinced that astrology is unscientific can function as a surprising validator, communicating to others her change of heart and the reasons for it.

If the results of the study by Lyons and collaborators are replicable and generalizable beyond GMOs, the skeptical community may want to consider harnessing conversion experiences as a persuasive strategy. Consider Janyce Boynton, once an ardent believer in the now convincingly debunked technique of facilitated communication (FC) for autism and other developmental disabilities. Boynton was once a trained facilitator; for readers who have seen the classic 1993 *Frontline* video, "Prisoners of Silence," Boynton was the facilitator in the Betsy Wheaton case, in which a sixteen-year-old girl with autism accused her father and brother of sexual abuse through FC. Following a series of controlled tests by Dr. Howard Shane of Boston Children's Hospital, which showed conclusively that she, not Wheaton, was authoring the messages, Boynton reluctantly decided to forgo FC and persuaded her school to stop using it. Two decades later, in a courageous 2012 article, Boynton wrote of her emotionally painful conversion experience, observing how difficult it was for her to surrender her initial beliefs in the wake of Shane's negative controlled tests:

> I felt such devastation, panic, pain, loneliness—a myriad of emotions difficult to put into words. The whole FC thing unraveled for me that day, and I did not have an explanation for any of it. Almost immediately, I started rationalizing away the truth. . . . I understand how difficult it may be for some facilitators to change their belief system. There is a lot at stake: people's careers, reputations, connections with their family member or client. Nonetheless, I urge practicing facilitators to take a long, hard look at their own behavior. (Boynton 2012, 11–12)

Despite overwhelming scientific evidence that FC does not work, the method remains alive and well, and even seems to be mounting a comeback in some quarters under the guise of "rapid prompting method" and allied techniques that are minor variants of FC (Lilienfeld et al. 2014). Concerted efforts by skeptics to stem the tide of FC's popularity appear to have failed. It would be interesting to see if presenting committed FC believers with Boynton's compelling narrative would shake their belief certainty. More broadly, it would be useful for the skeptical community to gather similar conversion stories and determine whether they can be harnessed in the service of beneficial attitude change.

Scott O. Lilienfeld, PhD, was a professor of psychology at Emory University, coeditor of the book *Science and Pseudoscience in Clinical Psychology, Second Edition* (2014), and author of several other books about science and pseudoscience in psychology.

References

Boynton, J. 2012. Facilitated communication—what harm it can do: Confessions of a former facilitator. *Evidence-Based Communication Assessment and Intervention* 6: 3–13.

Glaeser E., and C. R. Sunstein. 2014. Does more speech correct falsehoods? *The Journal of Legal Studies* 43: 65–93.

Kahan, D. M., E. Peters, M. Wittlin, et al. 2012. The polarizing impact of science literacy and numeracy on perceived climate change risks. *Nature Climate Change* 2: 732–735.

Lewandowsky, S., U. K. Ecker, C. M. Seifert, et al. 2012. Misinformation and its correction: Continued influence and successful debiasing. *Psychological Science in the Public Interest* 13: 106–131.

Lilienfeld, S. O., J. Marshall, J. T. Todd, et al. 2014. The persistence of fad interventions in the face of negative scientific evidence: Facilitated communication for autism as a case example. *Evidence-Based Communication Assessment and Intervention* 8: 62–101.

Lyons, B. A., A. Hasell, M. Tallapragada, et al. 2019. Conversion messages and attitude change: Strong arguments, not costly signals. *Public Understanding of Science* 0963662518821017.

Pratkanis, A. R. 2007. Social influence analysis: An index of tactics. In A.R. Pratkanis (ed.), *The Science of Social Influence: Advances and Future Progress*. Philadelphia, PA: Psychology Press, 17–82.

Vyse, S. 2019. An artist with a science-based mission. *Skeptical Inquirer* 43(2) (March/April).

CHAPTER 27

In Troubled Times, This Is What We Do

Kendrick Frazier

Vol. 42, No. 2
March/April 2018

I have often written in the *Skeptical Inquirer* about how what we do is a communal activity. There is a dynamic interaction between our authors/investigators who prepare our articles, reports, critiques, and reviews and our intelligent and curious readers, supporters, and conference attendees who provide moral (and financial!) support, information, ideas, and informed feedback. This is one of the decidedly cool things about the skeptical community. Everyone can contribute in some way.

And what is it we all do? Well, to quote the short version of the mission statement of SI and our Committee for Skeptical Inquiry that appears in every SI: We promote "scientific inquiry, critical investigation, and the use of reason in examining controversial and extraordinary claims." That's all! Yes, that mission is rather broad. And that is exactly what we try to do. We bring all the tools of evidence-based critical inquiry to popular questions and urgent issues that fascinate, mystify, confuse, and befuddle people. We seek scientifically validated information about issues and assertions and then provide a clear evaluation of those claims.

We call this activity "scientific skepticism." I often think of it as a field of intellectual and scientific inquiry that I call "Science & Skepticism." It is highly interdisciplinary. It draws upon all scientific fields. It also draws upon everything we know about human behavior, individually and in groups. It draws upon everything we know about how we think and how our brains work. It draws upon the great traditions of philosophy, beginning with the ancient Greek philosophers who founded rationalism (a purely scientific inquiry into the nature of man and the universe), humanism, and the concept of the individual. And it also embraces history and the humanities.

Our quest seeks to understand not only the external world of nature out there but our own selves, what makes us human—wonderful and creative,

212

flawed and exasperating. If we were an academic unit—say the [insert university of your choice] Institute of Science & Skepticism—we would have faculty from virtually every academic department including the schools of medicine, engineering, and law. But we aren't just an academic enterprise. We incorporate nonacademic traditions such as magicians' specialized knowledge of deception, investigative journalists' tools for getting at the truth, science communicators' skills in explaining complex scientific ideas, and skeptical investigators' blending of all these skills. We do all this in the quest to find out what is true and not true about the real world—including ourselves. And then we present those insights to the public in an appealing, understandable way.

What could be more important? Especially at this troubled time in our political and cultural history

- when fact and fiction are being blended at the highest levels of government;
- where beliefs and opinions are accorded greater sway than facts and evidence;
- where important science-oriented federal agencies are now headed by people who are not only scientifically uninformed but are defunding and in some cases even dismantling key parts of their agencies' scientific missions;
- where our political system is corrupted by "conspiracy theories and outright fabrication"; (Lest you think these remarks are partisan—our effort is decidedly nonpartisan—I point out that that last item is a quote from former Republican President George W. Bush's remarkable speech in New York on October 19 about internal threats to American democracy.)
- where longtime, legitimate, responsible, independent, mainstream news organizations are labeled "fake" and where scurrilous online "news" operations that really *are* fake disseminate intentional disinformation that too often gets accepted as true;
- where Russian meddling in our elections and in our social media causes further confusion and damage to our democracies;
- where pseudoscientific medical concepts and techniques have made deep inroads into our medical schools and universities and to enable that to happen proponents undermine the very idea of science;
- where genetically modified foods and organisms that can alleviate terrible diseases and help feed malnourished people in poor parts of the world are opposed by well-funded groups and well-off celebrities who think "organic" and "natural" foods are somehow better and not the product of a giant marketing industry;
- where religiously motivated leaders in our states, communities, and school boards continually try to sneak pseudoscientific, creationist ideas into public school curricula and try to prevent teaching evolution or even the age of the Earth;

- where a new flock of predatory journals that don't bother with the conventions of scientific integrity openly publish nonsensical and pseudoscientific papers in the guise of science; and
- where, overall, a kind of Orwellian, *1984*, Bizarro parallel world in which up is down and in is out afflicts our senses and deeply troubles our psyches.

We need independent, evidence-based, science-based critical investigation and inquiry now perhaps more than at any other time in our history. And that's what we do. That's what all of us in the skeptical community do.

We all must support critical inquiry and evidence-based thinking. We must honor those who do it, often at considerable sacrifice to themselves. We must gain a better understanding of how to encourage science-based thinking in others. We must help create a better informed and more enlightened nation and world.

Not just for us, but for the younger generations who succeed us. Let's leave this world better than it is now.

It is the challenge of our lifetime. Let's get to it.

Kendrick Frazier was the longtime editor of the *Skeptical Inquirer* **and a fellow of the American Association for the Advancement of Science. He was editor of several anthologies including** *Science under Siege: Defending Science, Exposing Pseudoscience.*

CHAPTER 28

Seven Big Misconceptions about Heredity

Carl Zimmer

Vol. 43, No. 3
May/June 2019

If someone says, "I guess it's in my DNA," you never hear people say, "DN—*what?*" We all know what DNA is, or at least think we do.

It's been seven decades since scientists demonstrated that DNA is the molecule of heredity. Since then, a steady stream of books, news programs, and episodes of *CSI* have made us comfortable with the notion that each of our cells contains three billion base pairs of DNA, which we inherited from our parents. But we've gotten comfortable without actually knowing much at all about our own genomes.

Indeed, if you had asked to look at your own genome twenty years ago, the question would have been absurd. It would have been as ridiculous as asking to go to the moon. When scientists unveiled the first rough draft of the human genome in the early 2000s, the final bill came to an Apollo-scale $2.7 billion.

Since then, advances in DNA sequencing and software for analyzing genetic data have steadily brought down the price tag. By 2006, it cost only $14 million to sequence a single human genome. Even at that drastically reduced price, though, only a few big labs with major financial support would dare take on such an expensive project. But in the years that followed, DNA sequencing continued its exponential cost crash, becoming cheap enough to turn into a consumer product.

If you want to get your entire genome sequenced—all three billion base pairs in your DNA—a company called Dante Labs will do it for $699. You don't need whole genome sequencing to learn a lot about your genes, however. The 20,000 genes that encode our proteins make up less than 2 percent of the human genome. That fraction of the genome—the "exome"—can be yours for just a few hundred dollars. The cheapest insights come from "genotyping"—in which scientists survey around a million spots in the genome known to vary a lot among people. Genotyping—offered by companies such as 23andMe and Ancestry—is typically available for under a hundred dollars.

Thanks to these falling prices, the number of people who are getting a glimpse at their own genes is skyrocketing. By 2019, over twenty-five million worldwide had gotten genotyped or had their DNA sequenced. At its current pace, the total may reach 100 million by 2020.

Future generations will look back at today as a pivotal moment in DNA's cultural history. People are no longer thinking of their DNA as a black box but as a database to be mined. They're learning that they have inherited mutations that raise their risk of certain diseases. They're getting estimates of their ancestry based on genetic markers that are common in certain parts of the world. They're merging their genetic information with genealogy to discover distant relatives. Some are also discovering some not-so-distant relatives that until now were family secrets.

There's a lot we can learn about ourselves in these test results. But there's also a huge opportunity to draw the wrong lessons.

Many people have misconceptions about heredity—how we are connected to our ancestors and how our inheritance from them shapes us. Rather than dispelling those misconceptions, our growing fascination with our DNA may only intensify them. A number of scientists have warned of a new threat they call "genetic astrology." It's vitally important to fight these misconceptions about heredity, just as we must fight misconceptions about other fields of science, such as global warming, vaccines, and evolution. Here are just a few examples.

Misconception #1: Finding a Special Ancestor Makes You Special

There are certain clubs to which ancestry is the key to admission. You can get into the Mayflower Society if you descend from the passengers of that famous ship. You can join the Order of the Crown of Charlemagne if you can prove that the Holy Roman Emperor is your ancestor. It's a thrill to discover we have a genealogical link to someone famous—perhaps because that link seems to make us special, too.

But that's an illusion. I could join the Mayflower Society, for example, because I'm descended from a servant aboard the ship named John Howland. Howland's one claim to fame is that he fell out of the *Mayflower*. Fortunately for me, he got fished out of the water and reached Massachusetts. But I'm not the only fortunate one; by one estimate, there are two million people who descend from him alone.

Mathematicians have analyzed the structure of family trees, and they've found that the further back in time you go, the more descendants people had. (This is only true of people who have any living descendants at all, it should be noted.) This finding has an astonishing implication. Since we know Charlemagne has living descendants (thank you, Order of the Crown!), he is likely the ancestor of every living person of European descent. And if you could get in a time machine and travel back a few thousand years, you could find someone who was a common ancestor of all living people on Earth.

Misconception #2: You Are Connected to All Your Ancestors by DNA

When you look at your family tree, you're looking at a series of branching lines that link you to your ancestors. What exactly flows down through those lines as they travel through time? A few centuries ago, people might say it was blood. In recent decades, blood has been replaced in our popular imagination with DNA. After all, our genes didn't come out of nowhere. We inherited them.

But genetics do not equal genealogy. It turns out that practically none of the Europeans who descend from Charlemagne inherited any of his DNA. All humans, in fact, have no genetic link to most of their direct ancestors.

The reason for this disconnect is the way that DNA gets passed down from one generation to the next. Every egg or sperm randomly ends up with one copy of each chromosome, coming either from a person's mother or father. As a result, we inherit about a quarter of our DNA from each grandparent—but only on average. Any one person may inherit extra DNA from one grandparent and less from another. If you go back to the next generation, you'll find that each great-grandparent contributed approximately an eighth of your DNA—but, again, that's only an average. Some of them may have contributed much more, others much less.

If you go back a few generations more, that contribution can drop all the way to zero. Graham Coop, a geneticist at the University of California, Davis, and his colleagues have calculated the odds of sharing no DNA with an ancestor as they moved back through the generations. If you go back ten generations, the odds of having DNA from any given ancestor drop to less than 50 percent. They go down even more as you push back further through your ancestry. While it is true that you inherit your DNA from your ancestors, that DNA is only a tiny sampling of the genes in your family tree.

Even without a genetic link, though, your ancestors remain your ancestors. They did indeed help shape who you are—not by giving you a gene for some particular trait, but by raising their own children, who then raised their own children in turn, passing down a cultural inheritance along with a genetic one.

Misconception #3: Ancestry Tests Are as Reliable as Medical Tests

Millions of people are getting ancestry reports based on their DNA. My own report informs me that I'm 43 percent Ashkenazi Jewish, 25 percent Northwestern European, 23 percent South/Central European, 6 percent Southwestern European, and 2.2 percent North Slavic. Those percentages sound impressive, even definitive. It's easy to conclude that ancestry reports are as reliable as stepping on a scale at the doctor's office to get your height and weight measured.

That is a mistake, and one that can cause a lot of heartbreak. To estimate ancestry, researchers compare each customer to a database of thousands of people from around the world. Those "reference populations" are typically selected because they have deep roots where they live. Some researchers select only people whose family has lived in the same place for three generations, for example. In each population, there are some genetic variants that are unusually common and others that are unusually rare. Researchers then look for these variants in a customer's DNA. They can identify stretches of DNA that are likely to have originated in a particular part of the world. While some matches are clear-cut, others are less so. As a result, ancestry estimates always have margins of error— which often go missing in the reports customers get.

To gain more certainty in their estimates, scientists are building up bigger databases. In 2018, Ancestry.com unveiled a new set of estimates for its customers. They got a lot of backlash. People who had initially been thrilled to discover a small portion of their ancestry came from Italy or Cameroon were devastated to now learn that they had no such link at all.

These estimates are going to get better with time, but there's a fundamental limit to what they can tell us about our ancestry. To say I am 43 percent Ashkenazi doesn't have the same timeless truth as saying I'm 43 percent carbon. Carbon has been carbon for billions of years. But the Ashkenazi people emerged through history. In the Roman Empire, people of Near East and European ancestries came together and started having children. In the Middle Age, Jews in northern and eastern Europe began to be persecuted and formed increasingly isolated communities. In these small groups, children increasingly inherited the same set of genetic variants. From an estimated population of just 350 ancestors, the Ashkenazi population has now reached ten million. They all share a number of distinctive genetic markers from that period of history. But their history reaches farther back in time, to older peoples.

Researchers are getting glimpses of those older peoples by retrieving DNA from ancient skeletons. And they're finding that our genetic history is far more tumultuous than previously thought. Time and again, researchers find that the people who have lived in a given place in recent centuries have little genetic connection to the people who lived there thousands of years ago. All over the world, populations have expanded and migrated, coming into contact with other populations. In Europe, for example, new waves of genetically distinct people have arrived from elsewhere every few thousand years, either replacing or interbreeding with the people who lived there before. Today, Europeans are genetically similar to each other, but only because the genes of their disparate ancestors—from places such as Africa, Turkey, and Russia—have been well mixed. If you want to find purity in your ancestry, you're on a fool's errand.

Misconception #4: There's a Gene for Every Trait You Inherit

When we learn about genetics in school, we learn about Gregor Mendel. In the 1850s, Mendel crossed lines of pea plants and discovered that their traits—such as the color of their flowers or the texture of their peas—were carried by invisible hereditary factors. Some factors were dominant, meaning that inheriting just one copy of them determined a trait. Other factors were recessive, meaning that they could shape a pea plant if it inherited two copies.

Mendel is a great place to start learning about heredity but a bad place to stop. There are some traits that are determined by a single gene. Whether Mendel's peas were smooth or wrinkled was determined by a gene called SBEI. Whether people develop sickle cell anemia or not comes down to a single gene called HBB. But many traits do not follow this so-called Mendelian pattern—even ones that we may have been told in school are Mendelian.

Consider your ear lobes. For decades, teachers taught that they could either hang free or be attached to the side of our heads. The sort of ear lobes you had was a Mendelian trait, determined by a single gene. In fact, our ear lobes typically fall somewhere between the two extremes of strongly attached to fully free. In 2017, a team of researchers compared the ear lobes of over 74,000 people to their DNA. They looked for genetic variants that were common in people at either end of the ear-lobe spectrum. They pinpointed forty-nine genes that appear to play a role in determining how attached they are to our heads. There well may be more waiting to be discovered.

None of those forty-nine genes is a gene "for" ear-lobe attachment. That language just doesn't make sense for the way most genes work. The genes that the scientists identified become active in many cells in an embryo. Some are active in skin cells across the body. Some are active in hair and sweat glands as well. Some help build the intricate anatomy of the inner ear. The attachment of our ear lobes is the result of a symphony performed by these players.

The genetics of ear lobes is actually very simple compared to other traits. Studying height, for example, scientists have identified thousands of genetic variants that appear to play a role. The same holds true for our risk of developing diabetes, heart disease, and other common disorders. We can't expect to find a single gene in our DNA tests that determines whether we'll die of a heart attack. Nor should we expect easy fixes for such complex diseases by repairing single genes.

Misconception #5: The Genes You Inherit Explain Exactly Who You Are

Throughout our lives—through our successes and failures, through our joys and suffering—we often wonder how things turned out the way they did. The more that scientists explore our DNA, the easier it is to shrug and say that it was all programmed in our genes.

Take, for example, a recent study on how long people stay in school. Researchers examined DNA from 1.1 million people and found over 1,200 genetic variants that were unusually common either in people who left school early or in people who went on to college or graduate school. They then used the genetic differences in their subjects to come up with a predictive score, which they then tried out on another group of subjects. They found that in the highest-scoring 20 percent of these subjects, 57 percent finished college. In the lowest-scoring 20 percent, only 12 percent did.

But these results don't mean that how long you stayed in school was determined before birth by your genes. Getting your children's DNA tested won't tell you if you should save up money for college tuition or not. Plenty of people in the educational attainment study who got high genetic scores dropped out of high school. Plenty of people who got low scores went on to get PhDs. And many more got an average amount of education in between those extremes. For any individual, these genetic scores make predictions that are barely better than guessing at random.

This confusing state of affairs is the result of how genes and the environment interact. Scientists call a trait such as how long people stay in school "moderately heritable." In other words, a modest amount of the variation in education attainment is due to genetic variation. Lots of other factors also matter, too—the neighborhoods where people live, the quality of their schools, the stability of their family life, their income, and so on. What's more, a gene that may have an influence on how long people stay in school in one environment may have no influence at all in another.

Misconception #6: You Have One Genome

In 2002, a woman named Lydia Fairchild applied for enforcement of child support when she separated from the father of her two children. The state of Washington required genetic testing to confirm his paternity. The tests showed he was indeed the father. But they also showed that Fairchild was not the mother.

State officials threatened to charge Fairchild with fraud, despite her protests that she had given birth to the children and the testimony of her mother, who had seen the birth of her grandchildren. When Fairchild went into a hospital to give birth to another child, a court official came to witness the delivery and watch the nurses draw blood for another DNA test. Once more, the test indicated that Fairchild was not the infant's mother.

This absurd situation arose because of the common assumption that each of us carries a single genome. According to this assumption, you will find an identical sequence of DNA in any cell you examine. But there are many ways in which we can end up with different genomes within our bodies.

Fairchild is known as a chimera. She developed inside her mother alongside a fraternal twin. That twin embryo died in the womb, but not before exchanging cells with Fairchild. Now her body was made up of two populations of cells, each of which multiplied and developed into different tissues. In Fairchild's case, her blood arose from one population, while her eggs arose from another.

Women can also become chimeras with their own children. During pregnancy, fetuses can shed cells that then circulate throughout a woman's body. In some cases they linger on after birth. They can then develop into muscle, breast tissue, and even neurons.

It's unclear how many people are chimeras. Once they were considered bizarre rarities. Scientists became aware of them only in cases such as Lydia Fairchild's, when their mixed identity made itself known. In recent years, researchers have been carrying out small-scale surveys that suggest that perhaps a few percent of twins are chimeras, but the true number could be higher. As for chimeric mothers, they may be the rule rather than the exception. In a 2017 study, researchers studied brain tumors taken from women who had sons. Eighty percent of them had Y-chromosome-bearing cells in their tumors.

Chimerism is not the only way we can end up with different genomes. Every time a cell in our body divides, there's a tiny chance that one of the daughter cells may gain a mutation. At first, these new aberrations—called somatic mutations—seemed important only for cancer. But that view has changed as new genome-sequencing technologies have made it possible for scientists to study somatic mutations in many healthy tissues. It now turns out that every person's body is a mosaic, made up of populations of cells with many different mutations.

Misconception #7: Genes Don't Matter Because of Epigenetics

The notion that our genes are our destiny can trigger an equally false backlash: that genes don't matter at all. And very often, those who push against the importance of genetics invoke a younger, more tantalizing field of research: epigenetics.

Ask five scientists to define *epigenetics* and you may get five different definitions. But they will all center on the fact that genes, on their own, do nothing. They simply store information that our cells can use as guides for building proteins or RNA molecules. But our cells only use genes in response to certain combinations of signals. It can be disastrous to use genes at the wrong time or in the wrong place. Genes involved in making enamel need to be switched on in developing teeth. But you wouldn't want your skin cells to make it too, trapping you in an enamel sarcophagus.

Our cells use many layers of control to make proper use of their genes. They can quickly turn some genes on and off in response to quick changes in their environment. But they can also silence genes for life. Women, for example, have two copies of the X chromosome, but in early development, each of their cells produces a swarm of RNA molecules and proteins that clamp down on one copy. The cell then only uses the other X chromosome. And if the cell divides, its daughter cells will silence the same copy again.

One of the most tantalizing possibilities scientists are now exploring is whether certain epigenetic "marks" can be inherited not just by daughter cells but by daughters—and sons. If people experience trauma in their lives and it leaves an epigenetic mark on their genes, for example, can they pass down those marks to future generations?

If you're a plant, the answer is definitely yes. Plants that endure droughts or attacks by insects can reprogram their seeds, and these epigenetic changes can get carried down several generations. The evidence from animals is, for now, still a mixed bag. In one intriguing experiment, researchers separated male mouse pups from their mothers from time to time, causing them stress. Later, they used sperm from those stressed mice to fertilize eggs, and some of their descendants proved to be unusually sensitive to stress. But skeptics have questioned how epigenetics can transmit these traits through the generations, suggesting that the results are just statistical flukes. That hasn't stopped a cottage industry of epigenetic self-help from springing up. You can join epigenetic yoga classes to rewrite your epigenetic marks or go to epigenetic psychotherapy sessions to overcome the epigenetic legacy you inherited from your grandparents.

You may feel more limber after your yoga class. And you may feel better after having talked about your anxiety. But your genes will still work much the same as they did before.

Carl Zimmer is an award-winning science writer and the author, most recently, of *She Has Her Mother's Laugh: The Powers, Perversions, and Potential of Heredity* (selected by *The Guardian* as the best science book of 2018). He writes the "Matter" column for *The New York Times*. His earlier books include *Evolution: Triumph of an Idea* (companion to the PBS Series), *The Tangled Bank: An Introduction to Evolution*, and *Microcosm: E. Coli and the New Science of Life*.

CHAPTER 29

Autism Wars

SCIENCE STRIKES BACK

Stuart Vyse

Vol. 42, No. 6
November/December 2018

In the field of autism treatment, the forces for science and evidence have won a few battles, and lost a few. Unfortunately, some of the most recent victories have been on the side of pseudoscientific and fad therapies, but a new army of researchers, practitioners, and advocates is fighting back.

Twenty years ago, it looked like facilitated communication (FC), a popular pseudoscientific treatment for autism, was dead. Proponents had suggested that many people with autism were trapped inside broken bodies. Autism was not a cognitive problem but a physical one. Inside these non-speaking people were intelligent, expressive minds, and if someone—a facilitator—just steadied their hands over a keyboard, FC could unlock the thoughts and feelings of the person within. Suddenly, with the help of their facilitators, people who were previously unable to speak were writing novels and poems and going off to college. The promise of FC was so miraculous that it spread like wildfire.

But, in the early 1990s, the first empirical tests of the technique began to appear, and the results were devastating. In virtually every case, controlled studies revealed that the facilitator—the autistic person's helper—was doing the typing, not the person with autism. It was a Ouija board-like phenomenon. The facilitators appeared to be entirely unaware that they were the authors of the words on the computer screen.

This research was a substantial blow to the proponents of FC, and a 1993 PBS *Frontline* episode, "Prisoners of Silence," was particularly effective in discrediting the technique (Palfreman 1993). Major professional organizations, including the American Psychological Association, the American Academy of Pediatrics, American Academy of Child and Adolescent Psychiatry, and the American Speech-Language-Hearing Association issued policy statements against the use of FC, and teachers and therapists went back to using more validated methods of educating people with autism. So, by the mid-1990s, it looked like the FC controversy was over, and science had won. Unfortunately, the story did not end there.

As readers of this column know, FC has surged back, and the autism wars have resumed on a number of new fronts. First, the pro-FC crowd criticized and denied the research. Syracuse University professor of education Douglas Biklen (2005), who is the primary promoter of FC in the United States, claimed the research methods used to evaluate FC were not suited to the particular needs of people with autism and had caused test anxiety. Many parents continued to believe their children were articulate writers whose mouths, hands, and arms did not work properly. Pro-FC researchers using less rigorous methods appeared to show some evidence of independent typing in a few individual students, promoting belief in the technique (Mostert 2010).

Second, new variations of FC were introduced that looked different but shared the same problems. The most popular of these new techniques is Rapid Prompting Method (RPM) developed by Soma Mukhopadhyay (2013). In this case, instead of hand-over-hand guidance on a keyboard, the person with autism points to a letter board, and a teacher or assistant reports the words tapped out. RPM is different from FC, but it has the same potential for unconscious prompting because the letter board is always held in the air by the assistant. As long as the method of communication involves the active participation of another person, the potential for unconscious guidance remains. Perhaps having learned a lesson from the FC episode of the 1990s, the proponents of RPM have avoided participating in research studies that might test its validity, and its popularity has grown. Furthermore, the original, thoroughly discredited version of FC continued to be promoted by a number of universities and professionals.

For those who lost track of the FC story back in the 1990s, the resurgence of pseudoscience was a surprise. Twenty years earlier, science had spoken, and that should have been that. But the promise of uncovering an intelligent, articulate child was just too appealing, and as a result, for some parents and professionals, FC became a system of belief.

Science Strikes Back

Although many of the recent battles have been won by the proponents of FC, RPM, and other related pseudoscientific therapies, in the years since the 1990s, the scientific viewpoint has scored some victories, particularly in the courtroom. Some of the earliest challenges to FC came when parents or others were falsely accused of sexual abuse through FC. Typically, a child with autism would type out a message describing acts of sexual abuse allegedly committed by a parent or someone else. When these cases went to trial, an essential question was, who was writing the abuse claims, the child with autism or the facilitator? In several cases, a simple double-blind test was able to show that the facilitator was the author of the abuse claims, not the person with autism. After these tests were conducted, defendants were usually acquitted or the charges were dropped.

Unfortunately, these victories came at a heavy cost, because the courts move slowly and people's lives can be damaged in the process. As recently as spring of 2018, a Hialeah, Florida, man was arrested and held in jail for thirty-five days while his case progressed (Ovalle and Gurney 2018). He was eventually released without any charges filed, but in addition to his incarceration, he had to endure the humiliation of having his picture, complete with prisoner's orange uniform, published in the *Miami Herald*—despite having done nothing wrong. No picture of the teacher who authored the false claims appeared in the paper. If this is a victory for science, it is an unsatisfying one.

Another development on the side of science has been the emergence of some very good writing on the topic. Much of this has come from national-level science journalists, including *Slate*'s David Auerbach (2015), *Forbes'* Steven Salzberg (2018), and Kavin Senapathy, who writes both for *Forbes* and here at *Skeptical Inquirer* (2018). In addition, Daniel Engber has been the leading reporter on the case of Anna Stubblefield, the Rutgers University philosophy professor who used FC to gain consent for a sexual relationship with a nonverbal man with cerebral palsy. Her first conviction and sentence of twelve years in prison for sexual abuse was thrown out on appeal (due to the judge's exclusion of some evidence in the first trial), but after serving less than two years of her sentence, Stubblefield avoided a second trial by pleading guilty to third-degree sexual assault, for which she was sentenced to time-served and fifteen years' probation (Napoliello 2018). Engber has covered this case at great length for both *Slate* and the *New York Times* (Engber 2017, 2018). Finally, Thomas E. Heizen, Scott O. Lilienfeld, and Susan A. Nolan (2015) have published a wonderful little book called *The Horse That Won't Go Away: Clever Hans, Facilitated Communication, and the Need for Clear Thinking*.

True Voices

Despite these few glimmers of light on the side of science and reason, the current state of the conflict appears to favor the forces of pseudoscience. RPM continues to spread, and it has been aided by a relatively new autism advocacy group, the Autism Self-Advocacy Network (ASAN), that frames people with autism as a minority whose rights to communication have not been respected (Autism Self-Advocacy Network n.d.). In the view of ASAN, access to RPM is part of the right to communication, and the organization advocates for its use.

For example, having published a policy statement about FC back in 1995, the American Speech-Language-Hearing Association (ASHA) recently posted a draft policy on RPM that includes a detailed comparison of RPM and FC. The main nugget of the policy is this statement:

> It is the position of the American Speech-Language-Hearing Association (ASHA) that use of the Rapid Prompting Method (RPM) is not recommended. . . . Furthermore, information obtained through the use of RPM should not be considered as the voice of the person with a disability. [Ad Hoc Committee on Facilitated Communication (FC) and the Rapid Prompting Method (RPM) 2018]

The ASAN advocacy group has issued a statement opposing this policy claiming that the ASHA's "blanket statement that specific forms of communication are *per se* inauthentic robs us of the right to communicate" (Crane 2018).

This is not ASAN's first statement promoting RPM. In March, 2016, ASAN filed a complaint with the U.S. Department of Justice in support of students in the Arlington (Virginia) Public Schools who were denied the use of RPM. RPM was not mentioned by name, but the complaint states that the students in question communicate best by "spelling words by pointing to letters on a letter-board held by a trained supporter" (Autism Self-Advocacy Network 2016).

Recognizing that the war was not yet won, a group of science-minded advocates, researchers, and professionals have launched a new effort. They call themselves "True Voices," and they are going on the offensive. Their first target was the University of Northern Iowa (UNI), which, in June of this year, sponsored the "Midwest Summer Institute: Inclusion and Communication," cosponsored by the Syracuse University Institute for Communication and Inclusion, which was formerly known as Facilitated Communication Institute. The conference schedule included a session called "Intro to Facilitated Communication Training," and offered college credit from the University of Northern Iowa to attendees. In the weeks before the UNI institute, scientists wrote an article in *The Conversation* that dramatically highlighted the kind of damage that can—and has—been done by FC (Helmsley et al. 2018). Next the group drafted a "Letter of Concern" about the UNI institute, signed by over thirty professionals and academics, and sent it to several officials at the university.

Attacking public institutions is an excellent strategy. Universities are supposed to be places of enlightenment and reason, and where public funds are involved, the promotion of discredited ideas is particularly controversial. Furthermore, attacking universities that lend credence to FC has proved to be an effective approach in the past. In May of 2015, I wrote an article on FC in which I mentioned that the University of New Hampshire Institute on Disability (IOD) regularly sponsored an "FC skill builders" group. Seven months later, I was contacted by the director of the institute who told me he had read my article and wanted me to know that, after a lengthy review process, the IOD had decided to cease all activities related to FC. I added an update about this development to the archived web version of the article.

It would be a mistake to think my article caused the change at IOD. I am sure they received criticism from a number of fronts. But the fact that the director contacted me suggests that he was interested in correcting the public record.

The True Voices effort at UNI has also begun to produce results. First, the episode provoked a flurry of bad publicity for the university. The previously mentioned *Forbes* article by Steven Salzberg was released the same week as the Midwest Summer Institute and mentioned it directly, asking why a university would offer college credit for instruction in a thoroughly discredited therapy. Other outlets publishing articles included *Inside Higher Ed* (Whitford 2018) and the Syracuse University student newspaper, the *Daily Orange* (Muller 2018). Syracuse University is the mecca of FC because Douglas Biklen, is an emeritus professor of education and the Institute for Communication and Inclusion is housed there (Syracuse University School of Education n.d.). Nonetheless, the *Daily Orange* has been a consistent critic, publishing a number of well-researched articles on FC.

Closer to home, two highly critical articles, quoting members of the True Voices group, appeared in the local Cedar Falls newspaper. The first article, "Facilitated Communication Conference Draws Fire at University of Northern Iowa," was published just prior to the conference (Miller 2018a).

The second article was released the same week as the conference, and after the university had received the letter of concern. It reported that the university would form a committee to look into the institute:

> We regularly evaluate UNI's sponsorship of conferences and events to ensure that we are supporting high-quality programming consistent with the mission of the university. . . . As part of this regular review, we will be convening a group of faculty experts from across campus to discuss the practices presented at this conference. (UNI spokesman Scott Ketelsen, quoted in Miller 2018b)

So, if nothing else, the True Voices offensive forced the University of Northern Iowa to endure some public criticism and prompted the administration to reevaluate their involvement with FC. If the experience at University of New Hampshire is any indication, UNI may choose to cut their ties to this discredited and dangerous technique. We can only hope.

Stuart Vyse is a psychologist and author of *Believing in Magic: The Psychology of Superstition*, which won the William James Book Award of the American

Psychological Association. He is also author of *Going Broke: Why Americans Can't Hold on to Their Money*. As an expert on irrational behavior, he is frequently quoted in the press and has made appearances on CNN International, the PBS NewsHour, and NPR's Science Friday.

References

Ad Hoc Committee on Facilitated Communication (FC) and the Rapid Prompting Method (RPM). 2018. "Rapid prompting method: Position statement." American Speech-Language-Hearing Association. Updated August 31. https://www.asha.org/policy/ps2018-00351/.

Auerbach, David. 2015. "Facilitated communication is a cult that won't die." *Slate*, November 12. https://slate.com/technology/2015/11/facilitated-communication-pseudo science-harms-people-with-disabilities.html.

Autism Self-Advocacy Network. N.d. Accessed June 12, 2023, https://autisticadvocacy.org/.

Autism Self-Advocacy Network. 2016. "ASAN files ADA complaint on communication access in schools." March 9. https://autisticadvocacy.org/2016/03/asan-files-ada -complainton-communication-access-in-schools/.

Biklen, Douglas. 2005. *Autism and the myth of the person alone*. New York: New York University Press.

Crane, Samantha. 2018. "ASAN letter to ASHA on the right to communicate." Autistic Self-Advocacy Network. July 2. https://autisticadvocacy.org/2018/07/asan-letter-to -asha-on-the-right-to-communicate/.

Engber, Daniel. 2017. "A second chance for Anna Stubblefield." *Slate*, June 14. https:// slate.com/technology/2017/06/the-conviction-in-the-anna-stubblefield-facilitated -communication-case-has-been-overturned.html.

Engber, Daniel. 2018. "The strange case of Anna Stubblefield, revisited." *New York Times Magazine*, April 5. https://www.nytimes.com/2018/04/05/magazine/the -strange-case-of-anna-stubblefield-revisited.html.

Heinzen, Thomas E., Scott O. Lilienfeld, and Susan A. Nolan. 2015. *The horse that won't go away: Clever Hans, facilitated communication, and the need for clear thinking*. New York: Worth.

Helmsley, Bronwyn, Howard Shane, James T. Todd, Ralf Schlosser, and Russell Lang. 2018. "It's time to stop exposing people to the dangers of facilitated communication." Conversation. May 22. https://theconversation.com/its-time-to-stop-exposing-people -to-the-dangers-of-facilitated-communication-95942.

Miller, Vanessa. 2018a. "'Facilitated communication' conference draws fire at University of Northern Iowa." *Gazette* [Cedar Rapids, IA], June 15. https://www.thegazette.com/education/facilitated-communication-conference-draws-fire-at-university-of-northern -iowa/.

Miller, Vanessa. 2018b. "University of Northern Iowa will review involvement in 'facilitated communication' conference." *Gazette* [Cedar Rapids, IA], June 19. https://www

.thegazette.com/education/university-of-northern-iowa-will-review-involvement-in-fa
cilitated-communication-conference/.

Mostert, Mark P. 2010. "Facilitated communication and its legitimacy—Twenty-first
century developments." *Exceptionality 18*, no. 1: 31–41.

Mukhopadhyay, Soma. 2013. *Developing communication for autism using rapid prompting
method: Guide for effective language.* Denver, CO: Outskirts Press.

Muller, Jordan. 2018. "Academics criticize conference co-hosted by SU." *Daily
Orange*, June 17. https://dailyorange.com/2018/06/su-co-host-conference-discredited
-communication-method-drawing-criticism-academics/.

Napoliello, Alex. 2018. "No more prison for ex-Rutgers professor who sexually assaulted
disabled student." *NJ.com.* May 11. Accessed July 22, 2018. https://www.nj.com/
essex/index.ssf/2018/05/anna_stubblefield_sentenced_for_second_time.html.

Ovalle, David, and Kyra Gurney. 2018. "Teacher's 'junk science' landed a Hialeahi dad
in jail on charge of raping his autistic son." *Miami Herald*, March 8. https://www
.miamiherald.com/news/local/crime/article204034704.html.

Palfreman, Jon, dir. 1993. "Prisoners of silence." *Frontline.* Aired October 19 on PBS.

Salzberg, Steven. 2018. "Facilitated communication has been called an abuse of human
rights. Why is it still around?" *Forbes*, June 18. https://www.forbes.com/sites/steven
salzberg/2018/06/18/facilitated-communication-may-be-an-abuse-of-human-rights
-why-is-a-university-teaching-it/?sh=3c87a5c329f3.

Senapathy, Kavin. 2018. "On unsubstantiated yet prevalent therapeutic interventions for
autism (part II)." *Skeptical Inquirer*, July 17. https://skepticalinquirer.org/exclusive/
on-unsubstantiated-yet-prevalent-therapeutic-interventions-for-autism-part/.

Syracuse University School of Education. N.d. "Douglas Biklen." Accessed June 12,
2023. https://soe.syr.edu/about/directory/douglas-biklen/.

Whitford, Emma. 2018. "Critics question conference on facilitated communication."
Inside Higher Ed, June 17. https://www.insidehighered.com/quicktakes/2018/06/18/
critics-question-conference-facilitated-communication.

CHAPTER 30

Skepticism and Pseudoexperiments

Benjamin Radford

Vol. 44, No. 5
September/October 2020

Pseudoscience is something that presents as, or is mistaken for, science—but is not. Pseudosciences are especially resilient and pernicious because they benefit from the hard-earned legitimacy of real science as arbiter of objective truth. From phrenology to astrology, homeopathy to facilitated communication, pseudosciences gain believers conditioned to (rightfully) respect science and its methods.

Pseudoexperiments are similar in that they typically present an experiment—a crucial tool in science—and apply it in a popular culture context (though without the input of actual scientists, who might needlessly complicate the attempt with seemingly frivolous trivialities such as control groups, blinding, research design, or, really, anything that might make it valid).

Pseudoexperiments are common in everyday life, though rarely identified as such. More often they're seen as rhetorical devices, entertainment, or simply news. They have a populist plausibility because they have a superficially valid lesson: To understand some phenomenon, you don't need to look at research design or boring statistics. You can "see for yourself" what's going on, what the truth is. They are especially effective because they give the illusion of objectivity. Pseudoexperiments, like scientific experiments, are usually portrayed as unbiased—though often activist—"real world" examples to prove a point. Like classic pseudoscientists, pseudoexperimenters want the *prestige* of science but don't want to put in the effort to make it meaningful or valid. Their goal is not the discovery of nuanced truth about the world but instead promoting an agenda by means of a simplified, memorable "gotcha" moment.

A pseudoexperiment is not necessarily the same as a *flawed* experiment. Conducting a perfect scientific experiment is very difficult (if not impossible), which is why p-values are factored into a study's results. Flawed research is a hallmark of many pseudosciences and paranormal topics, from ghosts to crop circles to cryptozoology. For pseudosciences such as psi, there is an inverse correlation between the quality of the research and the amount of good evidence obtained; the better the controls, the less empirical evidence appears (as Jim Alcock, Stuart Vyse, Scott Lilienfeld, and others have noted in this magazine). There are several types of pseudoexperiments.

Pseudoexperiments as Television Fodder

Most pseudoexperiments are done for a viewing audience, often for journalistic, entertainment, or commercial purposes. In January 2006, security specialist Bill Stanton appeared on *The Today Show* posing an alarming and provocative question: If you saw an innocent child being kidnapped by a stranger, would you help?

There are several approaches that the producers could take to answer the question, including interviewing a social psychologist about research on bystander effects. But the producers chose a more dramatic angle: using hidden cameras to record a seven-year-old named Rachelle faking being abducted on a city street. Rachelle's mother watched from a surveillance van as Stanton approached the girl, who stood alone in the middle of a sidewalk playing a video game. Stanton walked up to Rachelle and took her by the arm, saying things such as, "There you are, young lady! You come with me," while Rachelle protested, "No, no . . . you're not my daddy!"

Stanton and Rachelle repeated the scenario several times, and rarely did bystanders intervene. Some kept walking, and others glanced briefly at the scene, but few approached. The *Today Show* anchors called the results "shocking," and to everyone involved (and probably the audience), this seemed a clear and sad case of people reluctant to help someone in need. "It's frightening that no one will help," Rachelle's mother said.

Yet there may be a good reason more people didn't get involved—one completely missed (or ignored) by Stanton and the *Today Show* producers: the bystanders didn't believe that they were actually seeing a child abduction. Because the "abduction" was poorly staged, it's more likely that those who witnessed the scene simply—and accurately—recognized that the child was not in real danger. From the footage, it's clear that Rachelle was not an actress and didn't act scared when Stanton approached her. Her protests sounded like a typical child's whines instead of panicked pleas for help. Stanton did not strike the child or hurt her in any way, and Rachelle didn't scream, kick at, or fight off the man supposedly trying to abduct her. In short, there was little that would convince the average person that she was genuinely in danger. The problem isn't the seven-year-old's acting; it's a poorly conducted pseudoexperiment by Stanton and NBC News. *The Today Show*'s hidden camera test would be valid only if the bystanders actually believed that the child was in danger; it's not clear that was the case.

The highly rated ABC News series *What Would You Do?*, hosted by John Quiñones, follows an identical format and is now in its fourteenth season. Hidden cameras record people in public places such as restaurants and involve social justice issues, including race, gender identity, disability, and parenting. Will strangers step in to confront a staged racist or sexist remark made between two actors? Will passersby act against an able-bodied person using a handicapped spot? Quiñones, who was inspired to go into journalism by Geraldo Rivera—himself no stranger to alarmist television, especially during the "Satanic Panic" scare of the 1980s—said that he believes his show "holds up a mirror" to America.

While some of these staged performances are inevitably better executed than others, it's not clear how valid or generalizable the results are. People may choose (or choose not to) publicly intervene in events nearby for any number of reasons unrelated to the hypothesis offered by the show (apathy, prejudice, etc.). Bystanders may not fully understand what's going on; they may have visual or auditory disabilities and sincerely not have noticed the seemingly alarming event; they may be socially anxious or fearful about confronting others; and so on— though often the broadcast reactions are precisely what the producers hoped they'd be. Like *Candid Camera* and countless other shows, these pseudoexperiments are voyeuristic entertainment, not science experiments or philosophical musings on responsibility such as seen in *Rashomon* (1950) or *Force Majeure* (2014).

In some cases, it's a physical pseudoexperiment. In 1992, *Dateline NBC* aired a program titled "Waiting to Explode," which examined claims that General Motors trucks exploded when involved in low-speed collisions due to faulty gas tank design. To demonstrate this, the program showed footage of a truck exploding when being hit from the side, ostensibly as part of a safety test. It was later revealed that the explosion had been staged. Not only had the truck's gas cap been intentionally loosened, but an explosive charge had ignited the fireball. GM sued the network, and the journalistic credibility of NBC News suffered for years because of this pseudoexperiment.

Pseudoexperiments as Social Activism

While ratings-driven journalism is a common source of pseudoexperiments, many laypeople use them to raise awareness of social issues. For example, in 2018 a San Antonio woman named Stacey Alderete recorded herself entering her child's school. She had just dropped her daughter off and returned to the school after grabbing a backpack. She's seen in a short cell phone video following others through the main entrance and walking through a hallway. She's seen by several people but not confronted by any security staff.

To Alderete, it was a bold experiment that exposed a glaring—and possibly deadly—flaw in school security, one that could be exploited by a would-be mass shooter. The day after posting the video to YouTube, Alderete appeared at a public school-board meeting and said, "I proved yesterday just how vulnerable to violence or school shootings our children are at the high school. My video speaks volumes." She later added, "If I was able to do this without being detected, ANYONE CAN!" That interpretation was embraced and widely shared on news and social media (Bates 2018).

A more parsimonious explanation is that she was recognized. Alderete did not slip into the school unnoticed with a suspicious item; she was seen and (correctly) recognized as a nonthreatening mother of a child attending there. A young male carrying a gun, or even a large backpack, would arouse more fear and suspicion than an adult woman who's been seen at the school many times before. The mere fact that she wasn't personally confronted by security, of course, doesn't mean that her presence was unnoticed. It's entirely possible that she was seen on surveillance cameras by security staff who recognized her as having dropped off her daughter earlier and assumed she was returning to give her child a forgotten item or message.

It was later revealed that Alderete had an ulterior motive for her pseudo-experiment; it wasn't conducted to objectively test school security but instead because she had lost a re-election campaign as a school board member, emphasizing security. "Ms. Alderete is simply a disgruntled former board member that lost her last election and has spent a lot of her time since then trying to create havoc for the school district," Superintendent Abelardo Saavedra said. He also noted that it was "totally ridiculous" for Alderete to claim that she had exposed security flaws, because as a former school board member, she is a "fairly high-profile parent" and widely recognized by school staff and security (Caruba 2018).

Celestia Ward, in an article on "outrage vampires," described a typical social media pseudoexperiment by YouTube influencer "Joey Salads" and his 2016 viral video exposing the scorn that breastfeeding women face:

> The so-called experiment contrasts a sexy woman in a low-cut top with a young breastfeeding mom, both sitting on a mall bench "for science!" We see the attractive model receive plenty of glances, and one fellow even sits down and hits on her. Then, after the title "Now lets see how people react to Breastfeeding" [sic] flashes, we see the young mother heckled over and over again. In this two-minute video, four separate people hurl the same word, "disgusting," right at her,

and each time she meekly apologizes "I'm sorry, he's hungry." All these disgusted mall denizens also had an approach I've never seen in real-life situations. Rather than just grumbling or expressing their distaste in whispers as they amble by, each person walks right up to the young mother, stands in a position easily captured by the camera, and delivers the insult clearly and loudly . . . as if they had been coached on what to do. It was only one of many similar faked videos by Salads, who, when asked, "Why do you feel the need to fake an experiment?" replied he was "just trying to make something shocking and create controversy." (Ward 2018)

Other similar stunts include videos by a man named Adam Saleh, who claimed he was removed from a Delta Airlines flight because he spoke Arabic and was accused of being a terrorist. Like Alderete, he recorded the incident and posted it online, generating news headlines and anti-racist outrage. However, other passengers disputed Saleh's claims, reporting that he had lied: Saleh was not merely speaking Arabic but shouting on the plane, being disruptive and confrontational—clearly hoping to provoke a response from ordinary passengers that he could falsely portray as racism that would be sensational enough to go viral. In fact, "Saleh is a serial prankster who frequently makes YouTube videos in which he pulls stunts on flights. His YouTube channel features films in which he conducts 'experiments' to determine how passengers will respond to travelling alongside Muslims and people speaking Arabic" (Boult 2016).

Another social justice pseudoexperiment made national headlines in 2014 when a viral video showed a woman walking along New York streets with some men catcalling behind her. The video, filmed and released by an organization called Hollaback, initially garnered widespread attention (with over 40 million views) and praise, hailed as an important social experiment revealing the scope of street harassment and catcalling. But it soon attracted controversy when its status as a pseudoexperiment was revealed.

People began taking a closer look at what the video did—and didn't—show, what it revealed, and what it concealed. Nearly all the men seen catcalling, whistling, and making propositions at the white, conventionally attractive actress were people of color, mostly African American men. The video was clearly edited; she allegedly walked for nearly ten hours, but the video was just two minutes long. The footage wasn't merely sped up; instead only a tiny percentage of it was seen. Few suspected that it was outright staged (as Joey Salads's and Adam Saleh's videos had done), but this raised the question: Was the segment shown really representative of her experience during the walk, or did the editors cut it to make it seem that blacks were disproportionately harassing the woman? Part of that answer was related to where she walked and when.

The social justice video itself soon smacked of racism, and experts took a closer look. Professor Zeynep Tufekci (2014) analyzed the video, noting:

> I've taught Introduction to Research Methods to undergraduate students for many years, and they would sometimes ask me why they should care about all this "method stuff," besides having a required class for a sociology major out of the way. I would always tell them, without understanding research methods, you cannot understand how to judge what you see. The Hollaback video shows us exactly why.

Tufekci's excellent analysis is too complex to treat here; see her article "Hollaback and Why Everyone Needs Better Research Methods."

Advertising Pseudoexperiments

Pseudoexperiments have been a staple of advertising for generations. Countless side-by-side comparisons have "proven" how much better the hero product is versus the leading brand of everything from removers for stubborn laundry stains to car waxes—while leaving out important caveats and differences between the two.

In recent years, companies have used social media to share advertising pseudoexperiments, often in the hope they will go viral, amplifying their reach. One famous example is a series of short videos created by Dove/Unilever, a 2013 campaign called Real Beauty Sketches. In it,

> Dove hired former police forensic artist Gil Zamora to illustrate some psychologically revealing sketches. In a campaign created by Ogilvy Brazil, a series of women described themselves to Zamora in minute detail, from behind a curtain. The artist in turn created composites as though trying to identify a criminal. Next, each participant was asked to describe another woman present. The results are dramatic and sort of moving. Viewing the two sketches side by side—one based on self-description, one from a friendly stranger—it's clear how unflattering the women's own self-assessments are. (Berkowitz 2013)

The campaign was enormously successful, generating nearly 70 million views on YouTube and praise as empowering and poignant from news media, media critics, and feminist organizations. Soon critics noted flaws in the pseudoexperiment, suggesting that the advertisement didn't quite "prove" anything.

For one thing, the positive descriptions of the women in the video were rooted in what's conventionally attractive: thin, mostly Caucasian, with blue eyes and smooth complexions. Whatever the intended message, the focus was still on appearance (which is to be expected from a cosmetics company). From a research design point of view, it's not clear why Dove assumed that self-descriptions would (or should) match descriptions by other people. This is due in part to *demand characteristics*: people tend to respond in socially appropriate ways that make them appear kind and positive. People—especially women—tend to err on the side of humility when describing themselves to others, so as not to appear vain. As Autumn Whitefield-Madrano wrote in her book *Face Value: The Hidden Ways Beauty Shapes Women's Lives*, "We [women] learn that women aren't supposed to refer to themselves with words bolder than *reasonably attractive*" lest they be seen as vain (Whitefield-Madrano 2016, 58).

Similarly, people tend to be kind when describing the appearance of others so as not to appear cruel or mean-spirited. When being asked by an interviewer—on camera no less—to describe another person's appearance, few of us would be eager to highlight their flaws. The sketch artist was also aware of the pseudoexperiment's purpose and may have consciously or otherwise guided his depictions according to the client's goals.

Thought Pseudoexperiments

Of course, not all experiments are done in the real world. Philosophers, social scientists, and others engage in what are called thought experiments, whereby using various premises, assumptions, and rules of logic they can imagine what would, or should, happen given certain premises or conditions. They are usually framed as a rhetorical question intended to expose a real or imagined hypocrisy or double standard. Thought pseudoexperiments are common on social media, in which, for example, a meme (often rhetorically) asks the reader to compare two people or situations and typically directs them to (an apparently self-evident) conclusion. These thought pseudoexperiments typically involve political, environmental, or social justice issues and are framed to provoke outrage or righteous indignation.

Examples include memes such as "Imagine if [Harvey] Weinstein was black" under a photo of Bill Cosby's mugshot, suggesting that if the Hollywood producer were black, he would already be arrested and convicted, even though the cases are quite different. There are countless variables in the two cases. Race is certainly one of them, but there's no reason to think it's the only—or even the most important—one. It's not a real experiment; it's just asking us if we can imagine person A swapped out for person B in situation X. Unless you have the imagination of a turnip, the answer is always yes. We *can imagine* a scenario in which the outcome is as described. But that doesn't mean it is likely or even probable (for other examples, see Radford 2019). Similar memes compare two newsworthy people who got unusually long (or short) sentences for (allegedly) similar crimes, but from different socioeconomic strata. It's rhetoric thinly disguised as a neutral, objective question, akin to a conspiracy theorist "just asking questions." But if those premises are biased, skewed, or wrong, then the conclusions drawn from them will often be flawed.

Experiments versus Pseudoexperiments

Flawed experiments are common, but they may not have the same potential to mislead that pseudoexperiments do. From skeptical, media literacy, and critical thinking points of view, there are several flaws inherent in pseudoexperiments.

Perhaps most obviously, they emphasize anecdote over data, and anecdotes about a social problem are not necessarily valid evidence about that topic. The question is not what happened to one person in one specific case but instead whether that case is representative of the larger group or population. Pseudoexperiments always have an obvious agenda, so in that regard it's not an experiment at all. In scientific experiments, manipulating the data is a grave transgression; in pseudoexperiments, manipulating the data is the whole point. There's also a filedrawer effect: if a person conducts a pseudoexperiment and it doesn't turn out how they wanted it to, nobody sees it—or they simply edit out the inconvenient bits.

A second fallacy is failing to consider alternative explanations. Legal outcomes in court cases are especially vulnerable to misrepresentation in pseudoexperiments. People can cherry-pick examples all day to suit their needs. If you want to find a case where a wealthy white male was accused of something horrible and "got away with it" (whatever that might mean in a given context— from charges being dropped to an acquittal at trial, a mistrial to probation) presumably due to status and privilege, you can do it. The same is true if you swap out any of the variables: rich or poor, male or female, gay or straight, black or white, and so on. In mere minutes, an internet search will turn up some case, somewhere, in which a given category of person did (or didn't) "get away" with some presumably outrageous crime or act, plausibly due to whichever factor the presenter wishes to highlight. Often the cases are notable and newsworthy *precisely because* they're exceptional and unusual. Ordinary people doing ordinary things—or receiving statistically average sentences for ordinary crimes—rarely make the news. Selection and confirmation biases direct us to the dramatic, unrepresentative outliers and make inherently flawed examples for comparison.

Pseudoexperiments point the audience to a single, seemingly self-evident, overriding factor to the exclusion of others. The factor being highlighted could certainly be the main reason for the disparity, or it may have played a part, or it could have been wholly irrelevant. There's usually no way to know without a much closer look at research and statistics. In many cases, the overall point or "larger truth" of the meme or experiment is valid, but the pseudoexperiment chosen to prove or demonstrate that claim is not.

Pseudoexperiments are powerful because they're simple, easily digested, and effective at pushing buttons and provoking anger and outrage. They can also cause harm. The examples discussed here may make parents and children feel less safe in schools and in public; make Arab Americans feel less welcome on passenger airlines (or more self-conscious about speaking Arabic); make nursing mothers feel uncomfortable breastfeeding their babies in public; reinforce racist stereotypes about catcalling black men; and so on. Some of these are attention-seeking hoaxes, while others are misguided social justice stunts. Regardless of the methods and motivations, pseudoexperiments mislead and misinform. The real world is complex and hard enough to understand without pseudoscientific pseudoexperiments.

Benjamin Radford, MPH, MEd, is a scientific paranormal investigator, a research fellow at the Committee for Skeptical Inquiry, deputy editor of the *Skeptical Inquirer*, and author, co-author, contributor, or editor of twenty books and over a thousand articles on skepticism, critical thinking, and science literacy.

References

Bates, Josiah. 2018. Mother walks through Texas high school without being stopped, calls for better security. ABC News (February 23). Available online at https://abcnews.go.com/US/mother-walks-texas-high-school-stopped-calls-security/story?id=53304996.

Berkowitz, Joe. 2013. Forensic artist proves women literally don't know their own beauty. FastCompany.com (April 16). Available online at https://www.fastcompany.com/1682797/forensic-artist-proves-women-literally-dont-know-their-own-beauty.

Boult, Adam. 2016. Prankster 'kicked off flight for speaking Arabic'—Delta Air Lines defends actions. *The Telegraph* (December 22). Available online at https://www.telegraph.co.uk/news/2016/12/22/prankster-kicked-flight-speaking-arabic-delta-airlines-defends/.

Caruba, Lauren. 2018. Former South San ISD trustee gets school trespass warning. My SanAntonio.com (February 24). Available online at https://www.mysanantonio.com/news/education/article/Former-South-San-ISD-trustee-gets-school-trespass-12704155.php.

Radford, Benjamin. 2019. Misleading memes as cloaked propaganda. Center for Inquiry blog (June 21). Available online at https://centerforinquiry.org/blog/misleading-memes-as-cloaked-propaganda.

Tufekci, Zenyep. 2014. Hollaback and why everyone needs better research methods. Medium (November 3). Available online at https://medium.com/message/that-catcalling-video-and-why-research-methods-is-such-an-exciting-topic-really-32223ac9c9e8.

Ward, Celestia. 2018. Outrage vampires versus breastfeeding moms. Center for Inquiry blog (July 17). Available online at https://centerforinquiry.org/blog/outrage-vampires-versus-breastfeeding-moms/.

Whitefield-Madrano, Autumn. 2016. *Face Value: The Hidden Ways Beauty Shapes Women's Lives.* New York: Simon and Schuster.

CHAPTER 31

Arthur J. Cramp

THE QUACKBUSTER WHO PROFESSIONALIZED AMERICAN MEDICINE

Mike Jarsulic and Robert Blaskiewicz

Vol. 42, No. 6
November/December 2018

In the first decade of the twentieth century, an enterprising man in Grand Rapids, Michigan, named A.W. Van Bysterveld claimed that he could "locate the cause of your aches and pains" for free via the mail, if only you would send him a vial of your urine. Van Bysterveld was a self-proclaimed "Expert Inspector of Urine" and claimed that he used a "secret process handed down generation after generation, and most carefully guarded by the old families of Europe." Van Bysterveld assured prospective clients that though his "secret methods are not taught in schools," he "examines on an average of 25,000 bottles of urine a year. This alone stamps him as an authority of exceptional qualifications" (Cramp 1911a, 56).

Between March and November 1910, the American Medical Association's (AMA) Propaganda for Reform Department, led by Dr. Arthur J. Cramp, investigated Van Bysterveld's claims. A sample (containing tap water, pepsin, aniline dye, and ammonia) prepared by the chemists at the AMA was sent to the Van Bysterveld Medicine Company. The diagnosis came back: "Careful examination of the urine shows there is too much acid in the blood, which will cause a rheumatic condition, the back is weak, and you will have a tired nervous feeling most of the time" (Cramp 1911a, 56). Meanwhile, correspondents working with Cramp in Iowa and Michigan contacted Van Bysterveld's company and sent in samples that were identical to the first vial from the AMA. This time, the mixture in one case indicated poor blood, a malfunctioning liver, gas, nervousness, and heart problems; in the other case, it indicated weakened kidneys, a "catarrhal condition [inflammation of mucosal membranes] of the stomach and bowels," and nervousness (Cramp 1911a, 56–57).

Lastly, two more samples were sent in from different addresses in Chicago. This time, both vials were filled with 95 percent tap water with 5 percent glucose. This would indicate that the patient was not merely diabetic but was urinating a substance with more than twice the sugar content of a Red Bull. Again, the diagnoses came back as either liver disease or a catarrhal digestive track. Cramp's conclusion was unequivocal: "The whole things shows conclusively that the 'examination' of the urine is a farce, the diagnosis is a fake, and taking the money from victims for the 'treatment' of a purely imaginary disease is a fraud and a swindle" (Cramp 1911a, 57). Cramp further condemned those who profited from this mail-order pseudomedical company: "Those publications which accept the advertisements of this concern are, wittingly or unwittingly, participating in the profits of scoundralism" (Cramp 1911a, 57).

Arthur J. Cramp, MD, played an outsized but oddly unacknowledged role in the professionalization of American medicine. American medicine after the Civil War was still unorganized and largely unregulated. Practitioners needed little formal education and no single commonsense, science-based standard of evidence for treatments applied. As a result, numerous competing schools of thought and private interests fought for the attention of the American consumer, who was barraged with advertisements for nostrums and cures on nearly every page of every popular periodical. The American Medical Association, founded in 1847, had always looked down on secret proprietary treatments, and it saw as an important part of its mission the suppression of quack medicine. For instance, Eric W. Boyle notes that the AMA reserved the right to refuse admittance of journals that advertised proprietary medicines into the Association of Medical Journal Editors, a group that had been established by the AMA to raise the standards of medical publications. Nonetheless, the revenue generated by advertising questionable remedies was so substantial that even the *Journal of the American Medical Association (JAMA)* failed to uphold its own standards until the turn of the twentieth century (Boyle 2013, 12–3). Dr. Cramp contributed to the professionalization of American medicine by identifying and exposing quackery; subjecting claimed remedies to scientific analysis; assisting government, public, and professional policing efforts directed at controlling health fraud; and pursuing an aggressive thirty-five–year public education program.

Arthur J. Cramp was born in London on September 10, 1872. According to British public records, his father was a blacksmith, and at the age of sixteen, Cramp was apparently an engineer's apprentice. He moved to the United States at twenty. According to an undated biographical blurb in the files of the AMA, Cramp "did farm work in Missouri for a few years, attended an academy in that state. And then taught in a country school, at the same time writing a weekly column in the local newspaper" (Cramp, Arthur J, Special Data, n.d.). He married a woman from Missouri named Lillian Torrey. Cramp's family eventually settled in Waukesha, Wisconsin, where he and his wife worked at the Wisconsin Industrial School for Boys, a reformatory high school. According to the 1900 census records, they lived on the campus where they worked. Long before he became a quackbuster, then, he was an educator, and that ethos informed his later work. While at the school, they had a daughter, Torrey, who died a little over a year later. According to her death record, Torrey died January 2, 1900, from seizures related to meningitis. According to a letter from a colleague at the AMA to the historian of American quackery, James Harvey Young, Torrey's death while being treated by a quack is what compelled Cramp to pursue medicine and expose quackery. (It should be noted, however, no effective treatments for communicable meningitis existed at the time.) In 1906, Cramp graduated from the Wisconsin College of Physicians and Surgeons and joined the American Medical Association (Cramp, Arthur J., Correspondence, n.d.) as a medical editor in December of that year.

Cramp joined the AMA at a particularly auspicious time for practitioners who sought to combat quackery, as the country was in the middle of a period of progressive reforms. In 1905, the year before Cramp was hired, the AMA established the Council on Pharmacy and Chemistry, an in-house lab that analyzed the contents of patent medicines. That same year, Samuel Hopkins Adams published a series in Collier's that would be collected in an edition called *The Great American Fraud, a monumental and damning exposé of modern medical charlatanry.* Adams named names of fake practitioners, revealed the alcohol and narcotic content of supposed cures, and laid bare the profitable codependence of nostrum advertisers and publishers. In 1906, Upton Sinclair published *The Jungle,* a novel based on his observations of appalling and dangerous conditions in Chicago's meatpacking industry.

All of this occurred at the same time a pure food and drugs bill was moving through the Congress. The Pure Food and Drugs Act of 1906 regulated claims that were made on drug packaging, but in 1911 it was interpreted by the Supreme Court very narrowly to mean only that the ingredients on the label needed to be accurate. Claims of effectiveness remained unregulated. The 1912 Sherley Amendment was introduced to close this loophole, but during the legislative negotiations language regulating "false or misleading" claims was replaced by "false and fraudulent," which introduced the question of intent into legal matters, a more difficult case for regulators to prove (Cramp 1924, 424–425).

In 1910, the Carnegie Foundation released what was known as the Flexner Report, a comprehensive survey of medical education in the United States and Canada. The report led to reforms that standardized medical education, established scientific medicine as the educational gold standard, and diverted educational resources and accreditation away from homeopathic, proprietary, botanical, and eclectic medical schools (Page and Baranchuk 2010, 74). All these developments informed and validated the AMA's work of shedding light on proprietary medicines and quackery through aggressive public education, a campaign in which Cramp would feature prominently.

Even though the Pure Food and Drugs Act staked out new territory in the government's interest in public health, few federal resources were available to regulate medicine; each state had (and still has) its own medical board and standards of practice. As such, the AMA as a national organization was one of the few institutions that could conceivably standardize the practice of medicine. Cramp thought the AMA was taking on a job that should in theory be done by the government; however,

> the exigencies of politics make it well-nigh impossible for health agencies to tell unpleasant truths when these involve huge vested interests. Nevertheless, if the public's health is to be served, these truths must be told. The medical profession of America, recognizing this fact has assumed this responsibility and is discharging it through the Bureau of Investigation (Cramp 1933, 54).

The American Medical Association, as part of its attempt to promote the interests of its members, became, in the words of Eric W. Boyle, "the primary arbiter in disputes over the legitimacy of medical therapies" (Boyle 2013, 62). This aspect of the professionalization of American medicine was achieved largely through the efforts of Arthur J. Cramp. As soon as Cramp was hired on as an editor of the *Journal of the American Medical Association,* he began contributing regular pieces about quackery to it. As part of the general push to improve the quality of medical information available to physicians and the public, the AMA established the Propaganda for Reform Department, which grew out of the Council on Pharmacy and Chemistry and was headed by Cramp (Cramp 1933, 51). Cramp believed in what he once called "public education through publicity" (Cramp 1920, 788), and he believed that "the best that the medical profession can do in protecting the public is to turn the light on the methods of the faddist and the quack, so that his ignorance or fraud becomes apparent" (Cramp 1927b, 727–728). This was the mission of the Propaganda for Reform Department, which became the Bureau of Investigation in 1925 and the Department of Investigation in 1958 (Hafner et al. 1992, viii).

The centerpiece of the Propaganda Department was a growing collection of materials related to quackery. The collection grew out of information from Cramp's prodigious correspondence with professionals and members of the public who queried the AMA for information about a variety of treatments, practices, and practitioners that they encountered. Between 1918 and 1930, the number of letters answered by Cramp's office ballooned from under two thousand inquiries per year to over twelve thousand a year (Boyle 2013, 79). A fairly typical example of how this collection grew bit by bit can be seen in the file for a treatment called "Spray-O-Zone." In 1925, Cramp received a query letter from Francis E. Fronczak, Buffalo, New York's Public Health Commissioner.

Fronczak asked for any information the AMA had on Spray-O-Zone, a night-time nasal spray that claimed to "protect your children from infantile paralysis [polio]," and he included a sample of the nostrum in its packaging. On October 9, 1925, Cramp replied to the query:

> Your letter of September 23 was received in due course as was also the specimen of "Spray-O-Zone." As we had no record of Spray-O-Zone and had received no inquiries whatever about it, we were not justified in going to any large amount of work in investigating the nostrum. We did, however, ask the A.M.A. Chemical Laboratory to make a preliminary examination of the product. (Spray-O-Zone, 1925–1927)

He quotes the lab report, in which the chemists suggest that Spray-O-Zone was borax and potassium nitrate dissolved in water. A synopsis of the cursory qualitative evaluation of Spray-O-Zone appeared in the February 12, 1927, edition of the *Journal of the American Medical Association* (Cramp 1927a, 501). All of these items, and the packaging material, are preserved in the Spray-O-Zone folder. Through thousands and thousands of similar exchanges with physicians, regulators, and members of the public, Cramp steadily amassed a vast arsenal of information to unleash against medical hucksters and their enablers. Within five years of arriving at the AMA, according to historian James Harvey Young, "Cramp's office contained over 12,000 cards in a 'Fake File,' listing products, firms, and names of promoters. His 'Testimonial File' held the names of over 13,000 American and 3,000 foreign doctors who had given testimonials for proprietary drugs" (Young 1967, 131–2). By 1937, this card catalogue had expanded to some 300,000 entries (Boyle 2013, 80). Eventually, the collection would grow to 3,500 files spread over some ten thousand folders, totalling 370 cubic feet of material, or, as the archivists who described and catalogued the entire collection put it in the 1990s, 185 standard file drawers (Hafner et al. 1992, x).

Cramp made extensive use of this collection during his time at the AMA. He wrote a weekly column in *JAMA*, where he put the extraordinary medical claims to scrutiny, and he contributed the occasional article to H. L. Mencken's *American Mercury* magazine and the AMA's popular health magazine, *Hygeia*, for which he was an advisor (Fishbein 1969, 132). Not only did he investigate the products and quacks, but he also looked into the histories of the people involved, gathered information about their business plan, fact-checked advertisements, discussed the reasons that quacks were so convincing, and generally pulled back the curtain of a type of medicine grounded in marketing rather than science.

Among Cramp's other duties, he answered thousands of letters written to the AMA by laymen, Better Business Bureaus, and the advertising managers of publications that vetted ads they printed (Cramp 1933, 52). His department also supplied services that members of the public sought from government agencies, for instance, handling queries about questionable medical claims, products, and services that had been redirected to the AMA by local, state, and federal authorities. The AMA answered questions from government authorities seeking information about questionable medical practices that had appeared in their jurisdictions, and, said Cramp, "[played] an important part in bringing to the attention of state and federal officials schemes and methods that seem to be a menace to the public health, a violation of the law, or both" (Cramp 1933, 53). Starting in 1912, the Bureau also published educational materials, including posters, slide shows, and offprints of Cramp's *JAMA* articles about different forms of quackery. Much of that material, in turn, was gathered into Cramp's three volumes of Nostrums and Quackery, which he described as "a veritable encyclopedia on the nostrum evil and quackery" (Cramp 1933, 51–53). These were published between 1911 and 1936. Cramp was so central to the fight against quackery in the early twentieth century, Young remarks, that "scarcely an investigation was launched by a regulatory agency but that an inquiry went to Cramp to see what information the AMA already had on hand" (Young 1967, 132).

Cramp lamented the passing of this older type of advertising, which merely listed available products, as well as the rise of "selling copy," which created a false need for medicine in prospective consumers. *Semi-weekly* Camden Journal.

One of the challenges that dogged Cramp throughout his career was lax advertising standards. Though the Pure Food and Drug Act regulated what could be said on drug packaging, it did nothing to regulate advertisements and marketing of drugs elsewhere. Cramp knew that he was fighting an uphill battle as long as patent medicine manufacturers were able to market directly to potential consumers. His critiques of nostrum advertisements tended to focus on a few themes. First is the use of "selling copy," which sought to plant a need for a product in the minds of a target audience rather than indicating where products for a preexisting need could be found.

In practice, ads for quack remedies were intended to convince members of the public that they were sick with a particular condition, a sort of unethical, induced hypochondria. "No man," Cramp argued, "has any moral right to so advertise as to make well persons think they are sick and sick persons to think they are very sick. Such advertising is an offence against the public health" (Cramp 1918, 757). He roundly condemned "such periodicals and newspapers as are not above sharing the blood-money" of quack remedies (Cramp 1911b, 134). He saw these periodicals as enabling especially cruel fraudsters, or as he described them, "The swindler who sells stock in bogus companies to presumably intelligent human beings is a gentleman compared with those scoundrels who lie to the sick, humbug the suffering and fraudulently take the money of the incapacitated" (Cramp 1911a, 56).

Second, while other forms of merchandise that did not perform their advertised function or purpose would quickly be found out, fake medicine conspired with nature to trick patients into believing that fake treatments worked:

> The healing power of nature . . . is such, fortunate for biologic perpetuity, that the general tendency of the disordered animal economy is to get well. . . . In probably eighty per cent of all human ailments the afflicted person gets well whether he does something for his indisposition or does nothing for it. Herein lies the opportunity of the quack and the nostrum vendor. (Cramp 1924, 423)

He lamented the same difficulty modern quackbusters have in convincing the already committed that they are wrong and notes the power of the post hoc fallacy, which allows patients to mistake "sequence for effect" (Cramp 1918, 756).

Cramp also despised the use of testimonials in advertising, knowing how misleading they could be. Cramp's fights against testimonials were some of his most devastating attacks on quacks' credibility. A 1913 investigation into a supposed tuberculosis cure called Pulmonol is a case in point.

The testimonial-givers are, as always, divided into two classes; those who really had tuberculosis and those who did not have it. As we have said many times, it is useless to investigate fresh testimonials. Most of them are written in good faith and not until the cases have progressed further are the victims undeceived as to the efficiency of the nostrum. It is therefore necessary to wait a year or two before looking into testimonials of this class. We then find, invariably, that the consumptive who had relied on the nostrum is dead. (Cramp 1913, 1998)

Cramp then lists nine patients who had provided testimony on behalf of Pulmonal that had appeared either in newspapers or in advertising pamphlets. Cramp had statements from patients' physicians and some of the patients themselves, and, as so often was the case, he often found that the patients had either never been sick (if they existed in the first place), had self-diagnosed their disease, were still sick, or had died of the ailment they thought they had beaten (Cramp 1913, 1998–1999). When possible, Cramp would juxtapose a patient's testimonial with their obituary, as was possible when "Sargon," a preparation that was 18 percent alcohol, placed an ad in the *Rochester Democrat and Chronicle* featuring JR Kimber's testimony on June 25, 1930, five days after the same paper had announced his death (Cramp 1930, 285). Especially vexing to Cramp was that the Pure Food and Drugs Act regulated claims made only on packaging, not the claims made elsewhere. He used this fact to suggest a clever way that the public could check the magnitude of nostrum makers' false claims: "One can with almost mathematical accuracy determine the falsity of modern 'patent medicine' advertising: Subtract the claims made on the trade package from those made elsewhere; what remains—and the residuum will be large—is falsehood!" (Cramp 1920, 788).

Another important aspect of Cramp's educational project at the AMA was to give public lectures to schools, professional groups, and civic organizations across the country. He gave talks on subjects such as "Patent Medicine and the Public Health," "Pink Pills and Panaceas," "Fighting Deafness Quackery," and "Objectionable Medical Advertising." The AMA did not charge for Cramp's appearances (they only asked that a projector be available for his slides). During the early 1930s, Cramp was giving dozens of talks a year.

In June 1934, Cramp was in Cleveland attending the AMA's annual conference when he suffered a debilitating heart attack from which he never fully recovered. He was bedridden for several weeks following the episode, and when he returned to work, he tired very easily and was unable to keep his former pace. As a result, he canceled numerous presentations and took a drastically reduced workload. As he explained in a letter dated January 1935, turning down an offer to give a presentation in St. Clair County, Missouri:

> Last June, on the day I reached Cleveland for the annual meeting, I came down with a coronary thrombosis that kept me in the hospital for nearly six weeks and kept me away from my work for over four months. I am still able to work only at a very greatly diminished tempo and as avoiding all evening talks. As a matter of fact, I turned down some talks right here in Chicago because they were to be given in the evening. ("Cramp, Arthur J., Correspondence")

By September of 1935, the frustrating limitations that Cramp's heart attack placed on him prompted him to retire prematurely. In a letter dated September 24, 1935, Cramp explained:

> I have found that it is impossible for me to keep up my work here at the A.M.A. headquarters, and I expect in a few weeks' time (the end of November) to give up my work entirely. I had hoped to be able to stick it out until the end of November of next year, when I should have completed thirty years of continuous service, but I find that is impossible. ("Cramp, Arthur J., Correspondence")

In his retirement, he moved to Fort Lauderdale and then to North Carolina. After he retired, he still corresponded with the AMA and his editor Morris Fishbein, and he completed and published the third and final volume of *Nostrums and Quackery*.

Only one detailed description of Cramp's personality seems to exist, from W.W. Bauer, Cramp's colleague at the AMA during his final years with the organization, who was responding to an inquiry from the historian James Harvey Young. Bauer, who directed the AMA's Bureau of Health and Public Instruction, gave details that one would scarcely be able to perceive in Cramp's publications and correspondence. Bauer reported that Cramp was "typical of the small, slight Englishman. He was slender, quick, and birdlike in his movements with ruddy cheeks, clear blue eyes, [had] a small clipped mustache, and an imperial. In his later years, his hair and whiskers grew white." Cramp was a "perfectionist," and this trait characterized both his work and his personal habits. "He was always perfectly groomed and meticulously dressed," said Bauer. "He never sat down on a chair without first running his fingers over it to be sure there was no dust. He was also particular about his food and might be called a true gourmet. He ordered the best of food and usually in considerable variety and fairly large quantity, but he ate only a little of each. He usually had a bottle of imported ale with a meal and only took a few sips." Bauer said Cramp's home life was "devoted to reading and study" (Cramp, Arthur J., Correspondence, n.d.). Cramp was a cultured man and was a member of the Chicago Literary Club and the Chicago Ornithology Society, as well as numerous medical associations. Cramp wrote with a sharp wit and keen sense of the absurd, and he kept a copy of Lewis Carroll's works by his desk to get in the proper mindset to write about the absurdities he was fighting (Young 1967, 132). In an article about quack treatments that had received U.S. patents, Cramp singled out a concoction patented in 1913 by (the late) Alois Viquerat that claimed:

> The present invention relates to a composition which is intended to be used internally and which confers to the organisms immunity against the following microbial infectious illnesses: diphtheria, pneumonia, typhus, scarlet fever, influenza, septic infections, cerebralspinal meningitis, syphilis, pest, cholera and tuberculosis; it is also effective in another kind of disease, viz. goiter. (Viquerat 1913)

Cramp wryly noted, "By the time the patent was granted the inventor was dead and his estate got it. Since, by the use of his own preparation he should have been immune to practically all diseases, he probably died of senility" (Cramp 1926, 191–192). His colleagues at the AMA joked that quacks "had been pursued into the beyond by Doctor Cramp and his unrelenting ridicule" (Cramp, Arthur J., Correspondence).

Dr. Cramp died at the age of seventy-nine, in Hendersonville, North Carolina, on November 25, 1951, according to his obituary in *JAMA*, of arteriosclerosis and urema. The editors remembered him as a "prolific and constant contributor to the Journal, and a pioneer in the fight against quackery and fraud in the healing arts" ("Deaths" 1951, 1773). The legacy he left behind was well expressed in an earlier review of his Nostrums and Quackery series in the *British Medical Journal*:

> All volumes have been compiled by Dr. Cramp, who for thirty years has led the struggle against heartless fraud. This is a fine record of courageous persistence in public service. The persistence is all the more remarkable because the struggle is endless. . . . Whenever a fraud is exposed half a dozen new ones spring up to take its place, but Dr. Cramp has never been disheartened by the unending nature of the task to which he devoted his life. ("Work" 1937, 565)

The AMA closed the Department of Investigation in 1975. The files that Cramp and his successors had gathered over the decades became known as the American Medical Association Health Fraud and Alternative Medicine Collection, which is the only AMA archive that is open to non-members.

Mike Jarsulic is the senior HPC administrator at the University of Chicago's Biological Sciences Division. His skeptical interests include cancer quackery, the history of skepticism, and philosophy of science.

Robert Blaskiewicz is assistant professor of critical thinking and first-year studies at Stockton University in Galloway, New Jersey. He studies medical quackery, conspiracy theory, and pseudoscience.

References

Boyle, Eric W. 2013. *Quack Medicine: A History of Combating Health Fraud in Twentieth-Century America.* Santa Barbara: Praeger.

"Cramp, Arthur J., Correspondence, May, 1934–August, 1963." Folder 0176-03. American Medical Association Historical Health Fraud and Alternative Medicine Collection. American Medical Association, Chicago.

"Cramp, Arthur J, Special Data." Folder 0175-10. American Medical Association Historical Health Fraud and Alternative Medicine Collection. American Medical Association, Chicago.

Cramp, Arthur J. 1911a. The Van Bysterveld Medicine Company. *Journal of the American Medical Association* 56(1) (January 7): 56–7.

———. 1911b. Lawrence Hill, A.M., D.D., M.D.: Another consumption cure fake in Jackson, Michigan. *Journal of the American Medical Association* 56(2) (January 14): 134–137.

———. 1913. Pulmonol. *Journal of the American Medical Association* 61(2) (November 29): 1998–1999.

———. 1918. Modern advertising and the nostrum evil. *American Journal of Public Health* 8(10) (October 1): 756–758.

———. 1920. Truth in advertising drug products. *American Journal of Public Health* 10(10) (October): 783–789.

———. 1924. Therapeutic thaumaturgy. *The American Mercury* (December): 723–30.

———. 1926. Patent office magic—medical. *The American Mercury* (June): 187–194.

———. 1927a. Some miscellaneous nostrums. *Journal of the American Medical Association* 88(7) (February 12): 501.

———. 1927b. Fakes and fads in deafness cures. *Bulletin of the New York Academy of Medicine* 2(12) (December 1): 726–728.

———. 1930. A posthumous testimonial. *Journal of the American Medical Association* 95(4) (July 26): 285.

———. 1933. The work of the Bureau of Investigation. *Law and Contemporary Problems* 1(1) (December): 51–54.

Deaths. 1951. *Journal of the American Medical Association* 147(18) (December 29): 1773–1774.

Fishbein, Morris. 1969. *Morris Fishbein, M.D.: An Autobiography.* Garden City, NY: Doubleday.

Fresh family medicines. 1851. *Semi-weekly Camden Journal* 2(90) (November 14): 1.

Hafner, Arthur W., James G. Carson, and John F. Zwicky, eds. 1992. *Guide to the American Medical Association Historical Health Fraud and Alternative Medicine Collection.* Chicago: American Medical Association.

Page, Douglas, and Adrian Baranchuk. 2010. The Flexner Report: 100 years later. *International Journal of Medical Education* 1: 74–75.

"Spray-O-Zone, 1925–1927" Folder 0818-14. American Medical Association Historical Health Fraud and Alternative Medicine Collection. American Medical Association, Chicago.

The work of Dr. Cramp. 1937. *British Medical Journal* 1(3975) (March): 565.

Viquerat, Alois. 1913. Medicinal composition. US 1081069 A. (December 9). Google Patents. Available online at https://patents.google.com/patent/US1081069.

Young, James Harvey. 1967. *The Medical Messiahs: A Social History of Health Quackery in Twentieth-Century America.* Princeton: Princeton University Press.

CHAPTER 32

Environmentalism and the Fringe

David Mountain

Vol. 45, No. 4
July/August 2021

Environmentalism is more popular than ever—and with good reason: never has our dependency on the natural world or our culpability in its ongoing destruction been clearer. I'll spare you the customary roll call of ecological crises that tends to open articles about the environment (for every person spurred into action, another is sunk into despondency), but suffice it to say that our planet is in serious trouble. If humanity is to survive this century with any semblance of the quality of life enjoyed today, all of us need to act quickly to limit and reverse anthropogenic climate change and environmental destruction.

It's therefore heartening to see enthusiasm for environmentalism on the rise. Across much of Europe, support for Green parties is increasing. In the United States, three-quarters of people think more should be done to end the country's dependence on fossil fuels. The 2019 climate strike was one of the largest protests ever staged with an estimated six million people around the world—from Antarctica to Zambia—protesting leaders' continued inaction over the destruction of the planet.

By itself, however, environmentalism doesn't necessarily translate into effective action. An appreciation of the natural world and a concern for its future can inspire us to act, but it can't tell us *how* to act. How do we develop renewable sources of energy? How do we best preserve remaining areas of wilderness? And how do we build sustainable agricultural systems capable of feeding a global population rapidly hurtling toward ten billion? To answer these vital questions, we need science.

Unfortunately, environmentalism and science are not always the same thing. Indeed, throughout its history, environmentalism has been shaped by a range of fringe beliefs that have nurtured a tradition of unscientific thinking about the natural world. As a result, many sincere and well-meaning environmentalists today are wary of pragmatic, science-based solutions to the climate crisis—the very solutions that can get us out of this mess. It's time for environmentalism to acknowledge, and renounce, its long dalliance with the fringe.

The Origins of Environmentalism

To understand why environmentalism has been so vulnerable to fringe influences, we need to travel back to the 1920s and the dawn of modern environmental awareness. In some ways, our planet was much healthier. The global population had yet to reach two billion. Atmospheric carbon dioxide was around 300 ppm, compared to 410 ppm today. Nevertheless, it was a world undergoing rapid and unsettling change. Cities were expanding at unprecedented speed. Industrialization was consuming natural resources at ever-greater rates. And it was a world still ringing with the echoes of the First World War.

The prospect of an increasingly mechanized and inhuman society heralded by these developments concerned many who lived through them. The ways in which people acted on these concerns, however, were very different. On the one hand, biologists and agronomists, worried about the rates of soil erosion and deforestation around the world, set about developing sustainable forestry practices, improved farming methods, and higher-yield crops, thereby laying the foundations for much of modern environmental and agricultural science (Barton 2018).

On the other hand, many artists and intellectuals, especially in Europe, sought not to address the challenges of the modern world but to reject modernity altogether. Fueled by nostalgia for a rural era that was rapidly being lost to the slums and smokestacks of urbanization, they attempted to recapture in some form a pre-industrial way of life. They championed traditional farming methods and outdoor living as the key to healthy, meaningful lives. They searched for worldviews that celebrated a spiritual connection with the natural world, from Hinduism to transcendentalism to paganism. And they constructed a farrago of unscientific philosophies they hoped would restore humanity's relationship with the environment.

Modern environmentalism is the heir to both these traditions: the scientific and pragmatic *and* the spiritual and nostalgic. However we choose to measure environmentalism's success—laws passed, acres protected, lives improved—the former has been to its merit; the latter, ultimately, has been to its detriment.

Life Force and Cow Dung

An early example of the spiritual approach to the environment was biodynamic farming, which dates back to 1924 when Austrian philosopher and self-declared psychic Rudolf Steiner gave a series of lectures in the German city of Breslau on a new discipline he called "anthroposophy." Steiner was interested in improving agricultural yields without damaging the environment. What he wasn't interested in was investigating this important cause with science or even basic logic. Anyone masochistic enough to read his lectures will find themselves swallowing an indigestible word salad. In a single sentence, Steiner appeals to astrology, radiation, ether, and the metabolism of cows to explain the workings of anthroposophy. Elsewhere, he discusses the imagined effects of cosmic forces, karma, and "moon rays" on the life force of plants and animals, which he believes are at the heart of environmental and agricultural processes. Ultimately, Steiner argues—without a shred of evidence—that the earth is home to a range of mystical and magical forces and that agriculture can be revolutionized by harnessing these forces (Steiner 2007).

Steiner's lectures were a hit. Although he died just a few months later, a small but dedicated following continued his teachings under the new name of *biodynamics*. People were attracted to it for various reasons. Those wary of modern agricultural methods were drawn to its disdain for industrialization. Those left cold by the sterility and order of the modern world found comfort in its pseudo-pagan mysticism. In truth, there was nothing venerable about biodynamic farming. Despite his frequent allusions to ancient wisdom, Steiner invented his anthroposophical worldview entirely from scratch in the 1920s.

So how does biodynamic farming work? Well, in short, it doesn't. Biodynamic farmers attempt to enhance a farm's life force by applying potions, known as "preparations," to the soil, either directly or by mixing it into manure before spreading. These preparations are invariably concocted by subjecting natural substances to a series of bizarre, lengthy processes. One requires cow manure to be stuffed into a cow horn and then buried fifty centimeters underground for an entire winter. Another asks for a deer bladder to be bunged with yarrow blossoms and left out in the summer sun. And even if the resulting potions did somehow work, biodynamic farming guarantees its own uselessness by insisting that preparations should be applied in homeopathic amounts, with some processes asking for as little as one gram of potion per ten tons of manure (Dunning 2007). As Steiner himself confessed: "To our modern way of thinking, this all sounds quite insane" (qtd. in Barton 2018).

Flower Power

In the decades that followed, biodynamics' influence spread throughout Europe and North America. At the same time, others with an interest in agriculture and the environment also looked to the past—or, at least, an idealized imagining of the past—for direction. In the 1920s, a Swiss teacher named Hans Müller created the Swiss Farmer's Movement for Native Rural Culture, through which he expounded the benefits of pre-industrial farming techniques and peasant lifestyles. In the 1940s, the English folklorist Rolf Gardiner (later a founding member of the Soil Association, a prominent organic advocacy group) established the Kinship in Husbandry, a secretive organization with the aim of restoring rural ways of life in Britain. His acquaintance Gerard Wallop ran the English Mistery, a hyper-conservative group that hoped to revive feudalism in England. These various groups were united not just by nostalgia but a belief that there was more to the natural world than science could explain: something intimate, mystical, and unquantifiable (Reed 2010).

It wasn't until the 1960s and 1970s, however, that eco-spiritualism really took off. After all, this was the age of counterculture, a time when people throughout the West were looking for new and exciting things to believe in. As biologist Arthur Galston noted in 1972, a desire to reject "the synthetic, plastic world" of the establishment encouraged many to turn to nature for meaning and authenticity. Even his own students at Yale were attracted to this "anti-intellectualism," he glumly observed (Galston 1972).

Those looking to escape the establishment sought out any number of alternative beliefs. Some looked to Hinduism, Buddhism, and the Far East in the hopes of finding environmental philosophies uncontaminated by Western industrialism. Others turned to indigenous peoples such as Native Americans, who were said to retain an ancient spiritual connection with the natural world. Then there were those who looked to paganism for answers, reviving and reimagining nature-worship religions such as Wicca and Gaianism. Like biodynamics before, the resulting smorgasbord of beliefs was characterized by a desire to retrieve an older, more meaningful relationship with the earth and its ecosystems. In the words of one New Ager, it was about "regaining the intimate connection and awareness of our place in nature" (qtd. in Ferguson 1981).

Believers were typically motivated by a sincere concern for the environment. And their love of the esoteric didn't necessarily prevent them from participating in the practicalities of environmentalism. They could be found out in the Pacific Ocean, disrupting whale hunts; sitting in trees slated to be bulldozed by developers; or rustling up support for Earth Day in 1970, which proved a galvanizing moment for the environmental movement both in the United States and around the world. All too frequently, however, the spiritual dimension to their environmentalism fomented distrust of science and technology. To many counterculture environmentalists, these were part of the problem, not the solution. They represented, in the words of writer Edward Abbey, "the ever-expanding industrial megamachine" (qtd. in Drake 2013).

Take the neo-pagan movement. The emphasis on freedom and individuality makes New Age beliefs notoriously hard to generalize, but a representative sample of neo-pagans in the 1970s might have spent their free time growing organic vegetables, immersing themselves in nature, and attending group "apologies," in which they showed contrition to the earth for humanity's environmental destruction. Guided by the conviction that our planet is sacred, they also raised awareness of various environmental causes, including recycling, renewable energy, and the protection of wildlife refuges (Bloch 1998).

Campaigning and awareness-raising are of course good things. However, neo-paganism's suspicion of modernity all too often soured into a hostility toward mainstream society. Cities in particular were condemned as representing everything wrong with twentieth-century civilization. "If you look at [Earth] from space," explained one neo-pagan, "you'll see giant cancer cells called cities across the face of the planet" (qtd. in Bloch 1998). As a result, many neo-pagans were—and still are—wary of mainstream environmental efforts that are, by definition, best placed to generate change.

Moreover, by combining environmental causes with spiritual beliefs, neo-pagans and their fellow New Agers tainted environmentalism for many in mainstream society who didn't share their interest in things such as astrology, reincarnation, and "sexual magic." The influence of fringe beliefs such as neo-paganism led to environmentalists, even those who approached the subject as a strictly scientific matter, being lampooned as tree-hugging hippies. This caricature not only dissuaded people from becoming environmentalists but allowed environmentalism itself to be dismissed as a fringe belief. It's taken decades for the movement to shake off the stigma and reassert itself as a global and urgent concern.

Rebellion, Inc.

Eventually, even tree-hugging hippies grow up, get jobs, and settle down. This is what happened to the baby boomers who had formed the vanguard of New Age environmentalism. By the 1990s, their generation had ascended to the positions of power they once scorned: they were now politicians, lawyers, and business owners. This isn't to say that they abandoned their beliefs once they joined the mainstream. On the contrary: although they no longer chained themselves to trees or boarded anti-whaling boats, they still carried their anti-establishment dreams with them. The only difference was that they now had the power and influence to enact them. Seemingly blind to the irony, counterculture became mainstream (Heath and Potter 2005).

This is why the 1990s mark the moment when previously radical environmental beliefs became widely accepted, further compounding environmentalism's relationship with the fringe. Organic produce began appearing on supermarket shelves, where it was marketed as the green choice for shoppers worried about the environment. Alternatively, the ethically minded could head to their local farmers' markets and buy locally grown produce. Even biodynamic farming achieved some mainstream acceptance: today biodynamic vineyards cover an estimated 11,000 hectares around the world. In terms of influencing contemporary environmentalism, the commercialization of fringe beliefs and practices in the 1990s was just as influential as the rise of counterculture movements some thirty years previously.

At the heart of this commercialization was the desire, undimmed since the 1920s, to escape the chemicals and corporations of modernity for an older, gentler relationship with the earth. Organic farms contrasted the human scale of their enterprises with the sprawling monocultures of industrial farming. Farmers' markets likewise peddled nostalgia for a time before sterile supermarkets and faceless production lines. It would be unfair and inaccurate to claim that organic farming or local produce are unscientific in the way that, say, biodynamics is. But their origins and subsequent success owe just as much, if not more, to the spiritual and nostalgic strain of environmentalism as they do to the scientific and pragmatic. Consequently, when environmentalists today encourage us to buy local and eat organic, they often find themselves at odds with more pragmatic and scientifically grounded approaches.

Let's stay with organics for a moment. In some ways, organic farming can indeed be very good for the environment. Organic farms can help maintain local biodiversity, for example, and their soils can store more carbon than those on non-organic farms. But organic farming isn't a viable solution to the problem of sustainable agriculture, at least for most people. This is because organic agriculture is less efficient than modern, industrial farming methods and therefore requires a greater amount of land—typically 40 percent more—to grow the same quantity of food as a non-organic farm. Remember that the global population is currently approaching eight billion and is forecast to reach ten billion by around 2060. There's simply not enough land on earth to feed that many people using organic methods—at least not without converting the planet's remaining wildernesses into farmland, with the loss of ever more habitats and species. Despite this, many environmentalists today continue to advocate organic produce as the only ethical choice for those concerned about the natural world.

What about locally produced food? The principle behind it seems like common sense: the fewer miles your food has traveled to reach your plate, the fewer greenhouse gases released during its transit. In some situations, this is indeed the case—but not always. When you factor in other variables affecting the environmental impact of food production—water use, growing methods, storage techniques, and so on—in many cases transport is no longer the most significant source of emissions. In 2006, a team from Lincoln University in New Zealand did just this and found that the dairy, lamb, and apple farming industries emit more carbon dioxide in the United Kingdom than in New Zealand. As bizarre as it may seem, for U.K.–based environmentalists like me, the green thing to do is therefore to buy apples imported from New Zealand rather than buy the locally grown alternatives (Saunders et al. 2006).

Perils of Nostalgia

It's time for environmentalists to reassess our relationship with the megamachine. The hope of escaping modernity for some rural idyll was a pipe dream even in the 1920s. Today it's deluded to the point of being dangerous. Like it or not, the world is either industrialized or industrializing and, short of something catastrophic, will remain so for the foreseeable future. Environmentalists, if we are serious about saving the planet, need to accept this. We need to abandon false hopes of subverting our industrialized, urbanized societies and instead think about how we can work with them to change them for the better. Earlier I described environmentalism as a movement of two halves: one scientific and pragmatic and the other spiritual and nostalgic. Over the past century, the latter has become increasingly self-defeating as it grows ever more out of touch with reality. The future of environmentalism must be grounded in science and pragmatism.

It's worth remembering that the megamachine doesn't just destroy the natural world. Indeed, as science and technology get to grips with the scale of environmental crisis facing us, they are coming up with many ingenious solutions. Genetically modified crops have been developed that use 25 percent less water than their unmodified counterparts. State-of-the-art herbicides such as glyphosate improve crop yields while reducing costs, all without harming humans. Even nuclear power plants, once the supervillain of the environmental movement, have dramatically improved in terms of safety and reduced waste. In fact, a 2017 study found that the world will probably fail to meet the emissions targets set by the Paris Agreement without relying on nuclear power to some extent (Peters et al. 2017).

And yet, time and time again, we have witnessed the embarrassing spectacle of hazmat-clad Greenpeace activists ripping up genetically modified crops or of countries outlawing glyphosate on the basis of unfounded environmental concerns or of environmentalists simultaneously demanding that countries dramatically cut emissions while abandoning nuclear power. We have to contend with green activists regarding organic food as a matter of principle and not privilege or insisting that the solution to the climate crisis lies in the wholesale "decommercialization" of the West's "toxic economic system." This is the legacy of the environmental fringe.

The saving grace in all this is that these people mean well. Most share a sincere concern for the natural world. But as long as they continue to advocate solutions rooted in the nostalgia and spiritualism of the environmental fringe, they are hindering their own cause.

So keep the passion. Keep the compassion. But, for the earth's sake, drop the disdain for science.

David Mountain is a freelance writer based in Edinburgh. He is the author of *Past Mistakes: How We Misinterpret History and Why It Matters* **(Icon Books, 2020).**

References

Barton, Gregory. 2018. *The Global History of Organic Farming*. Oxford: Oxford University Press.

Bloch, Jon. 1998. Alternative spirituality and environmentalism. *Review of Religious Research* 40(1): 55–73.

Drake, Brian. 2013. *Loving Nature, Fearing the State: Environmentalism and Antigovernment Politics Before Reagan*. Seattle: University of Washington Press.

Dunning, Brian. 2007. Biodynamic agriculture. *Skeptoid* (January 18). Available online at https://skeptoid.com/episodes/4026.

Ferguson, Marilyn. 1981. *The Aquarian Conspiracy: Personal and Social Transformation in the 1980s*. London: Routledge and Kegan Paul Ltd.

Galston, Arthur. 1972. The organic farmer and anti-intellectualism. *Natural History* 81(5): 24–8.

Heath, Joseph, and Andrew Potter. 2005. *The Rebel Sell: How the Counterculture Became Consumer Culture*. Chichester: Capstone Publishing Ltd.

Peters, Glen, Robbie Andrew, Josep Canadell, et al. 2017. Key indicators to track current progress and future ambition of the Paris Agreement. *Nature Climate Change* 7(2): 118–122.

Reed, Matthew. 2010. *Rebels for the Soil: The Rise of the Global Organic Food and Farming Movement*. Abingdon, U.K.: Earthscan.

Saunders, Caroline, A. Barber, and Gregory Taylor. 2006. Food miles-comparative energy/emissions performance of New Zealand's agriculture industry. *Lincoln University Research Report No. 285* (January 18). Available online at https://researcharchive.lincoln.ac.nz/handle/10182/125.

Steiner, Rudolf. 2007. The agricultural course. *Rudolf Steiner Archive* (January 16). Available online at https://wn.rsarchive.org/Lectures/GA327/English/BDA1958/Ag1958_index.html.

CHAPTER 33

A Skeptic's Guide to Ethical and Effective Curse Removal

Benjamin Radford

Vol. 39, No. 4
July/August 2015

> My name is Chelsea, I have a curse that was cast on me, my son, and my boyfriend. Would you know of anyone that can help us with this situation? It's only getting worse as the days go by, it is frustrating, very stressful and is affecting our lives in major ways. I had ten or more different psychic readings and was told by all that someone has been working against me. They all gave the same exact stories of who and what has been happening. I have witnessed all the changes, the distance between us, confusions, arguments and tears in all who are involved. We were very close and happy, now fights out of nowhere just keep popping up. Also our happiness only lasts for 48 hrs each time then the fights start again.
>
> —Chelsea C.

I get e-mails like this a few times a year, and they are among the most difficult and heartbreaking pieces of correspondence I see. These requests for help require a delicate balance of several factors. I've struggled with these cases for years: What should I do? Reply with a polite e-mail saying, in effect, "Curses don't exist, don't worry about it"? Reply with something along the lines of, "I'm glad you reached out to me, I have greater magic than the person who cursed you. I will remove the curse, just light three candles in your bedroom window an hour before bed and in the morning you'll be fine"?

In theory either of these approaches could conceivably work; in practice neither is likely to help Chelsea or her family. As skeptics we often deal with people who hold false beliefs, and one of the most important lessons is that usually they are not stupid nor crazy but merely human, succumbing to cognitive and perceptual errors (or simply incomplete information) that plague us all. Our main goal as skeptics, as I see it, is to educate the public where we can—about psychology, science, skepticism, confirmation bias, critical thinking, and so on—but also to help people.

People who are scared and desperate (to find or contact missing loved ones, in grief, etc.) often seek out psychics for comfort and information. These people don't need didactic discourse on skepticism or logical fallacies. They don't need scorn, mocking, or condescending comments about the absurdity of believing in psychics; they need sensitivity, compassion, and help.

My main job is an editor, writer, and investigator, not a psychologist. But as an investigative skeptic with a background in psychology and years of experience in dealing with others in similar situations, the truth is that I can often help. I can't justify donating and devoting many hours of my time to any stranger seeking assistance, but I also don't feel it is ethical to simply ignore a plea for help from someone in genuine distress. Nor is a short boilerplate e-mail with links to information on curses and psychics on the CSI web site or Bob Carroll's SkepticsDictionary.com going to help this woman.

I didn't know exactly why Chelsea believed she was cursed, but I've spoken with others like her and I can imagine how genuinely terrifying it must be to perceive any accident, discord, or bad luck is a malicious sign of evil intent, a reminder that you are powerless in the face of an unseen supernatural force capable of tearing apart your family, harming your health or sanity, or even killing you. I have an unwritten personal and professional obligation to help if I can. I responded to Chelsea later that day with the following response. A one-page e-mail is no substitute for a family counselor, and I can't solve all her problems but I could do my best to address her immediate situation and offer a skeptical framework for understanding her situation based not on fear or magical thinking but science and psychology. For more on the sometimes delicate balance between strict skepticism and offering psychological help to people in need, see my article "Playing Witch Doctor: Hidden Ethics in Skeptical Ghost Investigations," in the Fall 2014 *Skeptical Briefs*. My reply to her is below, with my annotations in brackets.

> Dear Chelsea,
>   I'm sorry to hear about your troubles. [I wanted to sympathize and assure her that I'm concerned about their welfare, and her as a person.] Obviously there is something going on, [I wanted to validate her interpretation that something is going on, though I carefully avoided confirming that a curse is causing disruption in her life. Any overt attempt to tell her that she was not experiencing a curse would likely have been met with resistance.] but I'm not sure that the psychics are correct that someone has placed a curse on you. What might

have happened is that the psychics you spoke to gave you some information and someone brought up the idea of a curse, and you (not they) suggested who it might be, and they agreed with it and said you were right about the details and the person who is out to get you.

[Here I address (and undermine) one of the key pieces of experiential evidence she interpreted as confirming the curse: the psychics' validation of her information. By planting a seed of doubt and offering an alternate, non-threatening explanation for what she experienced I hope to subtly suggest that her interpretation of other curse-related phenomena might also have another, rational explanation.]

Obviously I wasn't there, and so I don't know what did or didn't happen, [I wanted to demonstrate open-mindedness about her situation, and not be aggressively skeptical or dismissive. Telling her "there's no such thing as curses, it's all in your head" would come off as rude and dismissive, and not be helpful to her. A psychologist must understand the worldview of the person he or she is trying to counsel and adapt their information and advice to best suit them] but I can tell you that you should be very careful because some psychics will tell a person who is having personal, financial, or relationship troubles that there is a curse on them, and offer to remove the curse for money. This is a scam, and many smart-but-desperate people have been conned out of money, and the psychics sent to jail. Like I said, don't know if that's what going on in your case but it might be, so please watch out for that and call the police if someone tries to do that because they are not trying to help you, they are trying to scam you. Some psychics are sincerely trying to help, but some aren't, and you often can't tell the difference. [Here I give her accurate information about psychic scams and leave the door open for her belief in psychics; if she has contacted ten or more psychics she believes in them, and this is not the situation for a dismissive or academic debunking of psychics or a discussion of cold reading techniques. I specifically describe her as smart, building her self-esteem a bit and reminding her that even smart people can misunderstand situations.]

I have studied and researched both psychics and curses, and one thing I can tell you for certain is that the power of curse is proportional to your belief in the curse. [My goal here is to establish my bona fides as an expert on the subject. If I want her to believe what I'm telling her and accept my advice to resolve this, she must find me credible. This is not a time for me to be wishy-washy or modest, it's a time for clear authority. Instead of directly debunking the existence of curses I focus on the psychology of curse beliefs and tell her that she has power over the curse.]

This is the same whether it's a voodoo curse or Gypsy curse or an African curse or any other kind: [Chelsea didn't tell me why she believed she was cursed, or by whom, or what type of curse it was.

By casting a wide net and explaining that all curses are fundamentally the same, I've covered my bases and assured that she will believe my advice applies to her situation.] The more you believe in its power over you, the more power it has over you. I've spoken to people whose lives have really been turned upside down by this sort of thing, but it was not so much the curse itself but the person's belief in the power of the curse. This is proven science and psychology (in addition to my books I have a Bachelor's degree in psychology and a Master's degree in science education, so I know what I'm talking about). [Again, establishing my bona fides as an authority on the subjects of curses, psychology, and science. I could have included folklore, of course, which would have been even more relevant, but it would not have carried the same significance. And I remind her that she is not helpless but can play an active role in removing the curse and improving her life.]

I obviously don't know all the details about what's going on in your specific case, [Again, demonstrating open-mindedness about her situation, trying not to overreach or deign to suggest that I know all about her or her situation. By admitting I'm fallible and limited by the information she gives me, it allows her to forgive any minor errors or mistakes that don't apply to her. It's similar to when a psychic reminds her audience that she's only human and not 100% perfect.] but from what you've told me it sounds very similar to others I've dealt with in the past, and I can tell you what has helped those people. [Establishing my experience on the subjects of curses. My message to her is: I understand your situation, I have experience in helping resolve it with other people, and I'm going to tell you how to address this.]

Here is what you need to do:

1) Stop going to psychics. They are not helping your situation. If consulting psychics or getting their information solved the problem, then you wouldn't be dealing with this right now. If you have been to at least 10 different psychics so far, then you could go to 10 or 20 more and there is no logical reason to think that they will help you or resolve the matter. All you are doing is spending your money and wasting your time (and probably making it worse). [It was clear that psychics were only exacerbating her problem; note that, as with curses, I avoided telling her that psychic power doesn't exist, because if she is a strong believer in psychics then any statements I made to that effect would put me (and therefore the perceived credibility of my advice) in conflict with her beliefs. Instead I took a different path and gave her a simple, practical reason not to continue seeing psychics: they all failed to help so far, which she could not deny.]

This is an issue that you can, and will need to, resolve by yourself. You are a strong and smart woman, I know that because you took the time and effort to find me and write to me. You can do this—in fact you're the only person who can. [I wanted to empower her to help herself, encourage her to see herself as the catalyst for change—not me, not psychics, and certainly not whoever she believed had cursed her.]

2) Remember that everyone has some of the troubles you're experiencing some of the time. People have confusions and fights and arguments and all that whether they are cursed or not, it's just part of life. Sometimes things are good, sometimes things are bad, but either way you need to try to address the root of the problem. A curse can make a problem much worse, from a psychological perspective, but it is not the cause of the fighting—that may be due to relationship issues, money problems, personality conflicts, or any number of other things. [I tried to frame the curse, or at least the harm that a curse can do, within the context of her belief system and a "psychological perspective" without flat-out denying the existence of a curse or malicious magic. If her e-mail had suggested that she had some doubt that she really was the victim of a curse, I would have been more skeptical about questioning that premise. Ridiculing or dismissing the idea that a curse may exist would have diminished my credibility in her eyes, and therefore sown doubt that I understood the situation and could resolve it through her.]

There may be some difficult dynamics between your son and your boyfriend, or you and your boyfriend that affect your son. . . . I don't know what they might be, but please look at those first, maybe even a friend or counselor might help. Fights and conflicts are caused by people, and they can be solved by people. [Here and elsewhere I personalized my advice as much as I could so that she would recognize how my directions and counsel applied to her and her specific relationship situation.]

3) Try not to focus on the negative things, which can bring negativity into your life and make the problem worse. [I almost rewrote this sentence, not wanting to invoke or give authority to the widely popular (and widely discredited) "law of attraction" as promoted in The Secret and elsewhere. However I decided that the statement was quite true from a psychological point of view, and in any event if Chelsea did happen to be an adherent to the law of attraction ideology I could use that to my advantage by pitting one blinkered worldview against a less toxic one.] You said that your happiness only lasts 48 hours, but don't get it in your head that that's part of a curse. [This is the last time in my response to Chelsea that I refer to a curse; by the end of the letter I want her focusing on the ways in which she can fix the problem through logic and psychology, not through consulting psychics or thinking of ways to remove a magical curse.]

There's nothing magical about 48 hours, it's just a pattern you've noticed. If you're doing okay for two days and then 48 hours approaches you may start to look for things to go wrong. Every day has good and bad things in them, and if you focus on the bad things, or let yourself magnify them, then you are creating your own self-defeating thought patterns, what in psychology is called a self-fulfilling prophecy. You need to avoid that. Focus on the good things in your life, the things that are going well. [I'm trying to give her some simple but sensible insight into the psychology of her beliefs, and confirmation bias.]

It won't happen overnight, but from my experience this is what's needed to solve your situation. It's not like suddenly everything is perfect, but what will happen is that if you follow these instructions and you are patient, 48 hours without a huge fight will stretch into three days, and then three days will become a week, and after a week or two it will be gone. Of course, like I said, we all have fights and conflicts, so if something comes up, it's not a setback, it doesn't mean it's not working, it just means that you're human! I know these steps are difficult and easier said than done, [Here I try to guard against the expectation that my advice will magically fix the situation immediately, and caution that it will be a gradual process. I close by reminding her that I am concerned about her, trying to help her, and offering sympathy and compassion.] but if you want to break free from this, this is what you should do. I hope this helps, and please get in touch with me in a few weeks to tell me how it's going.

All best,

Ben Radford

Research Fellow, Center for Inquiry

I'm sure other skeptics may have handled the situation differently, or better than I did. I'm not suggesting that this is the only, or the best, way to handle this situation, but I did the best I could under the circumstances. I did indeed get a reply back from Chelsea:

Hi Mr. Radford, Thank you so much for that advice, maybe that is what I needed and you are absolutely right! I will definitely work on that and get back to you with an update. Best advice I've gotten yet. I feel better already! Thanks again. Best regards, Chelsea.

Part IV

SCIENCE AND PSEUDOSCIENCE

DUBIOUS CLAIMS

CHAPTER 34

Vaccines, Autism, and the Promotion of Irrelevant Research

A SCIENCE-PSEUDOSCIENCE ANALYSIS

Craig A. Foster and Sarenna M. Ortiz

Vol. 41, No. 3
May/June 2017

Larry Kusche's review of the Bermuda Triangle mystery (1986; 2015) provided one of the clearest victories for reason over rumor. His method remains convincing because it was so straightforward. Kusche demonstrated that many disappearances had been wrongly attributed to the Bermuda Triangle; the actual events either did not happen as reported or likely occurred outside of the infamous area. Kusche also demonstrated that the rate of actual disappearances within the triangle did not seem to differ meaningfully from the rate of disappearances in other parts of the ocean. In so doing, Kusche highlighted two issues that occur frequently in pseudoscience. The first involves reporting events inaccurately, and the second involves relying on handpicked observations rather than a representative set of observations (Hansson 2013).

Vaccinations and Autism: Inaccuracies and Anecdotes

Much of the debate surrounding vaccines and autism has been similarly based on straightforward considerations about whether the reported evidence is accurate and whether a representative set of evidence supports the vaccines-cause-autism claim. This debate was triggered by Andrew Wakefield and others' (1998—Retracted) research involving twelve children who were referred to a pediatric gastroenterology unit. Wakefield et al. reported that all twelve children had experienced developmental problems at varying intervals after exposure to the measles, mumps, and rubella (MMR) vaccine; nine of these developmental disorders were identified as autism with a tenth being questionably identified as autism. Wakefield et al. used these results to suggest that the MMR vaccine could contribute to a syndrome involving gastrointestinal problems and regressive autism. This research fueled the well-known concern about childhood vaccination and its alleged contribution to autism—a genie that shows no sign of returning to its bottle.

Even if the Wakefield et al. results had been legitimate, the small sample size and selective nature of the sample should have encouraged a cautious interpretation. Nevertheless, the evidence was at least broadly interpretable: a small group of children received the MMR vaccine and subsequently experienced developmental problems. Of course, this evidence was not accurate. Wakefield created fraudulent results presumably for financial reasons (Deer 2011). Subsequent research revealed no evidence that childhood vaccine administration elevated the rate of developing autism (Taylor et al. 2014). In the absence of any scientific connection between vaccinations and autism, promoters of the vaccines-autism link could still handpick observations. Anti-vaccination proponents pointed to the many vaccinated children who later developed autism (noted by The Logic of Science 2016). The best-known example has probably been actress Jenny McCarthy's son. In a strange twist, some have raised the possibility that McCarthy's son did not actually develop autism (Rubin 2008).

We view the handpicking of confirmatory observations with understanding. It is natural for humans to seek explanations for difficult events, and the temporal proximity between a vaccination and the diagnosis of autism could certainly feel causal. Nevertheless, this aggregation of anecdotes, no matter how broad, is not science. Most objective observers can understand that if millions of children are vaccinated and a fraction of children develop autism, then many children who have been vaccinated will subsequently develop autism, even in the complete absence of any cause-effect link. Obviously, many people still believe that vaccinations cause autism, but at least the science and pseudoscience surrounding this debate remains relatively clear. The scientifically minded can point to the absence of any correlation between childhood vaccination and the development of autism. Proponents of the vaccines-autism link can point to the number of children who were vaccinated and later developed autism.

Vaccines and Autism: The Promotion of Irrelevant Research

This context is important for understanding an interesting tactic that subsequently developed regarding vaccinations and autism. The supposed science behind the vaccine-autism link came back. The vaccines-cause-autism community began offering several non-Wakefield studies as evidence of supportive science. As far as we can tell, this tactic was popularized by blogger Ginger Taylor in 2007 when she published a list of "just over a dozen studies" supporting the link between vaccines and autism (see Taylor 2013). This list has since expanded to well over 100 studies. Taylor's work is paralleled by other lists to include Walia's (2013) list of twenty-two studies, Adl-Tabatabai's (2015) list of thirty studies, and Anti-Vaccine Scientific Support Arsenal's (2015) list of twenty-six studies, the latter containing mostly perspectives and reviews. To illustrate the dissemination of this argument, the Activist Post's Facebook page, where Walia's list was posted on August 24, 2015, appears to have more than 500,000 followers presently. We do not need to address these studies specifically, as the collective implication of these studies has been debunked sufficiently by scientific research (e.g., see Taylor et al. 2014) and by scholars who have reviewed the concerns associated with many of these studies (e.g., Ditz 2013; The Logic of Science 2016).

Instead, our purpose is to highlight the promotion of irrelevant research as its own developing characteristic of pseudoscience. This tactic shares a connection with the old pseudoscientific tactic of using scientific-sounding language (Shermer 2002) because both tactics might make a community appear more scientific. Nevertheless, the promotion of irrelevant research goes much further by pointing to purportedly important scientific findings.

Radner and Radner (1982, 36) gave a nod toward the promotion of irrelevant research when they described the "grab-bag approach to evidence." According to Radner and Radner, pseudoscience will use quantity of evidence (the grab bag) over quality of evidence in an attempt to wear down opponents. Radner and Radner's grab bag focused mostly on the continued offering of confirming observations (e.g., Bermuda Triangle disappearances) or questionable pieces of evidence (e.g., an old jet-shaped figurine as evidence of ancient aliens). Radner and Radner also included the misuse of research findings in their grab bag after explaining that pseudoscientists are reluctant to "weed out" bad evidence from a scientific debate. To illustrate, they noted that parapsychologists continued to use the results of a flawed research design as evidence for psi.

At the same time, Radner and Radner did not really describe the promotion of irrelevant research as a specific method for making pseudoscience look like science. This tactic, as it has been used in the vaccines-autism domain, involves much more than the inclusion of a dubious study or two that lie at or near the center of a science-pseudoscience debate. *The promotion of irrelevant research is an active aggregation of several questionable or peripherally related research studies in an attempt to justify the science underlying a questionable claim* (see also Barrett [2008], who mentioned this tactic briefly). It includes, among other things, (a) results that have dubious legitimacy, (b) results that possibly occurred due to chance, (c) results based on inappropriate statistical procedures that create a false perception of a relationship, (d) results coming from poorly controlled research, (e) results where the supposedly harmful aspects of vaccines were manipulated at much stronger levels than is actually present in a vaccine, (f) results where the dependent variable was tangentially related to autism but was not autism (e.g., gastrointestinal problems), and (g) results containing multiple explanations due to confounding variables (see Ditz 2013; The Logic of Science 2016).

The promotion of irrelevant research potentially changes the landscape of the overall discussion surrounding vaccines and their relation to autism. It can shift the argument away from the misuse of handpicked examples or a manageable number of fallible research designs (all the while ignoring the broader absence of any systematic link between vaccinations and autism). It can move the argument to an assortment of scientific findings with questionable relevance and legitimacy (all the while ignoring the broader absence of any systematic link between vaccinations and autism). By doing so, the vaccines-cause-autism community no longer ignores the entirety of the science regarding vaccines and autism. They have instead developed and promoted a scientific debate that does not actually exist in science. Instead, this bogus scientific debate takes place in an electronic world between people who are usually consumers of science rather than being scientists themselves. Among actual scientists, this does not seem to be an issue. Taylor et al. (2014), in their statistical integration of several studies that examined vaccination and autism development, referred to almost none of the studies provided in the lists of research supposedly supporting the vaccination-autism link. Other scientists who have conducted research involving vaccination and autism have demonstrated similar levels of disregard (e.g., Uno et al. 2012).

We can illustrate the promotion of irrelevant research by considering this development relative to other questionable health-related claims. A Google search for "studies that show that vaccines cause autism" revealed the four aforementioned lists within the first thirty-two non-advertisement results. A Google search for "studies that show that new age crystals" did not immediately reveal similar lists within the first thirty results. A Google search for "studies that show that chiropractic" did reveal lists of research supporting chiropractic medicine that appear similar to the lists used to promote the vaccines-autism link. Hall (2014) reviewed the top chiropractic studies of 2013 (Luck 2013) and offered several methodological concerns. There is, however, a critically important distinction between this debate and the bogus scientific debate in the vaccine-autism domain. Hall reviewed the top chiropractic studies as reported by a chiropractic website. This stands in obvious contrast to proponents of the vaccines-autism link pointing to a set of research findings that appear to be generally irrelevant to the actual scientific debate.

Why has the promotion of irrelevant research occurred so prominently in the vaccines-autism domain? The anti-vaccination movement boomed after the publication of what appeared to be legitimate scientific research (Wakefield et al. 1998—Retracted). This caused many individuals to become firmly entrenched in the vaccines-cause-autism view. This initial entrenchment was surely reinforced by not vaccinating children, encouraging others to avoid vaccinating children, or both. It is difficult to admit wrongdoing in this regard because the consequences involve the well-being of children. The debate is also interdependent because this issue involves the well-being of unvaccinated *and* vaccinated children. This interdependence likely gave vaccination supporters additional initiative to press the anti-vaccination community for scientific justifications.

The vaccines-autism link has also been more neatly debunked than other forms of health-based pseudoscience. The vaccines-autism link is more easily examined because the results are not complicated by psychological rather than physical explanations. For instance, chiropractic medicine is difficult to investigate because it is seemingly impossible to create a proper control condition where participants believe they received chiropractic treatment when they did not. The vaccines-autism debate is not complicated by a placebo effect in this way. In fact, the vaccines-cause-autism theory could subtly encourage autism spectrum disorder diagnoses in vaccinated children; of course, this pattern of results has not been revealed in extensive scientific examinations (e.g., Taylor et al. 2014).

We also believe that the development of the Internet and social media has enabled this pseudoscientific tactic (see, e.g., Kata 2012). This promotion of irrelevant research has occurred overwhelmingly—possibly exclusively—in this context. The Internet provides quick access to a wealth of scientific and pseudoscientific information that allows groups on both sides of this debate to disseminate information swiftly and widely. This seems to have enhanced the ability of the scientifically minded to press supporters of the vaccine-autism link for the science behind their claims. Similarly, the Internet surely enhanced the ability for supporters of the vaccine-autism link to find and list several research studies that look, at first glance, as if they provide a noteworthy scientific argument. Taylor's (2013) original list was clearly intended to address concerns that there is no scientific evidence for the vaccine-autism link. If the Internet and social media hastened the promotion of irrelevant research, it explains why the Radner and Radner (1982) grab bag focused primarily on confirmatory observations and poor examples of evidence. It would have been difficult for nonscientists in 1982 to quickly piece together and share lists of any research that has some remote possibility of supporting a pseudoscientific claim. The continued ability for individuals to do so suggests that the promotion of irrelevant research is likely to continue as a pseudoscientific tactic.

Pyrrhic Victories and Practical Implications

Unfortunately, the promotion of irrelevant research is intentionally or unintentionally clever because it can obscure the distinction between science and pseudoscience. Science advocates can explain fairly easily why handpicked observations or particular research designs might be misleading. It is more difficult for science advocates to address an extensive list of research findings that have dubious relevance and legitimacy. Yet if these lists are not addressed, newcomers might initially see science that purportedly promotes both points of view, which can create a sense of false equivalence (Skeptical Raptor 2015).

At the same time, entering this debate could create a series of Pyrrhic victories. Scientifically savvy individuals who critique the promotion of irrelevant research might win several battles while experiencing a setback in their overall campaign—a campaign that is waged with arguments rather than soldiers in an effort to promote reason rather than empires. Those who truly love science will be tempted to enter this debate knowing that they ultimately have a winning hand, but the vaccines-cause-autism community can offer a long list of weak counterarguments in the form of fraudulent research, poor research, tangentially related research, and alleged pro-vaccination conspiracies. Sadly, by the time the arguments about the science underlying vaccinations and autism are hashed out, newcomers to this issue might be understandably fatigued and confused about what they should believe (Radner and Radner 1982; Skeptical Raptor 2015). This can force science to rely on authority rather than a digestible explanation of the existing scientific evidence. Ironically, the reliance on authority to confirm belief is another tactic commonly associated with pseudoscience (e.g., Hansson 2013).

It might be wise for those who wish to promote science and reason to steer clear of the specifics associated with handpicked irrelevant research. Getting into the weeds of this debate might encourage people to believe that all scientific research endeavors are equally compelling and that the group with the greater number of supposedly supportive studies is the winner. It is probably more effective to keep the focus on the basic scientific principles that are more easily understood. If vaccines cause autism, then vaccinated children should be developing autism at an elevated rate compared to non-vaccinated children. There is no evidence of this pattern (see Taylor et al. 2014 for an introduction). The tangential findings provided by several questionable and legitimate studies consequently lack any substantial relevance. Only those who are really interested in the details of these generally irrelevant studies should bother with them. They can be directed to thoughtful reviews (e.g., Ditz 2013; The Logic of Science 2016).

Perhaps more important, the promotion of irrelevant research is, oddly enough, an acknowledgement that science matters. If science did not matter, there would be no reason to offer lists of supposedly supportive scientific studies. This creates two contradictory stances that the vaccines-cause-autism community can be asked to clarify. First, should decisions about vaccines and autism be based on the scientific evidence? It seems that some in this community believe that science matters (e.g., Taylor 2013) whereas others question the utility of science (e.g., Jameson 2015). Second, is there a conspiracy that stifles research supportive of a vaccine-autism link (e.g., Olmsted 2014)? If so, why are there so many published studies that supposedly support that link?

The answers to these questions need to be established before, not after, a debate ensues about whether science supports the vaccines-autism link. Clarifying the ground rules is clearly necessary because the vaccines-cause-autism community can claim that their science is legitimate but that any science discrediting the vaccine-autism link is biased by some type of nefarious pro-vaccination agenda (see, e.g., Jameson 2015). This community can also argue that there would be additional research supportive of the vaccine-autism link if it had not been suppressed by a nefarious pro-vaccination agenda (see, e.g., Olmsted 2014; Mikkelson 2015). If the vaccines-cause-autism community wants to have a legitimate scientific debate, we are confident that members of the scientific community would be happy to entertain it. It just needs to come with a commitment that scientific findings cannot be omitted for undocumented impropriety and that suppositious research findings that have been suppressed by some unsubstantiated pro-vaccination agenda would not be considered.

Conclusion

The promotion of irrelevant research reveals a fundamental contradiction. It acknowledges the importance of science but disregards the most informative scientific studies and the general consensus of the scientific community. Our science versus pseudoscience analysis of this development serves two purposes. First, we hope that it encourages an effective response to the promotion of irrelevant research in the vaccines-autism domain. Second, we want to highlight the promotion of irrelevant research as an important pseudoscientific tactic in its own right. We believe that the promotion of irrelevant research will expand as a pseudoscientific tactic, and promoters of science and reason should therefore be prepared to identify and address it.

The views expressed in this article are those of the authors and do not necessarily reflect the official policy or position of the United States Air Force Academy, the Air Force, the Department of Defense, or the U.S. Government.

Craig A. Foster is a CSI fellow and former professor in the Psychology Department at SUNY Cortland where he taught a course titled *Psychology of Pseudoscience*. He also conducts research in the areas of critical thinking and pseudoscience.

Second Lieutenant Sarenna M. Ortiz is an admissions advisor at the U.S. Air Force Academy. She is a graduate of the U.S. Air Force Academy, where she majored in Legal Studies and Behavioral Sciences and Leadership.

References

Adl-Tabatabai, Sean. 2015. 30 solid scientific studies that prove that vaccines cause autism. *YourNewsWire.com* (December 9). Available online at http://yournewswire .com/30-solid-scientific-studies-that-prove-vaccines-cause-autism/.

Anti-Vaccine Scientific Support Arsenal. 2015. Vaccines do cause autism—undeniable scientific proof (blog entry). *Anti-Vaccine Scientific Support Arsenal* (April 29). Available online at https://avscientificsupportarsenal.wordpress.com/.

Barrett, Stephen. 2008. "Research" associated with the promotion of questionable theories, products, and services. *Quackwatch* (November 26). Available online at https:// www.quackwatch.org/06ResearchProjects/ploys.html.

Deer, Brian. 2011. How the vaccine crisis was meant to make money. *British Medical Journal* 342: c5258.

Ditz, Liz. 2013. Those lists of papers claiming that vaccines cause autism: They don't show what they claim (Part 1) (blog entry). *I Speak of Dreams* (August 23). Available online at http://lizditz.typepad.com/i_speak_of_dreams/2013/08/-those-lists-of -papers-that-claim-vaccines-cause-autism-part-1.html.

Hall, Harriet. 2014. The top 10 chiropractic studies of 2013 (blog entry). *Science-Based Medicine* (January 21). Available online at https://www.sciencebasedmedicine.org/top -10-chiropractic-studies-of-2013/.

Hansson, Sven O. 2013. Defining pseudoscience and science. In M. Pigliucci & M. Boudry (Eds.), *Philosophy of Pseudoscience: Reconsidering the Demarcation Problem* (pp. 61–77). University of Chicago Press.

Jameson, Cathy. 2015. Vaccines: The science has spoken (blog entry). *Age of Autism* (June 7). Available online at http://www.ageofautism.com/2015/06/vaccines-the -science-has-spoken.html.

Kata, Anna. 2012. Anti-vaccine activists, web 2.0, and the postmodern paradigm—an overview of tactics and tropes used online by the anti-vaccination movement. *Vaccine* 30(25): 3778–3789.

Kusche, Larry. 1986. *The Bermuda Triangle Mystery Solved*. New York, NY: Galahad Books. Reprinted 1995.

———. 2015. The Bermuda Triangle delusion: Looking back after forty years. *Skeptical Inquirer* 39(6) (November/December): 28–37.

Luck, Marissa. 2013. The top 10 chiropractic studies of 2013 (blog entry). *Chiro Nexus* (December 30). Available online at https://www.chironexus.net/2013/12/top-10-chiropractic-studies-2013/.

Mikkelson, David. 2015. Rumor: Data suppressed by the CDC proved that the MMR vaccine produces a 340% risk of autism in African-American boys. *Snopes.com* (February 3). Available online at http://www.snopes.com/medical/disease/cdcwhistleblower.asp.

Olmsted, Dan. 2014. Midweek mashup: Autism research suppression (Web log post). *Age of Autism* (November 4). Available online at http://www.ageofautism.com/2014/11/midweek-mashup-autism-research-suppression.html.

Radner, Daisie, and Michael Radner. 1982. *Science & Unreason*. Belmont, CA: Wadsworth.

Rubin, Daniel B. 2008. Fanning the vaccine-autism link. *Neurology Today* 8(15): 3.

Shermer, Michael. 2002. *Why People Believe Weird Things: Pseudoscience, Superstition, and Other Confusions of Our Time* (Revised and Expanded). New York, NY: St. Martin's Griffin.

Skeptical Raptor. 2015. Science deniers use false equivalence to create fake debates (blog entry). *Skeptical Raptor* (December 29). Available online at http://www.skepticalraptor.com/skepticalraptorblog.php/science-deniers-false-equivalency-pretend-debate/.

Taylor, Ginger. 2013. No evidence of any link (blog entry). *Adventures in Autism* (July 24). Available online at http://adventuresinautism.blogspot.com/2007/06/no-evidence-of-any-link.html. To be clear, this appears to be a post from June 14, 2007, that was updated on July 24, 2013.

Taylor, Luke E., Amy L. Swerdfeger, and Guy D. Eslick. 2014. Vaccines are not associated with autism: An evidence-based meta-analysis of case-control and cohort studies. *Vaccine* 32(29): 3623–3629.

The Logic of Science. 2016. Vaccines and autism: A thorough review of the evidence (blog entry). *The Logic of Science* (April 28). Available online at https://thelogicofscience.com/2016/04/28/vaccines-and-autism-a-thorough-review-of-the-evidence/.

Uno, Yota, Tokio Uchiyama, Michiko Kurosawa, et al. 2012. The combined measles, mumps, and rubella vaccines and the total number of vaccines are not associated with development of autism spectrum disorder: The first case–control study in Asia. *Vaccine* 30(28): 4292–4298.

Wakefield, Andrew J., Simon H. Murch, Andrew Anthony, et al. 1998. RETRACTED: Ileal-lymphoid-nodular hyperplasia, non-specific colitis, and pervasive developmental disorder in children. *The Lancet* 351(9103): 637–641.

Walia, Arjun. 2013. 22 medical studies that show vaccines can cause autism (blog entry). *Activist Post* (September 12). Available online at http://www.activistpost.com/2013/09/22-medical-studies-that-show-vaccines.html.

CHAPTER 35

Illusions of Memory

Elizabeth F. Loftus

Vol. 40, No. 1
January/February 2016

The honorary doctorate awarded to me by Goldsmiths is deeply meaningful at this particular time in my life. It gives me a chance to talk with you about my work on illusions of memory—or the memories that people sometimes have of seeing things or doing things that they never saw or did.

When I began my life work on illusions of memory, I had no idea it would one day prove to be such a socially relevant and politically explosive topic. Of course, couples and siblings quarrel endlessly about whose memory of past events is right—that is the amusing and infuriating *Rashomon* aspect of every family's life. But who could foresee, in the late twentieth century, "recovered memory therapy" or that people would come to believe with all their hearts that they remembered being abducted by aliens or Satanic cults? Who knew that by the first decade of the twenty-first century we would find hundreds of individuals in prison who were innocent—proven innocent by DNA analysis—and the major cause for their wrongful convictions was faulty human memory?

And so, as my research on memory evolved, its findings became ever more applicable in the service of justice.

To briefly summarize, in that research I and my collaborators showed that you could alter people's memories for crimes, accidents, and other events that they had recently witnessed. You could pretty easily make someone believe that a car was going faster than it really was or that the bad guy had curly hair instead of straight hair. Later we would show that you could plant entire events into the minds of ordinary healthy people, letting them believe that they had experiences that they never ever had—even experiences that would have been pretty traumatic had they actually happened.

So you can see how these findings might be applicable in the service of justice. They help us understand how improper handling of eyewitness testimony can lead to false memories and the conviction of innocent people. They help us understand how suggestive or coercive therapy can lead people to develop memories of being abused in a Satanic cult, accusations that can cause untold misery for innocent people and their families.

At the same time, this research became emotionally controversial and the focus of terrible hostility among those who could not accept its findings or its implications for the real world. In my case, disgruntled people have written countless threatening letters. They have tried generating letter-writing campaigns to the chair of my former academic department, the president of the university, and even the governor of the state to get me fired from my academic job. They have threatened violence at places where I've been asked to speak, prompting universities on several occasions to provide armed and unarmed guards to accompany me during those speeches. People spread defamatory insults in their own writings, in newspaper columns, and, of course, on the Internet. One person even sued me in court when I published an article questioning the veracity of a psychiatrist's case study of a young woman's recovered memories of maternal sexual abuse. That litigation dragged on for almost five years before it ended.

Through these experiences, I've learned firsthand that science is never dispassionate, at least not if you are studying anything that has political, emotional, or financial implications for people's lives: child testimony, sex, the unreliability of projective tests such as the Rorschach, or, in my case, illusions of memory. I could have chosen to study memory in the sea slug—hardly anyone would get exercised about that. But I chose to study human memory, eyewitness testimony, false memories, confabulation in adults and children, and harmful therapeutic methods. And big trouble came my way.

But I'm proud of the work I've been able to accomplish as a psychological scientist and proud of the people I've had a chance to help along the way. I've learned to accept the hassles as the price that all scientists pay for doing research that matters or that threatens deeply held beliefs.

And this brings me back to Goldsmiths—where we are today: I am especially grateful for the honorary doctorate from Goldsmiths—where a number of superb academics are doing research that threatens deeply held beliefs, and also matters to people. That is why this honor holds such extra-special, extra-poignant, meaning for me.

Elizabeth F. Loftus is Distinguished Professor of Social Ecology, and Professor of Law, and Cognitive Science in the Psychology and Social Behavior and Criminology, Law and Society departments in the School of Social Ecology at University of California, Irvine.

CHAPTER 36

The "Lie Detector" Test Revisited

A GREAT EXAMPLE OF JUNK SCIENCE

Morton E. Tavel

Vol. 40, No. 1
January/February 2016

Recently I came across one of television's true crime programs that presented a provocative example from actual case files: A woman had been brutally murdered in her apartment. Her former boyfriend with whom she had recently had a vigorous altercation became the leading murder suspect. During the investigation, this man was "offered" a polygraph (lie detector) test in the attempt to establish his likely guilt or innocence. He agreed to be tested and was found to have "failed," thus presumably indicating his likely guilt. Despite this result, the evidence presented at the trial was deemed insufficient for a guilty verdict, and he was acquitted. Convinced that the polygraph test was accurate, his local community made him a pariah; he was shunned and even threatened with bodily harm. Several months later, however, another man, the actual murderer, was apprehended and convicted. Thus at least for this first suspect, despite his ordeal, the story had a satisfactory ending, and the lie detector test itself proved inaccurate and misleading. This outcome leads to an obvious question: How often does "lie detection by machine" prove false?

Much of the American public seems to be convinced that the "lie detector" is valid, as indicated by its ubiquitous use in "whodunit" literature and on television crime, psychology, talk, and news shows. After all, faced with such an avalanche of widespread approbation, who could doubt the validity of such a test? Supporting this illusion is the fact that federal, state, and local police departments and law enforcement agencies across the United States are generally avid proponents of this method.

But let's take a closer look at this subject. Questions about the accuracy of this test should be amenable to modern scientific methods. Interestingly, this challenge is strikingly similar to those we face regularly as medical researchers and practitioners when we evaluate various tests in the attempt to establish the presence or absence of many diseases. Through this lens, therefore, I can provide some insight on a method that is often uncritically analyzed.

The Procedure and Its History

The "lie detector" test has been used for nearly a century, and it employs a "polygraph," which, during questioning, continuously records an examinee's blood pressure, respiration, pulse rate, and skin resistance (an indirect measure of perspiration).

The usual format compares responses to "relevant" questions with those of "control" questions. The control questions are designed to control for the effect of the generally threatening nature of relevant questions. Control questions concern misdeeds that are similar to those being investigated, but refer to the subject's past and are usually broad in scope; for example, "Have you ever betrayed anyone who trusted you?"

A person who is telling the truth is assumed to fear control questions more than relevant questions. This is because control questions are designed to arouse a subject's concern about their past truthfulness, while relevant questions ask about a crime of which they are suspected. A pattern of greater physiological response to relevant questions than to control questions leads to a diagnosis of "deception." Greater response to control questions leads to a judgment of no deception. If no difference is found between relevant and control questions, the test result is considered inconclusive.

In the effort to improve the test's accuracy, alternative means of questioning have been suggested, such as the "guilty knowledge test" (Ruscio 2005). Rather than attempting to determine the truthfulness of an examinee's responses to relevant items, this technique aims to expose "hot" responses to questions only a guilty individual could display. This is done with a series of multiple-choice questions, one of which contains the incriminating information. This type of interrogation is limited only to instances of specific wrongdoing, but it has not been sufficiently investigated and is quite likely to suffer from the same short-comings as the conventional procedure.

The test records the activity of the sympathetic branch of the autonomic (involuntary) nervous system that influences heart rate, respiratory rate, blood pressure, and perspiration. Although this part of the nervous system is active at all times, it increases during excitement, rage, anxiety, fear, or fright, any of which could be caused by lying. But deception is a cognitive function that defies direct measurement. Indeed, throughout the entire history of medical science, there have been no scientific studies that have shown that the emotional response linked to lying could be measured. Moreover, reactions associated with lying and any other assumed emotional stresses can be quite variable. Some people may stay calm with a gun at their head. By contrast, others may respond excessively with heart thumping and sweaty palms at simply shaking someone's hand. The polygraph examination itself often causes fear and anxiety, and if such responses are excessive in response to a given question, then one may be deemed to have failed that question by a polygraph examiner.

The Evidence

Because of this obvious biologic improbability in this era of evidence-based medical science, the premise of lie detection by polygraph has fallen under a cloud of skepticism, and it is justly considered pseudoscience by most of the scientific community (Iacono 2001).

The American Polygraph Association (APA), a professional organization for polygraph examiners, predictably has complete faith in the accuracy of the test. They have licensing procedures in twenty-eight states and their own trade journal, *Polygraph*, in which they report scientifically questionable studies and provide anecdotes of the accuracy of their trade. The majority of these members complete a six-week to six-month post–high school training course in the art of polygraphy. They are not required to have formal training in medicine, psychology, physiology, or behavior—the very disciplines on which the testing is based. The majority of their members cater to the legal system upon which their economic livelihood depends, thus creating the background for a clear conflict of interest.

As expected, polygraph examiners will usually confidently state that the exam is highly accurate, around 95 percent. This implies that if 100 guilty suspects are given a polygraph exam, ninety-five of them will be detected through the test, meaning that only five of these 100 will be a false negative and not be detected by this method. On the other hand, they will state that if you are telling the truth, you have almost a 100 percent chance of being cleared by the test (Reid and Inbau 1977).

However, to be acceptable to modern scientific standards, the two characteristics that must be established are the *sensitivity* and *specificity* of this test: Sensitivity is that percentage of positive test results when lying is known to exist. Specificity is the percentage of negative results in the absence of prevarication. Under both scenarios, to establish the test's accuracy, the presence of both lying and honesty must be determined independently of the test procedure itself. True accuracy must also be derived from real-life conditions because, for obvious reasons, it cannot be derived from volunteers in laboratory settings that lack the emotional pressures of real suspicion. To accomplish this, a group without known prevarication is tested to assess the results. But absolute proof of honesty is evasive in such individuals, because even if lying is absent, the anxiety associated with the test may cause false positive test results, reducing test specificity, which is a major confounder.

But what about test results in those who are actually guilty? We have little knowledge of the frequency with which liars are judged to be truthful. Moreover, guilty subjects—and many others—can purposely control their reactions using what are called "countermeasures," sufficient to confuse the results enough to produce a "false negative." One outstanding example of a false negative was that of Aldrich Ames, who in 1995 had successfully passed five polygraphs during his long career in intelligence, and, despite this, he was subsequently arrested and convicted of spying. After Sandia National Labs Senior Scientist Alan P. Zelicoff published a strong commentary in *Skeptical Inquirer* calling polygraphs "a dangerous ruse" (Zelicoff 2001), Ames himself wrote a remarkable letter to the editor from federal prison confirming Zelicoff's points, adding that polygraphs are indeed "junk science," "a superstition," and "a refuge from responsibility." "Like handing fate to the stars, entrails on the rock, bureaucrats can abandon their duties and responsibilities to junk scientists and interrogators masquerading as technicians," Ames wrote (Ames 2001). In 2003, another example was provided by Gary Ridgway, who eventually was found to be the Green River Killer, having murdered forty-nine women in the Seattle area. Ironically, Ridgway had passed a lie detector test in 1987, while another man—who was proved to be innocent—failed. Although such isolated examples are admittedly anecdotal, they raise suspicions that the test itself may be flawed and should be carefully scrutinized through critical scientific means that employ large numbers of tests, as described below.

The polygraph test was not subjected to modern scientific investigation until three decades ago (Saxe et al. 1983; Lykken 1981). Since then there have been several studies employing improved methodology. Despite the impossibility of achieving a completely satisfactory research design, these tests clearly refute the high accuracy previously claimed. These studies have appeared in reputable peer-reviewed journals (Horvath 1977). They generally report a sensitivity ranging around 76 percent. This means that out of 100 liars only seventy-six of them will be detected by the polygraph. But adding doubt about this apparent test sensitivity is the fact that establishing true "guilt" often cannot be dissociated from prior polygraph testing, which may have contributed to confessions and/or guilty verdicts in court, skewing the data toward a falsely high sensitivity value of the test. The testing examiner also may have prior suspicions about the honesty of the examinee, which can in turn cause bias in the test's interpretation and further skew the data. The skill of the examiners also is associated with considerable variability of results, creating another major source of inaccuracy. Some smaller police departments with limited budgets may designate a police officer to be their department examiner instead of using an outside professional examiner. The designated testing officer may have little or no formal training, other than the limited training provided by the company selling the instrument to the police department. This can lead to situational bias based upon an officer's predisposition to believe most suspects are generally guilty. As a result, the officer may categorize an inconclusive result as untruthful. If the testing officer is aware of the circumstances of the individual being tested, and the weight and extent of the evidence indicating the individual being tested is the likely perpetrator of the crime, unbiased testing is virtually impossible. Officers are also generally trained to ask questions using phrasing that may cause someone to answer in a manner that may appear to be an admission of guilt. This type of questioning format would be inappropriate in a polygraph examination. For these various reasons given above, establishing true sensitivity of this test is therefore unlikely ever to be achieved, but the estimations above are likely overstated.

Even more damning—but unsurprising—is that studies report an average specificity of 52 percent, meaning that out of 100 people who are not lying, only fifty-two will be identified as telling the truth while forty-eight of these honest individuals will be branded as liars. These odds are similar to a coin toss, which would have a specificity of 50 percent. Other studies (Brett et al. 1986; Kleinmuntz and Szucko 1984; Lykken 1981) have shown even lower specificity values, indicating that "positive" test results are virtually useless. In 2003, the National Academy of Sciences, after a comprehensive review issued a report, *The Polygraph and Lie Detection* (National Academy of Sciences 2003), stating that the majority of polygraph research was "unreliable, unscientific and biased" and concluding that fifty-seven of approximately eighty research studies—upon which the American Polygraph Association relies—were significantly flawed. It concluded that, although the test performed better than chance in catching lies—far from perfect—perhaps most important, the test produced too many false positives.

If anything, there is perhaps one minor advantage to subjecting suspected felons to such testing (Lykken 1981; Lykken 1991): 25 to 50 percent of exam- inees will, under the intense psychological pressure of the exam, confess to the misdeed at hand, having been persuaded that they have been proven dishonest by "scientific" means. It is usual for the polygraph examiners to interrogate the subjects whom they believe have "failed" the test. Examiners may state that there is no way now to deny the objective guilt demonstrated by this "impar- tial" scientific device, and that the only available option is to confess, which they often do. But, while perhaps effective, this in no way exonerates the test itself. Perhaps various forms of torture such as water boarding might be just as effective. And one might ask further how often this form of interrogation might lead innocent persons to falsely admit, out of fear or other threats, to some form of wrongdoing.

Justification for Continued Use?

For all these reasons, the continued use of polygraph lie detection has the potential to cause much harm to those many innocents who are falsely judged dishonest by its results. A single failure could conceivably ruin one's life. Since 1923, polygraph evidence has not been admissible in federal court cases because the test was deemed to lack scientific validity. Sadly, however, it is still used widely by the court systems of many states. Moreover, suspects are frequently "offered" this test prior to criminal proceedings, but if, for any reason, they decline to be tested, this refusal alone may cause them to be presumed guilty. Conversely, if they consent to be tested, they are risking the commonly occurring false positive outcome, which in the view of many prosecutors and juries, supports a guilty verdict. Thus from the standpoint of the accused, he/she is caught in a catch 22 situation.

Even more regrettable is the attempted application of this test for pre-employment or security clearances. In this context, testing large groups with a low base rate of dishonesty will disclose a large number of false positive responses, as encompassed in the mathematical principles of Bayes theorem (Tavel 2012). Understanding these concepts, the American Medical Association's Council on Scientific Affairs (1986) recommended that the polygraph not be used in pre-employment screening and security clearance, with which I fully concur.

Adding clear restrictions to this testing, The Federal Employee Polygraph Protection Act, passed in 1988, virtually outlawed using polygraphs in connection with employment. That law covers all private employers in interstate commerce, which includes almost every private company that uses a computer, the U.S. mail, or a telephone system to send messages to someone in another state.

Under the Act, it is illegal for all private companies to:

1. Require, request, suggest, or cause any employee or job applicant to submit to a lie detector test;
2. Use, accept, refer to, or inquire about the results of any lie detector test conducted on an employee or job applicant;
3. Dismiss, discipline, discriminate against, or even threaten to take action against any employee or job applicant who refuses to take a lie detector test.

Federal applicants and employees are also generally protected from lie detector tests by civil service rules. Despite all these apparent safeguards, they often must submit to a polygraph examination in the quest of a coveted security clearance for federal employment or to retain such a job. The Employee Polygraph Protection Act allows polygraph tests to be used in connection with jobs in security and handling drugs, or in investigating a specific theft or other suspected crimes.

As a result, these examinations continue to be used by federal agencies such as the FBI, CIA, and National Security Agency, where they are commonplace in decisions concerning employment. Applicants might expect them as a form of initial screening even before they start working at the agency. They may also be required to take follow-up polygraph tests from time to time. Prior to a likely job offer, the polygraph test is often the last remaining hurdle. Under these circumstances, the need to pass can be a very nervous event for anyone—especially those who have not been subjected to a polygraph before, and of course, as explained, this can easily trigger a false positive response, resulting in an unjustified rejection. Thus use of such a test for this purpose is very difficult to defend.

Given the fallacies of such testing, one would assume that enterprising individuals would provide instruction to subjects on how to pass such a test (through "countermeasures"), thus leading to the conclusion—rightly or wrongly—that one is being truthful. Moreover, the prior administration of certain drugs to block the sympathetic nervous system could also be expected to impair test accuracy. Actually, instruction on passing these tests is easily available on some Internet sites. Therefore it is difficult to understand why anyone providing personal instruction to this end would actually be prosecuted as if this were a felony. This is exactly what occurred recently (Taylor 2013) when an Indiana man was accused of "threatening national security" by teaching government job applicants and others how to pass lie-detector tests. He was sentenced to eight months in prison after federal agents had targeted him in an undercover sting. At the time of this writing, at least one additional case is pending on similar grounds. Although legal intricacies extend beyond this discussion, this issue appears to threaten First Amendment rights. Moreover, that such instruction is even possible exposes the flaws of a procedure largely regarded as pseudoscience by the scientific community. To exemplify this point further, let us present an intriguing—although farcical—analogy: Suppose we discovered that, through coaching, one could teach persons how to transfer firearms past metal detectors at air terminals without triggering alarms? Such a revelation would logically cause one of two reactions by the supervising authorities: 1) Eliminate such a flawed test in favor of one that is accurate; or 2) Attempt to silence those coaching this "deception" with threatened legal penalties, including incarceration. The obvious answer to this question—number one—requires little thought. But with regard to polygraph testing, this answer has thus far not been chosen by our authorities, for they have exercised the second option. The mere fact that authorities have chosen to suppress information of this type might, in itself, be considered as a tacit admission that such testing is fatally flawed.

Conclusions

In summarizing its many pitfalls, Iacono (2001) concluded in an article titled "Forensic Lie Detection: Procedures without Scientific Basis," the following, which I paraphrase for clarity: Although this form of testing may be useful as an investigative aid and tool to induce confessions, it does not pass muster as a scientifically credible test. Its theory is based on naive, implausible assumptions about its accuracy. It is biased against innocent individuals and can be beaten simply by artificially augmenting responses to control questions. Although it is not possible to adequately assess the error rate of this test, both of these conclusions are supported by published research findings in the best social science journals.

Given such overwhelming evidence of inaccuracy as I have presented here, how can we, as a society, react to such a perversion of science? The logical solution is to completely abandon this method of testing. The most urgent necessity is the complete elimination of this testing in connection with employment. All remaining state and federal laws that allow use of the polygraph, in or outside of court settings, should be abandoned. Moreover, as long as it remains in use, there is simply no justification for prosecuting those who provide information about how to "pass" such a test.

Unfortunately, various state and national polygraph certifying and licensing organizations—whose livelihood depends upon their own continued existence—are well entrenched in our society. Their provision of services to most law enforcement agencies creates a symbiotic relationship that is difficult to overcome. Nevertheless, to eradicate this blight the scientific community, as well as others who understand these concepts, must educate the public and relentlessly urge the responsible authorities to discontinue the present unsatisfactory status quo.

Morton E. Tavel, MD, Clinical Professor Emeritus, Indiana University School of Medicine, is a physician specialist in internal medicine and cardiovascular disease. He is author of *Snake Oil Is Alive and Well: The Clash Between Myths and Reality* (2012) and *Health Tips, Myths and Tricks: A Physician's Advice* (2015). His website is http://www.mortontavel.com/.

References

American Medical Association Council on Scientific Affairs. 1986. Polygraph. *Journal of the American Medical Association* 256: 1172–75.

Ames, A. 2001. The polygraphs controversy (letter to the editor). *Skeptical Inquirer* 25(6): 72–73.

Brett, A. S., M. Phillips, and J. F. Beary. 1986. Predictive power of the polygraph: Can the "lie detector" really detect liars? *The Lancet.* 1: 544–547.

Horvath, F. 1977. The effect of selected variables on interpretation of polygraph records. *Journal of Applied Psychology* 62: 127–136.

Iacono, W. G. 2001. Forensic "lie detection": Procedures without scientific basis. *Journal of Forensic Psychology Practice* 1: 75–86.

Kleinmuntz, B., and J. Szucko. 1984. A field study of the fallibility of polygraphic lie detection. *Nature* 308: 449–450.

Lykken, D. T. 1981. *A Tremor in the Blood: Uses and Abuses of the Lie Detector.* McGraw-Hill, New York.

———. 1991. Why (some) Americans believe in the lie detector while others believe in the guilty knowledge test. *Integrative Physiological and Behavioral Science* 26: 214–222.

National Academy of Sciences. 2003. *The Polygraph and Lie Detection, Committee to Review the Scientific Evidence on the Polygraph.* The National Academies Press, Washington, D.C.

Reid, J. E., and F. E. Inbau. 1977. *Truth and Deception: The Polygraph ("Lie Detector") Technique.* Williams & Wilkins, Baltimore.

Ruscio, J. 2005. Exploring controversies in the art and science of polygraph testing. *Skeptical Inquirer* 29(1): 34–39.

Saxe, L., D. Dougherty, and T. Crosse. 1983. Scientific validity of polygraph testing: A research review and evaluation. *Conference: OTA-TM.* U.S. Congress Office of Technology Assessment.

Tavel, M. E. 2012. *Snake Oil Is Alive and Well: The Clash between Myths and Reality. Reflections of a Physician.* Chandler, AZ: Brighton Press, 51.

Taylor, M. 2013. Indiana man gets 8 months for lie-detector fraud. *Seattle Times* (September 6).

Zelicoff, A. P. 2001. Polygraphs and the national labs: Dangerous ruse undermines national security. *Skeptical Inquirer* (July/August). Online at http://www.csicop.org/si/show/polygraphs_and_the_national_labs_dangerous_ruse_undermines_national_security/.

CHAPTER 37

Does Astrology Need to Be True?

A THIRTY-YEAR UPDATE

Geoffrey A. Dean

Vol. 40, No. 4
July/August 2016

The original article in *Skeptical Inquirer* Winter 1986–87 had a relaxed cover picture by artist Ron Chironna and this introduction by Editor Kendrick Frazier: "We begin publication in this issue of Geoffrey Dean's two-part 'Does Astrology Need to Be True?' a comprehensive investigation of the claims of *serious* astrology as defined by 'serious' astrologers. Although we are striving for shorter articles, so that we can cover a wider range of interests, we publish this lengthy inquiry because of its special significance. As one of our reviewers of Dean's manuscript wrote, 'It is without doubt the best article on astrology I have ever seen.'"

Other than its length, the original article had three claims on your rapt attention. It was the result of much recycling among colleagues and noted skeptics, including Susan Blackmore, Ray Hyman, Ivan Kelly, Andrew Neher, and Marcello Truzzi, whose critical comments kept it on the rails. It ignored the nonsense of sun sign astrology and focused on the *real thing* as used in consulting rooms, on why people believe in it (because it seems to work), and on the results of tests (astrology stops working when cognitive artifacts such as confirmation bias are controlled). And finally it asked, not is astrology true, but does it *need* to be true? A change that in one hit ended a centuries-old shouting match over claims of truth. The answer was no.

The real thing was not hard to find. In Western countries, it was the subject of roughly 100 periodicals and 1,000 books *in print* and was practiced or studied by roughly one person in 10,000. (The proportion today has been affected by astrology on the Internet but is probably not much different.) My conclusion in the original article was:

In the last ten years various studies have addressed astrology (the real thing, not popular nonsense) on the astrologer's terms. The results of these studies are in agreement, and their implications are clear: Astrology does not need to be true in order to work, and contrary to the claims of astrologers authentic birth charts are not essential. What matters is that astrology is *believed* to be true, and that authentic birth charts are believed to be essential.

Business as Usual

Astrologers replied in their usual way to criticism, dismissing it as biased and ignorant. Their repeated claim—that their daily experience confirms their fundamental premise *as above so below*—is still heard from the rooftops. They still misinterpret cognitive artifacts in a chart reading as evidence of links with the heavens. And they still explain away all failures by the same old excuses, such as stars incline and do not compel; another factor is interfering (there is *always* another factor), and astrologers are not infallible. Astrology is thus made nonfalsifiable, whereupon belief and paying clients follow automatically. It then gets worse.

Unwelcome evidence is dismissed because, they say, research is biased; astrology is too nuanced to be testable by science, and (the ultimate clincher) research funding is nonexistent. Yet astrologers insist that looking at birth charts will convince us that astrology works. Just try it and our eyes will be opened at last! But they cannot have it both ways. Astrology cannot simultaneously be difficult to test and yet easy to prove. Their response to this contradiction is usually a scornful silence.

Nevertheless, the past thirty years have seen big advances in research design, the availability of data, and the use of computers to break the calculation barrier. At one time, astrologers using logarithms could take many hours to calculate a comprehensive birth chart; a home computer can do the same for dozens of charts while you cough.

The result has been hundreds of controlled tests of astrology by both believers and critics. Most studies are little-known; so for forty years, my colleagues and I have been visiting astrology collections and searching academic databases for every useful study ever made. We have published the results in *Tests of Astrology: A Critical Review of Hundreds of Studies* (Dean et al. 2016). Some of the notable studies we found are outlined below. They show how skeptical inquiry has advanced on astrological claims during the last thirty years.

Every astrological consultation involves feedback (as shown above) to help the astrologer pick chart factors that fit the situation. But how accurate are their meanings? The late Dr. Andrew Patterson lectured in engineering at the University of Witwatersrand. His interest in astrology began in the 1960s, and for many years he was a teacher and invigilator in South Africa for the U.K. Faculty of Astrological Studies. His scientific background resulted in that rarest of combinations—a fine critical sense plus an encyclopedic grasp of astrology—which he applies below to the challenge of learning astrology. As you read his account (abridged from 1991), remember he is a teacher of astrology, not a debunking skeptic.

Astrology is more difficult to learn than anyone realizes. Probably we have all had much the same experience. You meet astrology via a friend and become hooked. You start studying. But after a while you grow uneasy. It is not clear how Sun in Leo (*must shine*) differs from Moon in Leo (*needs to shine*). When asked to describe Saturn (*restriction*) in 8th house (*death*) you are not sure where to start. All you can say about a hard aspect is that it represents a challenge, whereas an easy aspect is, well, easy. As for a quincunx, you struggle just to pronounce it.

To clear up your confusion, you buy every recommended book. But they just make your confusion worse. Consider the interpretations given in those books. They are either *all the same* so they blur into one another as with Leo above. Or they are *all different*, thus Sun square Saturn varies from "a life of hardship" to "loss of father." Or they are *all useless*, being either amazingly general or amazingly specific, thus Mars in Libra varies from "lack of commitment" to "passion for sword-dancing." Or they are *all evasive* as in "Neptune dissolves," which conveys nothing while pretending to convey everything.

Patterson concludes by pointing out that truth in astrology is tested by how well it matches the symbolism. Anything that passes this test is seen as true, not because it is actually true but because it could be true. Being able to say that the truth (whatever it is) is consistent with the symbolism is not terribly useful. Which is why astrology is so hard to learn (Patterson 1991).

Which Zodiac to Use?

Western astrologers use the tropical zodiac tied to the seasons, while Eastern astrologers use the sidereal zodiac tied to the stars. Around 200 AD, the two zodiacs coincided, but today precession has put sidereal signs almost one sign ahead of tropical signs. So have their meanings changed?

British astrologer J. E. Sunley spent ten years comparing meanings between tropical sign X and sidereal sign X as given in astrology books. In principle, their meanings should be mostly different, but he found they were mostly similar—which is consistent with signs having no meaning at all except in the minds of astrologers. It explains why tens of thousands of Western tropical astrologers can agree that in their experience Scorpio is intense, while hundreds of thousands of Eastern sidereal astrologers can look at much the same piece of sky—which they call Libra—and agree that in their experience it is not intense but relaxed. So much for experience.

But if *relative* sign meanings are okay, as in Leos get on well with Sagittarians, what is there to worry about?

Sun Signs for Lonely Hearts

Sun sign compatibility was explored by Manchester University's David Voas (2007) using data gathered for the 2001 census in England and Wales. Traditionally, favorable angles between any two sun signs are said to be the conjunction 0° (Leo and Leo), sextile 60° (Leo and Libra), and the legendary trine 120° (Leo and Sagittarius). Despite possible conflict with other factors in the two charts (among sun sign astrologers this is the default explanation for awkward findings), if the claim is true then it should show up in a large enough sample: ten million marriages, for example.

Voas notes that completion errors are problematic. Census forms are typically completed by one member of the household, who for some reason may enter their own birthday for that of their spouse. Others may enter January 1 or July 1 if an exact birthday is unknown, which is sometimes the case in old people's homes and for people born overseas. If dates of birth are illegible or missing (about 0.5 percent of all responses), the census office enters the day as the first of the month and assigns the months in rotation. Voas carefully removed all such artifacts but was unable to find evidence for useful sun sign effects.

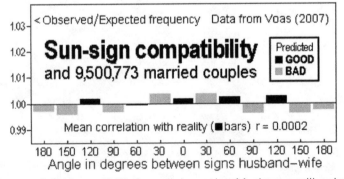

Astrology predicts compatibility for certain angles. Most are positive, but none are useful or statistically significant and all are explained by knowledge of sun signs biasing the outcome, as explained in the next section. *Chart created by Geoffrey A. Dean.*

Thanks to his enormous sample, Voas's test was the most sensitive test of sun signs ever made. But none of the 144 possible sun sign pairings differed significantly from chance alone. In terms of predicting compatibility, sun signs absolutely did not work. You will not find this result in astrology books.

Experience 1, Science 0

British astrologer and former journalist Dennis Elwell (1930–2014) was noted for his eloquence. In an article in the *Astrological Journal* (1991), he restated the faith of astrologers in their experience as follows: "Like many others, I persevere with astrology because experience has shown that by and large its basic assumptions are correct. . . . If some piece of research proves a dead end, I do not question the authenticity of my experience, I question the competence of the research, or its underlying assumptions."

He held that failures to verify astrological claims were caused by the wrong approach because the right approach always worked. One of his favorite examples was how the birth chart for the Declaration of Independence on July 4, 1776, showed strong links with the Statue of Liberty. Thus the statue is big (Jupiter), made of copper (Venus), has a female form (Venus), and appears in the birth chart as Venus conjunct Jupiter exact within 3°. And so on through dozens of events and associated people. It was "the kind of evidence that astrologers recognise and respect," and it convinced Elwell (as claimed in his 1987 book *Cosmic Loom: The New Science of Astrology*, which contains no science despite its title) that "science will eventually be obliged to embrace the astrological if it is to unify its picture of the universe." So please test astrology by case studies, not by statistical studies of groups.

Okay, let's do it. Suppose we've been told that the above chart has its Sun conjunct Uranus, whose meaning "very frequently indicative of great talent" could hardly be more apt—is astrology already discernible? Indeed, the statue is an innovative (Uranus) national monument resting on Sun-ruled granite and lit by electricity (Uranus). It is 151 feet high (equals 1° Leo, which sign is ruled by the Sun) on an eleven-pointed island (obviously the eleventh sign Aquarius, ruled by Uranus). Everywhere we look we find the predicted Sun-Uranus links. Yes, it's amazing!

Since this is "the kind of evidence that astrologers recognise and respect," we now have good reason to believe in astrology—except the chart has no actual Sun-Uranus conjunction (Uranus is 40° from the Sun, not 0° or any other aspect within traditional limits). Elwell's respected evidence is no evidence at all.

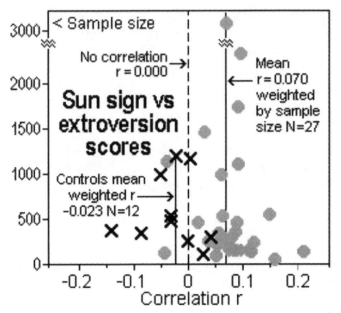

The correlation between sun sign and extroversion scores is usually weakly positive and was once hailed as proof of astrology. But it disappears (crosses) if subjects don't know sun sign meanings or are tested against their moon sign (which few people are aware of). So the effect is in fact an artifact of sun sign knowledge. *Chart created by Geoffrey A. Dean.*

Sun Signs and Self-Image

Odd-numbered signs from Aries onward are said to be extroverted. The rest are said to be introverted. Ask Sagittarians (odd-numbered and said to be sociable and outgoing) a question related to extroversion (such as "Do you like parties?"), and knowledge of astrology might tip their answer in favor of yes rather than no. In fact, this answer-tipping can be detected if people know their sun sign but not if they don't. When taken together with opinion polls, the results suggest that one in three people believes sufficiently in sun signs to measurably shift their self-image in the believed direction—of which a tiny fraction may believe sufficiently to bias their choice of partner as in the previous section.

Astrologers Put to the Test

Charles Carter, the leading British astrologer of the 1930s, was noted for exceptional clarity of expression. Here is an example from his book *The Principles of Astrology* (1925, 14): "Practical experiment will soon convince the most sceptical that the bodies of the solar system indicate, if they do not actually produce, changes in: (1) Our minds. (2) Our feelings and emotions. (3) Our physical bodies. (4) Our external . . . affairs and relationships with the world at large."

Thirty years ago, such claims began to be tested by jumbling up birth charts with things such as their owner's case histories and personality traits. Could astrologers match them correctly? The outcome was maybe yes but mostly no. Since then, more tests have been made that bring the total to sixty-nine, and new ways have been developed to analyze the results. For example, the correlation between a reading and reality can be plotted against sample size to clarify what is happening. The plots in the figure below show how it works.

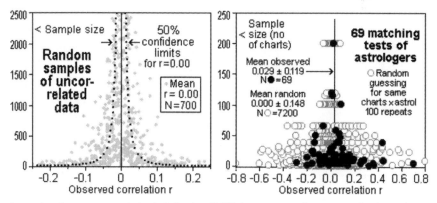

Samples from uncorrelated data (r = 0.00) have sampling errors that produce non-zero correlations (r ≠ 0.00) especially for small samples (common in astrology). Right: If astrologers could accurately match birth charts to their owners then the black dots would peak on the right. But they peak close to r = 0.0, or zero accuracy, and are skewed to the right indicating the presence of publication bias against negative results, hence the slightly positive mean (r = 0.029). Here, all means are weighted by sample size. Discarding tests with small sample sizes or less familiar criteria makes no difference. *Chart created by Geoffrey A. Dean.*

The studies in this figure are too numerous and too consistent with hundreds of other studies to be easily dismissed. Also, their subsequent meta-analysis shows that the differences between results are entirely explained by sampling errors, which leaves nothing for astrology and astrologers to explain; to paraphrase Pierre-Simon Laplace, we have no need of such hypotheses. But for completeness, we should still look at some of those other studies as shown in the figure below.

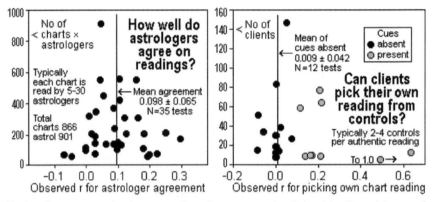

Tests of agreement when reading the same chart avoid all problems of judging accuracy, such as errors in birth time. The agreement should be high because astrologers tend to read the same books, but it is only weakly positive (r = 0.098) and is nowhere near the 0.8 generally required for tests applied to individuals (as astrology is). What is astrology worth if astrologers cannot even agree on what a chart means? Right: Clients given several chart readings cannot pick their own reading unless it contains cues (as required by the experimental design), such as the name of their sun sign, in which case they do quite well. *Chart created by Geoffrey A. Dean.*

The power and sensitivity of our tests so far are beyond anything the ancients could have dreamed of. But astrologers airily dismiss the results because, as one put it, "We have enough cumulative experience to know that it [astrology] works, whether the computer studies and the scientists agree with us or not" (Alexander 1983, xii).

Claims Tested on 3,290 People

For his PhD in psychology, German astrologer and psychotherapist Peter Niehenke (1984) circulated copies of a 425-item questionnaire for testing astrological claims. It was advertised in two newspapers and a New Age magazine and by notices at Freiburg University. He duly received 3,498 responses (requiring more than 110 reams of paper), of which 3,290 provided usable birth data, of which 62 percent were from birth certificates. The questions had been tested in a pilot study to make sure they were free of problems. Each was relevant to a given factor (planet, sign, house, or aspect) to see if the subjects identified with that factor regardless of whether it was actually in their birth chart. Thus Sun-Saturn aspects were explored by questions involving their supposed meanings such as disappointment, misfortune, pessimism, and guilt feelings.

Overall, no result was consistently in support of astrology. For example, subjects with four Saturn aspects (said to indicate heavy responsibility and depression) felt no more depressed than those with no Saturn aspects and showed no correlation with depression scores. Subjects with good trines to Jupiter (said to indicate optimism and good fortune) felt no sunnier than those with none. Aspects between aggressive Mars and the Sun, Moon, or Ascendant showed no correlation with aggressiveness scores. Responses to the question "I am unlucky in love: yes/no?" showed no correlation with aspects to Venus from Jupiter or Saturn, or with the house position of Saturn, all of which are said to be highly relevant. In the end, Niehenke decided there was more to astrology than being true or false: "a world in which astrology exists is surely a more enjoyable world than one without it. The need that astrology be a reality is much stronger than all the rational demonstrations against it" (1984, 15).

300,000 Chart Factors

In 1996, U.S. database engineer Mark McDonough wrote software to store and deliver the 30,000 birth data in AstroDataBank, the world's second largest collection of timed birth data. After several years of work, he could automatically analyze any subset of data for 300,000 chart factors (that's not a misprint; the large number is due to fashionable ideas such as asteroids and planetary nodes) taken individually or in combination and identify which factors differed the most from controls. But when applied to actual birth data grouped by, say, occupation or events, the results if positive (which was not often) failed to replicate. There was no evidence that astrological claims were valid: nothing actually worked. He asked for an explanation, but nobody had a clue. So he abandoned astrology to follow other interests.

Wrong Charts Make No Difference

Do astrologers get right answers from wrong charts? If they do, then their fundamental premise as above so below is disconfirmed. The idea might seem difficult to test—what astrologer wants to read *wrong* charts?—but it happens purely by accident and is surprisingly common. The astrologer gives a reading that satisfies the client *but the wrong chart has been used*. It makes no difference how wrong it is—by hours, days, or years—the chart still works. Astrologers recognize this but see it as some occult property of astrology that puts it beyond human understanding. Skeptics may disagree.

Les Gauquelins et leur Héroïsme

The most heroic studies in astrology were made by French psychologists Michel Gauquelin (1928–1991) and his wife Françoise (1929–2007). They used statistical testing and large samples mostly from the nineteenth century. Their results for traditional astrology (signs, aspects, transits) were consistently negative. Nothing worked. Therefore they were surprised to obtain positive results for what was later called the Mars effect (and, later still, planetary effects because the Moon, Venus, Jupiter, and Saturn were also involved): the tendency for eminent professionals to be born when the planet matching their occupation (such as Mars for sports champions, Jupiter for actors) had just risen or culminated. Planetary effects were new in that, unlike previous factors, they were critically dependent on the *hour* of birth.

Statistically, the effects were often very significant, which to astrologers meant strength. But their effect *sizes*, which for over thirty years nobody bothered to calculate, averaged a tiny $r = 0.04$ ignoring direction. So the effects were actually weak and were significant only because large samples were tested (typically more than 1,000). Indeed, the effects were so weak that if applied to 100 clients, on average only two would get readings more accurate than tossing a coin—and even then only if they were among the one in 20,000 who were eminent. Yet the effects replicated and were not explainable by faulty procedures (see the figure below).

Ironically, planetary effects created baffling puzzles even for astrology. Why only five planets? Why no effect for the sun or for signs and aspects? Why occupation and not personality? Why contrary to all expectations are planetary effects *larger* for *less-precise* birth times? And why are there such strange effects in the first place?

For forty years, nobody had a clue. Astrologers predictably saw the effect as proof of the higher realities in which astrology is said to operate. But after eight years of work, I uncovered a new artifact capable of explaining all the puzzles—namely the misreporting of birth times to match the pop astrology of the day (Dean 2002). The level of misreporting was very small, but then again so were the planetary effect sizes—and as opportunities for misreporting disappeared, so did planetary effects. Nobody knows if planetary effects still apply today, but that's only because privacy laws make new data hard to find. In any case, planetary effects are far too weak to be of practical use to astrologers.

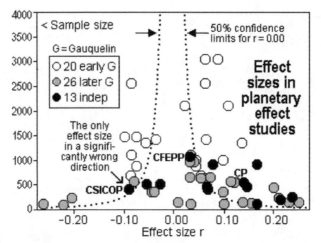

Effect sizes for the fifty-nine known studies based on computer-calculated data spread over five planets—Moon 12 percent, Venus 4 percent, Mars 38 percent, Jupiter 25 percent, Saturn 21 percent. Some results have negative effect sizes (e.g., Saturn for painters), so they do not cancel out positive effect sizes. What matters is the proportion of effect sizes that lie outside the 50 percent confidence limits. If significantly more than 50 percent, as is the case here, then planetary effects seem to be real. *Chart created by Geoffrey A. Dean.*

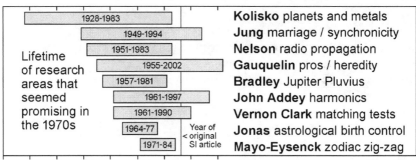

1920 1930 1940 1950 1960 1970 1980 1990 2000 Year

Last of the Astro Mohicans: Gauquelin planetary effects were the last of the astrological areas that had seemed promising in the 1970s and which still remained when my original *SI* article was published in 1986. The indicated end year is when the promise was lost due to the discovery of artifacts—poor control of chemistry (Kolisko), biased samples (Jung), inappropriate tests (Nelson), social artifacts (Gauquelin), astronomical artifacts (Bradley), sampling errors (Addey and Clark), unavailable data (Jonas), and sun sign knowledge (Mayo-Eysenck). *Chart created by Geoffrey A. Dean.*

But might consolation be found in Indian astrology, claimed by Indian astrologers to be vastly better than anything available in the West?

An Indian Test of Indian Astrology

Indian astrology is hugely different from Western astrology. It is more complex, uses the sidereal zodiac, and fortune-telling is the norm. The scientific revolution that eroded astrology in seventeenth century Europe did not happen in India, so it has had a free run ever since. Today it is firmly entrenched at all levels of Indian society. But no controlled test had been made in India until the one by Jayant Narlikar and colleagues at the Inter-University Centre for Astronomy and Astrophysics in Pune (Narlikar 2013).

They gave each of twenty-seven volunteering Indian astrologers (mean experience fourteen years) a different set of forty timed charts each, and a team of astrologers 200 timed charts (a larger number than in any Western test), to see if they could tell bright children from mentally retarded children. This is a commonly accepted claim in India, but neither group outperformed tossing a coin.

Nightmare on Time-Twin Street

"Time twins" are people born close enough in time and geography to have similar birth charts. At a given moment, the birth chart supposedly indicates trait X, the next moment it is trait Y, and so on. So time twins should be more alike in X than expected by chance, which makes them *the* definitive test of astrology, since all confounding reading artifacts are avoided.

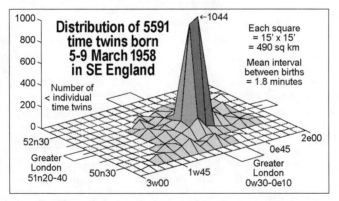

Distribution of 5,591 NCDS births in Southeast England during March 3–9, 1958 with the expected peak in the Greater London area. *Chart created by Geoffrey A. Dean.*

In a city of one million people, more than 2 percent will have a time twin born within one minute, about the same proportion as people with an ordinary twin, and about 20 percent will have a time twin born within ten minutes. The numbers increase very rapidly with time difference and city size. Indeed, the number of time twins in Western history is so enormous (hundreds of millions) that many similarities in personality and events will occur by chance alone. So the handful of cases routinely cited by astrologers cannot hope to be convincing.

The systematic testing of time twins was explored by Ivan Kelly and me (2003) using cohorts from the National Child Development Study (NCDS) of 16,000 children born in the United Kingdom during March 3–9, 1958. To minimize variations in birth place, we analyzed only those born in Greater London. Birth times for 92 percent of cases were reported to the nearest five minutes, and the rest to the nearest minute. For each person, we selected a total of 110 variables measured at ages eleven, sixteen, and twenty-three that were said to be shown in the birth chart such as ability, accident proneness, behavior, occupation, personality, and physical data such as height, weight, vision, and hearing. Data collection had required whole armies of researchers well beyond anything astrologers could achieve. For the purposes of testing astrology, this database was a dream come true.

But for astrology itself the results were a nightmare: support for astrological claims was nowhere in sight. For example, Saturn sets every day momentarily

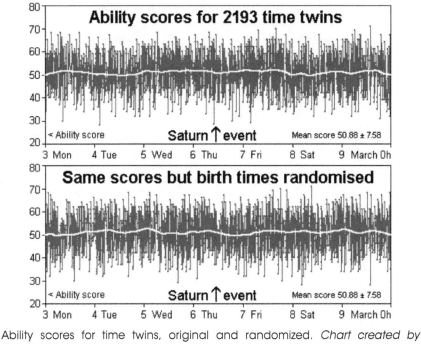

Ability scores for time twins, original and randomized. *Chart created by Geoffrey A. Dean.*

exactly on the horizon, a position traditionally held to greatly boost its strength. At that time in London on March 6, 1958, it was also square the Moon within 0.1°, which is also held to boost its strength. It was not just a strong Saturn event; it was also the strongest Saturn event for the entire week. Saturn is held to indicate *restriction* and *limitation*, so its effect should show up as a dip in measures of ability. But it did not.

A Strong Saturn Fails to Show Saturn Effects

Ability scores (a composite of fifteen tests such as intelligence, reading, and mathematics) plotted against time of birth for 2,193 births (see the figure above) shows no discernible effect from the Saturn event, no daily rhythm that might coincide with rising or culminating planets as in Gauquelin's planetary effects, and no clear difference from the same data when the birth times are randomized (lower plot). The white lines are forty-one-point moving averages (forty-one points is about three hours). None of the other 110 variables fared any better when analyzed by a battery of tests. But can we be sure that the test is really appropriate? It may be that ability is too broad a measure to show Saturn effects, in which case we need something such as extroversion that is more definitely linked to Saturn (*caution, reserve*). Perhaps Saturn effects are too focused to be discernible during seven days, in which case we need a smaller time frame. All as in the next test (see the figure below).

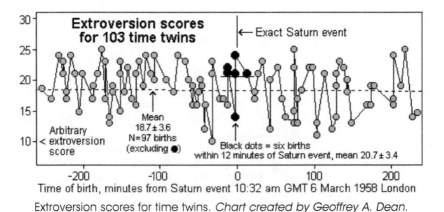

Extroversion scores for time twins. *Chart created by Geoffrey A. Dean.*

No Saturn Effect on Extroversion

Astrology predicts a drop in extroversion scores (here based on ratings on thirteen relevant scales such as impulsive–cautious) below the mean during the Saturn event (black dots in the figure above). But if anything they increase, albeit not significantly (p by a t-test is 0.22). The extroversion scores show no tendency to group together. Enlarging the Saturn event window makes no difference, so the time twin similarities predicted by astrology are not detectable.

Could Lack of Resources Be the Problem?

At this point, the last hope for astrology's factual validity seems to disappear, but there are still straws to clutch at. Astrologers claim that with enough funding, research facilities, and right-minded researchers, astrology would soon regain its rightful place as queen of modern worldviews. This belief has been put to the test in each of more than a dozen PhD theses that have involved tests of astrology.

It did not work for Niehenke's PhD thesis. So let's look at the PhD thesis of Pat Harris, a British astrologer whose website offers you a £30 ($45) Astro Fashion Profile based on sun signs. Earlier, during twenty years of professional practice, one of her clients had conceived via IVF (in vitro fertilization) under astrological conditions that were absent during seven failed attempts. This seemed to suggest that astrology could improve the IVF success rate—an idea she explores in her thesis (Harris 2005).

Later in an article in the *Journal of Sexuality, Reproduction & Menopause* (2008), Harris claimed that "attempts to conceive during [astrologically] optimal times have an increased likelihood of success," even though an editorial note advised that her results were not statistically significant. The May 2009 issue contained a letter from Jacky Boivin, professor of health psychology at Cardiff University, who noted that Harris's two samples of twenty-seven and twenty-eight women were too small to escape sampling artifacts (for which about 400 would be needed), thus her claim "is completely unwarranted."

Unusually, her thesis was under an embargo (normally granted only if it contains commercially sensitive material) that for five years prevented its release. In due course, I found it to contain no birth data, no proper controls, no expectancies, no details that would allow an independent check, and success rates inflated in much the same way as predicting dryness in arid areas—exactly the sort of errors and omissions that in my day would get your thesis rejected. What was her university thinking?

Like Niehenke, Harris did not let her results influence her belief in astrology. And here we encounter astrology's dark side—on another website she now offers for £90 ($135) the best astrological dates for achieving conception plus a £148 ($220) telephone analysis of your birth chart to optimize fertility. In her 2009 letter, Professor Boivin had commented that Harris's paper "should not have been published because it falls short of the scientific standards adopted to create the evidence base for interventions in fertility. . . . People with fertility problems are willing to try anything to achieve pregnancy, and giving them false hopes is yet another way of taking advantage of this vulnerability."

This of course calls into serious question the scientific and ethical standards of Harris's actions. So let's try one more time with the PhD thesis of Keith Burke (2012), a former U.S. astrologer who went further than most by cofounding a for-profit institute for teaching New Age topics. He taught astrology classes and held workshops through the institute, wrote astrology articles and a textbook, and lectured at national conferences. Verily the definitive right-minded researcher!

He had noted that the Moon is generally held to be as important as the Sun but had received little attention by researchers. There was also a clear similarity between the Moon's meaning in each of the four astrological elements and four of the Big Five personality dimensions. So this became the subject of his PhD thesis at the Pacifica Graduate Institute, an accredited clinical training graduate school in California that was even better suited to astrology than a pure research school. According to astrologers, the results ought to support astrology. But the effect sizes for 192 subjects with timed births, mean age forty-nine, were not only at chance level, but three were in the wrong direction (see table below).

Element	Big Five	r	p
Fire	Extroversion	−0.082	0.49
Earth	Conscientious	−0.006	0.27
Air	Intellect	−0.074	0.31
Water	Neuroticism	0.050	0.94

Timed birth charts against the Four Astrological Elements and four of the Big Five personality dimensions. *From data collected by Keith Burke.*

The funding, research facilities, and right-mindedness (to say nothing of a promising hypothesis) had been to no avail. Unlike Niehenke and Harris, Burke had already stopped reading charts for clients, a decision helped by his concerns about people looking not for counseling but for major life answers that a chart cannot give. He is now a clinical psychologist and a professor of behavioral sciences, and he does not use astrology in his profession or personal life (Burke 2015).

The case for and against astrology can now be briefly stated. Since thirty years ago, the case against has become stronger. The case for remains unchanged.

Cases for and against Astrology

Astrology is among the most enduring of human beliefs and has undisputed historical importance. A warm and sympathetic astrologer can provide wisdom and therapy by conversation with great commitment that in today's society can be hard to find. To many people, astrology is a wonderful thing: a complex and beautiful construct that draws their attention to the heavens, making them feel they are an important part of the universe. However, to their discredit, astrologers fail to recognize astrology's many problems. They refuse to accept that experience can be unreliable; they brush aside negative evidence; and they dismiss critics as close-minded by definition. As a result, astrologers are promoting both an illusion and a deceit. They are astrology's own worst enemies. Ultimately, the issue is a personal one—whether factual truth is to be more important than personal meaning. Skeptics will no doubt have thoughtful responses to that one.

Answer to the Title Question

The tests outlined here lead to the same answer as do hundreds of other tests. They confirm that nothing in a birth chart is sufficiently true to support the meanings claimed by astrologers. Their books, classes, and conferences are not built on evidence but on opinions based on opinions based on opinions, thus perpetuating the seeing of faces in clouds. Millennia have not wearied them.

So the answer to the thirty-year-old question in the title remains the same. No, astrology does not need to be true in order to seem to work. It is simply a time-honored cover for artifacts that better explain the outcomes. Astrologers have had ample opportunity to prove otherwise by controlled tests but have not done so, a failure most easily explained by their being unable to do so. As a consequence, astrologers should not be surprised if they find themselves disqualified from positions of credibility in Western society.

Nevertheless, depending on who we are, we can still see astrology as beautiful, spiritual, helpful, controlling, lucrative, great fun, or simply stupid. But one final question.

How to React?

French social scientist Laurent Puech (2003, 267), in a book-length study of the pretensions of astrology, suggests that the best reaction to astrology lies in the provision of reliable information and critical tools:

> Whether we like it or not, astrology and recourse to astrologers is here to stay. I think they will never disappear because they fill a need. They will be simply more or less important according to the times. How to react? . . . [It is] not a question of censuring astrology but of helping people to find reliable information about it, and also to find the minimum critical tools for evaluating it.

The problem for astrologers who wish to promote their invalid views of astrology is how to stop people from finding out the truth, even though some may see astrology as having more to it than being true or false.

Geoffrey Dean, PhD, is a British technical editor living in Perth, Western Australia. He is a fellow of the Committee for Skeptical Inquiry and a contributor to four Prometheus Books, including Skeptical Odysseys. He and his associates are best known for critical studies into astrology since the 1970s and the skeptical website www.astrology-and-science.com.

Postscript

TOWARDS UNDERSTANDING ASTROLOGY

Testing astrology is one thing. Understanding it is something else because people need to know they are getting *all* the evidence from both sides with nothing awkward left out. If you have no understanding of how astrology can offer what science does not—a psychologically meaningful link between individuals and the cosmos (or so astrologers claim)—you are open to being led astray. So in October 2022 the authors of *Tests of Astrology* published *Understanding Astrology*, see References. Among other things it increased the number of studies reviewed to more than a thousand. It could also be downloaded for free. But did it change the previous conclusion? Here we need to start at the beginning.

Nearly forty years ago the British parapsychologist Carl Sargent (1986) described the then state of astrological research as "a nightmare. It consists almost entirely of one-off studies. . . . Almost nothing can be concluded from this research, since independent replications with standardized procedures are wholly lacking." Today the message from *Understanding Astrology* is that little has changed, And there are still many reasons why people believe in astrology and why it does not need to be true.

RESEARCH METHODS HAVE CONTINUED TO ADVANCE

Here are two remarkable examples. (For full details see *Understanding Astrology* pages 259–260 and 749–752.) First, a US professor of information science analysed many thousands of Twitter and Facebook data to see if sun signs diffentiated between people, all without assuming sign meanings or asking astrology-related questions (which even ten years ago would have been impossible). The results showed convincingly that sun signs do not differentiate people by personality, which again makes sun signs the most disconfirmed factor in astrology.

Second, in India a team of information technologists spent ten years comparing large samples of exactly opposite cases, each with a decisive impact on life (Cancer vs No Cancer, Divorced vs Long Married, Retarded vs Intelligent, Celebrity vs Ordinary). Each case was described by 136 Indian chart factors said to be specific to each case. Although each case was claimed to be easily visible in Indian birth charts, and exactly opposite cases could hardly be more different, the difference (mean effect size –0.000 sd 0.028) wasn't even slightly better than tossing a coin. For an area so far without hard evidence, it was a giant leap forward.

A CONTINUING PUZZLE

Nevertheless research astrologers continue to report effect sizes that, although small (typically < 0.1), are often statistically highly significant. It implies that some small parts of astrology are genuinely true. The puzzle is how can critics honestly imply otherwise.

Sometimes the answer is obvious. For example astrologer Ken McRitchie (2022) in *Journal of Scientific Exploration* gives a meta-analysis claiming to show that astrology is producing significant results of high quality. Except it is not a genuine meta-analysis (one that uses all relevant studies and a secondary statistical analysis) but is merely a list of ten cherry-picked studies. Hardly a *scientific* exploration.

Otherwise the answer is surprisingly simple. Methods of control in astrology—such as shuffling birth data—cannot tell the difference between effects due to *astrology* and those due to *prior knowledge* of astrology. If your birth chart contains X and your behavior is X-ish, is it because your birth chart contains X (hooray for astrology) or because you know it contains X (hooray for role-playing)? Either way neither you nor tests can tell who the winner is. And it doesn't end there.

PRIOR KNOWLEDGE CANNOT BE DISMISSED AS UNLIKELY

Today "there is no area of human existence to which astrology cannot be applied" (Parker and Parker, 1975). Astrology is also sufficiently popular for its effect on self-image to be a confound in every population used to test it. You don't even need to be aware of your prior knowledge because forgotten images can still operate subconciously to cue your behaviour (Bargh, 2017). So prior knowledge cannot be dismissed as unlikely.

However, given that astrology has no known physical explanation (gravity, magnetism, radiation, quantum effects), and that prior knowledge is more plausible than astrology (which is effectively mysterious principles unknown to science), it follows that effect sizes in controlled tests of astrology should be about the same as those in tests of self-image. No formal large-scale comparison has been made, but their similarity is apparent in the many effect sizes reported in *Understanding Astrology*.

For example pages 712 and 726 report ten effect sizes averaging 0.020, maximum 0.030, for sample sizes averaging 20,000, range 4000 to 65,000, all controlled by shuffling, all with p-values of 0.05 through 0.001 and therefore all seen by astrologers as evidence for astrology. But all are less than the effect size of 0.07 for prior knowledge based on the meta-analysis of sun signs vs extroversion, so all are more plausibly seen as evidence for prior knowledge. Although the effect size for prior knowledge is likely to depend on situation—immediately after reading sun sign descriptions 0.32 was recorded by 55 students (Delaney and Woodyard, 1974)—its decisive importance has not been recognised until now.

Conclusion

The previous conclusion (astrology doesn't need to be true) still holds but with an important qualifier. Unless prior knowledge of astrology has been conclusively shown to be absent (and no amount of standard controls or shuffling will do this), claims that positive effect sizes reflect genuine astrological effects will be premature. Positive effect sizes are more plausibly explained by prior knowledge.

Nevertheless astrology still deserves a place in our future. Even though some may see astrology as having more to it than being true or false, in terms of longevity and ongoing popularity it has a clear edge over other questionable beliefs. For every student of pseudoscience astrology would seem to be a good place to start.

References

Alexander, R. 1983. *The Astrology of Choice*. New York, NY: Weiser.

Burke, K. 2012. Big five personality traits and astrology: The relationship between the moon variable and the NEO PI-R [Big five personality inventory]. PhD thesis, Pacifica Graduate Institute, California.

————. 2015. Personal communication with the author.

Carter, C. 1925. *The Principles of Astrology*. London: Theosophical Publishing House.

Dean, G. 2002. Is the Mars effect a social effect? *Skeptical Inquirer* 26(3) (May/June): 33–38.

Dean, G., and I. W. Kelly. 2003. Is astrology relevant to consciousness and psi? *Journal of Consciousness Studies* 10(6, 7): 175–198.

Dean, G., A. Mather, D. Nias, et al. 2016. *Tests of Astrology: A Critical Review of Hundreds of Studies*. Amsterdam: AinO Publications. For details, visit http://www.astrology-and-science.com.

Elwell, D. 1987. *Cosmic Loom: The New Science of Astrology*. London: Unwin Hyman

————. 1991. Can astrology win through? *Astrological Journal* 33(1): 49–50.

Harris, P. 2005. Applications of astrology to health psychology. PhD thesis, School of Social Sciences, University of Southampton.

————. 2008. Managing fertility treatments and stress with astrology. *Journal of Reproduction, Sexuality, and Menopause* 6(3): 43–44.

Narlikar, J. V. 2013. An Indian test of Indian astrology. *Skeptical Inquirer* 37(2) (March/April): 45–49.

Niehenke, P. 1984. The validity of astrological aspects: An empirical inquiry. *Astro-Psychological Problems* 2(3): 10–15.

————.1987. *Kritische Astrologie, 194*. Freiburg: Aurum.

Patterson, A. 1991. Why astrology is so hard to learn. *Considerations* 6(3), 5–13.

Puech, L. 2003. *Astrologie: Derrière les mots* [*Astrology: Behind the Words*]. Sophia Antipolis: Éditions book-e-book.

Voas, D. 2007. Ten Million Marriages: A Test of Astrological Love Signs. https://astrologiaexperimental.files.wordpress.com/2014/05/voas astrology.pdf. A version without tables is available in *Skeptical Inquirer* 32(2): 52–55, Ten million marriages: An astrological detective story.

Postscript References

Bargh, J. 2017. *Before You Know It: The Unconscious Reasons We Do What We Do.* New York: Touchstone. The result of more than thirty years of research.

Dean, G., A. Mather, D. Nias, and R. Smit. 2022. *Understanding Astrology: A Critical Review of a Thousand Empirical Studies 1900–2020.* Amsterdam: AinO Publications, 948 pages, nearly 1500 illustrations, 4000 references, Open Access, hardcover copies with sewn binding (lies flat) are available only from the publisher. Or download it for free from www.astrology-and-science.com/U-aino2.html/.

Delaney, J. G., and H. D. Woodyard. 1974. Effects of reading an astrological description on responding to a personality inventory. *Psychological Reports* 34, 1214.

McRitchie, K. 2022. How to think about the astrology research program: An essay considering emergent effects. *Journal of Scientific Exploration* 36(4), 706–716.

Parker, J., and D. Parker. 1975. *The Compleat Astrologer.* London: Mitchell Beazley, 81. A classic best-selling textbook

Sargent, C. 1986. In Modgil, S. and C. (eds.), *Hans Eysenck: Consensus and Controversy.* Philadelphia and London: Falmer, 347–348.

CHAPTER 38

Hans J. Eysenck

THE DOWNFALL OF A CHARLATAN

David F. Marks

Vol. 44, No. 6
November/December 2020

I write this article as a long-term investigator into psychology, health-related behavior, and claims of the paranormal. The article concerns an orchestrated saga of intellectual dishonesty by Professor Hans J. Eysenck, late of King's College London, and two of his acolytes: medical sociologist and therapist Ronald Grossarth-Maticek and the late Cambridge University psychologist Carl L. Sargent. This story will send shudders down the spines of many other Eysenck acolytes, but it is a story that must be told. The scientific record needs to be corrected and Eysenck's false claims excised.

Dramatis Personae

At the time of his death in 1997, Eysenck was the third most cited psychologist in the world—only slightly behind Sigmund Freud and Jean Piaget. Eysenck's career was mired in controversy for his racist views on the genetics of intelligence, his acceptance of tobacco company funding to support his claim that personality is a more significant cause of cancer and heart disease than smoking, his advocacy of the tenuous scientific evidence on astrology, and his one-sided, credulous defense of parapsychology.

The three parties to this story practiced wishful thinking, bias, and willful deceit in equal measure. The main actor, Hans J. Eysenck, today stands accused of dozens of "unsafe" publications (more on that presently) and of fake science on an industrial scale. The second, Ronald Grossarth-Maticek, served as a kind of "sorceror's apprentice," providing Eysenck with copious quantities of impossible data; he remains in practice in Heidelberg, Germany. In return, Eysenck provided Grossarth-Maticek with an unvalidated certificate of qualification in behavior therapy (see https://www.ruprecht.de/2020/01/28/documents-from-the-property-of-ronald-grossarth-maticek/). The third, Carl L. Sargent, hero-worshipped Eysenck but, following credible accusations of fraud, had to leave his post at Cambridge University to seek another career in role-playing computer games. Meanwhile, Sargent had written four coffee-table books with Eysenck acclaiming the wonders of parapsychology and psychic powers.

Before we get to Eysenck's and Sargent's support of parapsychology, a much greater scandal has hit the headlines: allegations of data manipulation and fake scientific and medical claims in multiple dozens of "unsafe" publications. The journal that I edit, the *Journal of Health Psychology*, published Anthony Pelosi's (2019) exposé of a series of impossible findings that had been reported by Eysenck in the 1980s. My supporting editorial included an Open Letter to the Principal of King's College London, Professor Byrne, seeking a full investigation into H. J. Eysenck (Marks 2019). King's College responded by running an enquiry that reached the conclusion that twenty-six of Eysenck's publications were "unsafe" (King's College London 2019). To date, there have been fourteen retractions and seventy-one "expressions of concern" on papers dating as far back as 1946 (for more on this, see the Retraction Watch database at retractiondatabase.org). News of the scandal broke in the *Guardian* (Boseley 2019) and in *Science* (O'Grady 2020). The story doesn't end here.

Intellectual Dishonesty

Hans Eysenck's intellectual honesty was a subject of concern over many decades since suspicions surfaced in the 1960s. In those days, Eysenck's book *Uses and Abuses of Psychology* was required reading for all psychology undergraduates, and it sold millions of copies. Little did we imagine that the great man himself would later be proved to be one of psychology's principal abusers! One of my lecturers, Dr. Vernon Hamilton, revealed something about Eysenck's laboratory ways.

Hamilton privately told me that his ex-boss, Eysenck, *had cheated doing his data analyses.* Hamilton didn't stay long and was asked to leave the institution. This ultimately led to public conflict with Hamilton (Eysenck 1959). Similar unsettling concerns were raised by others working for or with Eysenck, and rumors circulated over several decades.

Anthony Pelosi (a psychiatrist at Priory Hospital Glasgow), with Louis Appleby, had critiqued Eysenck's research in the early 1990s, but no action to sanction Eysenck was taken. The only society of professional psychologists in Britain, the British Psychological Society, washed its hands of Eysenck. The Society had got its fingers burned investigating the earlier massive data fraud by Sir Cyril Burt, and it was unwilling to risk another scandal by investigating Eysenck—who purely coincidentally(?) was Burt's most famous student. Additionally, the 2019 King's College investigation was half-baked. The unnamed investigating committee ignored Eysenck's multiple single-authored publications even though they were based on the very same datasets supplied by Grossarth-Maticek, craftily diverting the blame away from Eysenck toward his ever-willing apprentice (Marks and Buchanan 2020).

Richard Smith, retired editor of the *British Medical Journal*, astutely remarked:

> When forensic accountants detect fraud they assume that everything else from that person may well be fraudulent. Scientists tend to do the opposite—assuming that everything is OK until proved to be fraudulent. But as proving fraud is hard, lots of highly questionable material remains untouched. . . . I think of the example of R. K. Chandra, who was eventually found guilty not only of research fraud but also of financial and business fraud. His first paper established to be fraudulent was in 1989. Why, I ask myself, would you start being honest after you'd practiced fraud—yet hundreds of his papers are left unremarked, including unfortunately some that have been shown to be fraudulent. (Smith 2020)

A reliable source and long-time colleague of Eysenck states: "Eysenck was a mendacious charlatan. I base that not so much on his published fiction but his denial of the link between smoking and cancer was pernicious. His espousal of the beliefs of the John Birch Society was egregious . . . a grant had to be withdrawn and several researchers dismissed." Not good.

Hans Eysenck's Books on Parapsychology

A profile of Hans Eysenck based on his biography by Rod Buchanan and also on his parapsychology books with Carl Sargent provides insights into Eysenck's intellectual values as a scientist and scholar. There were four books with Sargent, all having Eysenck as first author: *Explaining the Unexplained: Mysteries of the Paranormal* (1982), *Know Your Own Psi-Q* (1983), *Explaining the Unexplained: Mysteries of the Paranormal* (2nd Ed.) (1993), and *Are You Psychic?* (1996). The collaboration between the two authors began in the early 1980s in Sargent's heyday at Cambridge and continued until 1996. These four books present a distorted and strongly biased view that psychic powers are scientifically proven, when in fact evidence suggests exactly the opposite (Marks 2020).

Eysenck's and Sargent's naiveté and credulity are everywhere apparent. In one of their books, Sargent and Eysenck argued that the experiments of William Crookes with the medium Daniel Dunglas Home were evidence of supernatural powers. Yet critics such as Victor Stenger noted that "Crookes gullibly swallowed ploys such as this and allowed Home to call the shots . . . his desire to believe blinded him to the chicanery of his psychic subjects" (Stenger 1990, 156–157).

Eysenck and Sargent present a totally one-sided view of the scientific evidence on psi. They affect the naive and frankly idiotic stance that fraud and trickery do not need to be considered. David Nias and Geoffrey Dean (1986) summarized their criticisms of the Eysenck/Sargent books thus: "the failure of Eysenck and Sargent's books to cover trickery and credulity is a serious deficiency" (368).

These books are among the most distorted and misleading accounts of parapsychological phenomena ever published by academic psychologists. The four books are a total disgrace, and how Eysenck had the gall to put his name to them—perhaps only to build his reputation as the fearless contrarian—is beyond imagination. In addition to the terrible scholarship, there is convincing evidence of scientific fraud by Sargent. How much Hans Eysenck knew about this, we will never know because Eysenck's wife destroyed his papers after his death. However, Susan Blackmore's report on Sargent's fraud became public knowledge several years into the collaboration and years before the third and fourth books with Eysenck were published.

If Sargent had kept his trickery hidden from Eysenck, then Eysenck could have been an innocent party. In a partnership built over more than fourteen years, surely there would have been a conversation that included a question of the kind, "Oh, I hear you left Cambridge, why was that?" If, as seems likely, Sargent admitted the occurrence of some kind of experimental "error," then Eysenck would have been party to covering up Sargent's deceit. Did Eysenck imagine nobody would notice? Or perhaps he simply did not care. After all, that great Cambridge genius, Isaac Newton, had done the same kind of thing, and Eysenck saw no problem with a bit of data fudging. According to Eysenck, a genius does whatever is necessary to prove his or her theories, as he stated in one of his many potboilers.

Chronology of Events

A chronology of events shows how the career of this aspiring parapsychologist unfolded:

- 1979: The University of Cambridge awards Sargent a PhD, which he claims was the first awarded to a parapsychologist by this university.
- 1979: The Society for Psychical Research provides a grant to Susan Blackmore enabling her to visit Sargent's lab at Cambridge in November. The original plan was to visit for a month. However, Blackmore was only able to stay

eight days from November 22–30, 1979. Blackmore summarized her visit to Sargent's lab this way:

> [Sargent's Ganzfeld] research was providing dramatically positive results for ESP in the GF and mine was not, so the idea was for me to learn from his methods in the hope of achieving similarly good results. . . . After watching several trials and studying the procedures carefully, I concluded that CS's [Carl Sargent's] experimental protocols were so well designed that the spectacular results I saw must either be evidence for ESP or for fraud. I then took various simple precautions and observed further trials during which it became clear that CS had deliberately violated his own protocols and in one trial had almost certainly cheated. I waited several years for him to respond to my claims and eventually they were published along with his denial. (Harley and Matthews 1987; Sargent 1987; see also Marks 2020).

- 1980: Sargent writes a monograph, *Exploring Psi in the Ganzfeld.* Parapsychological Monographs No 17.
- 1980: Sargent, Harley, Lane, and Radcliffe publish "Ganzfeld Psi Optimization in Relation to Session Duration" in *Research in Parapsychology.*
- 1981: Sargent and Matthews publish "Ganzfeld GESP Performance in Variable Duration Testing" in *Journal of Parapsychology.*
- 1982: Eysenck and Sargent publish their first book together, *Explaining the Unexplained: Mysteries of the Paranormal.*
- 1983: Eysenck and Sargent publish their second book, *Know Your Own PSI-Q.*

The Inevitable Downfall

- 1984: The Parapsychological Association Council asked Martin Johnson to head a committee to investigate Blackmore's accusation of fraud by Sargent. My book *Psychology and the Paranormal* describes what happened next: "The Parapsychological Association (PA) invited CS to provide an account of the 'errors' that SB [Susan Blackmore] had reported, but he declined to offer any explanation. The PA President, Stanley Krippner, wrote to CS at four different addresses, but still received no reply. The PA's 'Sargent Case Report' dated 10 December 1986 found that, in spite of strong reservations about CS's randomisation technique, there was insufficient evidence that CS had used unethical procedures."

- Sargent was "reproved" for failing to respond to the Parapsychological Association's request for information. However, Sargent had allowed his membership to lapse through non-payment of dues and was informed that, should he wish to renew his membership, his application would be considered with "extreme prejudice," i.e., Sargent would be unlikely to be readmitted as a member.
- The final report of this committee reprimanded Sargent for failing to respond to their request for information within a reasonable time.
- 1985: Sargent leaves Cambridge University and the parapsychology field (stated in the 2nd edition of *Explaining the Unexplained: Mysteries of the Paranormal*, 1993). Sargent moves into full-time authoring of game-books.
- 1987: Susan Blackmore's 1979 "A Report of a Visit to Carl Sargent's Laboratory" is finally published in *Journal of the Society for Psychical Research*.
- 1993: Undeterred by the report of cheating, Eysenck and Sargent publish their third book, *Explaining the Unexplained: Mysteries of the Paranormal*. (2nd ed.)
- 1996: Eysenck and Sargent publish their fourth book, *Are You Psychic?: Tests & Games to Measure Your Powers*, a revised version of *Know Your Own Psi-Q*.

The Hidden Truth

Two editions of the book by H. J. Eysenck and Sargent (1982, 1993) raise questions about how much Eysenck knew of the fraud accusations against Sargent in Blackmore's Society for Psychical Research report of 1979. In the 1982 edition of the first book, the procedural problems with Sargent's Ganzfeld research are not even mentioned. In the 1993 edition, the authors refer to "spirited exchanges on GF research" between Blackmore, Sargent and Harley (189).

However, the Ganzfeld evidence of psi is described by them as "very, very powerful indeed." They do not mention the accusations of fraud against Sargent, his departure from Cambridge University, or his repeated noncooperation with the Parapsychology Association enquiry. I obtained an update from Susan Blackmore on her current thinking about her thirty-year-old allegation of fraud by Sargent and on psi research more generally, which I reproduce below. Here are Susan Blackmore's answers to a few specific questions:

> DM: Do you think, in the light of everything that has come to light, CS committed fraud at Cambridge? (Ideally, a yes or a no).

> SB: Yes, at least on one specific trial.

DM: Do you think CS knowingly deceived anybody (including possibly himself) or was he simply a victim of confirmation bias/ subjective validation?

SB: The former.

DM: Is there anything else you would like to say about research on psi?

SB: In the light of my decades of research on psi, and especially because of my experiences with the GF, I now believe that the possibility of psi existing is vanishingly small, though not zero. I am glad other people continue to study the subject because it would be so important to science if psi did exist. But for myself, I think doing any further psi research would be a complete waste of time. I would not expect to find any phenomena to study, let alone any that could lead us to an explanatory theory. I may yet be proved wrong of course. (Blackmore 2019)

Conclusions

With this history, we can establish the following facts and conclusions.

1. A pattern of data manipulation in Hans J. Eysenck's and two of his collaborators' research is evident over several decades. The earliest paper of concern was published in the *Proceedings of the Royal Society of Medicine* in 1946. Following a recent exposure in the *Journal of Health Psychology* by Pelosi (2019) and an ensuing enquiry by King's College London (2019), journals have retracted fourteen of Eysenck's papers and published seventy-one expressions of concern.
2. A reliable source accused Eysenck of cheating with his data analyses as early as the 1960s. Eysenck's colleagues and PhD students at that time publicly critiqued Eysenck's laboratory methods, but no action was taken.
3. In the late 1970s and early 1980s, Eysenck formed a long-term collaboration with Cambridge parapsychologist Carl Sargent. Although Carl Sargent had been accused of fraud in 1979, Eysenck's publications with Sargent occurred over the period 1982–1993. These books grossly distorted the scientific evidence of the paranormal.
4. Anthony Pelosi and Louis Appleby (1992; 1993) and others raised serious questions about publications by Eysenck with R. Grossarth-Maticek. The authorities failed to respond.

5. Anthony Pelosi (2019) again voiced his concerns. My own editorial (Marks 2019) appealing to Kings College London to open an enquiry finally led to concrete action. Twenty-five publications by H.J. Eysenck and R. Grossarth-Maticek were deemed by Kings College London to be unsafe, yet the true number of unsafe publications is most likely well over 100.

6. As suspicions strengthened over a seventy-five-year period from the mid-1940s, torpor and complacency in the academic system enabled research malpractice to continue, not only Eysenck's, Grossarth-Maticek's, and Sargent's, but across the board.

7. The currently available systems for regulating research integrity and malpractice in the United Kingdom are an abject failure. A totally new approach is required. An Independent National Research Integrity Ombudsperson needs to be established to significantly improve the governance of academic research. Perhaps then, future Hans Eysencks can be stopped in their tracks.

David F. Marks is a professor of psychology at City University London.

References

Blackmore, S. 2019. Personal communication with the author (August 1).

Boseley, S. 2019. Work of renowned U.K. psychologist Hans Eysenck ruled 'unsafe.' *The Guardian* (October 11). Available online at https://www.theguardian.com/science/2019/oct/11/work-of-renowned-uk-psychologist-hans-eysenck-ruled-unsafe

Eysenck, H. J. 1959. Anxiety and hysteria—a reply to Vernon Hamilton. *British Journal of Psychology* 50(1): 64–69.

Harley, T., and G. Matthews. 1987. Cheating, psi, and the appliance of science: A reply to Blackmore. *Journal of the Society for Psychical Research* 54(808): 199–207.

King's College London. 2019. King's College London enquiry into publications authored by Professor Hans Eysenck with Professor Ronald Grossarth-Maticek. Available online at https://www.kcl.ac.uk/news/statements/hans-eysenck.

Marks, D. F. 2019. The Hans Eysenck affair: Time to correct the scientific record. *Journal of Health Psychology* 24(4): 409–420.

———. 2020. *Psychology and the Paranormal. Exploring Anomalous Experience.* London: Sage Publications.

Marks, D. F., and R. D. Buchanan. 2020. King's College London's enquiry into Hans J Eysenck's 'unsafe' publications must be properly completed. *Journal of Health Psychology* 25(1): 3–6.

Nias, D. K. B., and G. A. Dean. 1986. Astrology and parapsychology. In S. Modgil and C. Modgil (eds.). 2012. *Hans Eysenck: Consensus and Controversy.* Abingdon, UK: Routledge, 357–371.

O'Grady, C. 2020. Famous psychologist faces posthumous reckoning. *Science* 369(6501): 233–234. Available online at https://science.sciencemag.org/content/369/6501/233.full.

Pelosi, A. J. 2019. Personality and fatal diseases: Revisiting a scientific scandal. *Journal of Health Psychology* 24(4): 421–439.

Pelosi, A. J., and L. Appleby. 1992. Psychological influences on cancer and ischaemic heart disease. *British Medical Journal* 304(6837): 1295.

———. 1993. Personality and fatal diseases. *British Medical Journal* 306: 1666–1667.

Sargent, C. L. 1987. Sceptical fairytales from Bristol. *Journal of the Society for Psychical Research* 54(808): 208–218.

Smith, R. 2020. Personal communication with the author.

Stenger, V. 1990. *Physics and Psychics: The Search for a World beyond the Senses.* Amherst, NY: Prometheus Books.

Related Articles in *Skeptical Inquirer*

These previous articles deal, in part, with Eysenck collaborator Carl Sargent's questionable papers:

- Susan Blackmore, "Another Scandal for Psychology: Daryl Bem's Data Massage," November/December 2019
- Susan Blackmore, "Daryl Bem and Psi in the Ganzfeld," January/February 2018
- Susan Blackmore, "Psi in Psychology," Summer 1994
- Susan Blackmore, "The Elusive Open Mind: Ten Years of Negative Research in Parapsychology," Spring 1987

CHAPTER 39

Beware the Naturopathic Cancer Quack

Britt M. Hermes

Vol. 44, No. 2
March/April 2020

Naturopathic medicine is not any kind of medicine, and its practitioners are nothing short of quacks. I know because I used to be one. Herein lies my story.

Six years ago, I was a quack.

From 2011 until 2014, I was a licensed naturopathic doctor in the United States. Thanks to poor legislation in many states, I could advertise as a primary care physician, write prescriptions, and bill insurance companies. I called myself "Dr. Britt." In Arizona, I worked outside the health insurance system and catered to a patient population interested in aligning chakras and detoxing with diets and enemas. While I was more drawn to using *real* medical treatments rather than esoteric therapies, I was happy to be employed by the renowned naturopathic cancer "specialist" who owned the clinic.

I likely would have gone the entirety of my career as a naturopath if I hadn't unexpectedly found myself in a nerve-rattling situation. I quit after discovering my boss had been importing a non–Food and Drug Administration (FDA) approved drug called Ukrain and administering it to cancer patients (Thielking 2016; Arizona Naturopathic Physicians Medical Board 2015). His patients would get regular intravenous injections of Ukrain despite, in some cases, being discharged by their medical oncologists due to the severity of their disease. Others had chosen him for first line treatment. They flocked to him because his credentials included FABNO (Fellow of the American Board of Naturopathic Oncology).[1]

It didn't strike me as unusual that the packages, containing dozens of old-fashioned glass ampules, were marked with foreign postage. I assumed patients had been given informed consent, as they were spending thousands of dollars on the treatments. We all trusted him. One day the shipments stopped. My boss said, "They were probably confiscated."

It is potentially a federal crime to give an unapproved medication to patients (Friedman 2018; FDA 2019). Under my boss's orders, I had assisted with the administration of Ukrain, which is advertised as a natural cancer drug that cures almost any type of cancer and does not harm healthy cells (Nowicky 2019). Although this type of marketing is generally a red flag for a bogus cure, Ukrain does contain compounds that seem to show anti-cancer activity (Ernst and Schmidt 2005). However, it is also associated with life-threatening side effects, such as liver failure, tumor bleeding, and bone marrow toxicity (Boehm et al. 2015; Gansauge et al. 2002; Stickel et al. 2001; Moro et al. 2009). During the time that our patients had been receiving intravenous injections of Ukrain, its manufacturer had been arrested in Austria for charges of fraud—he had been relabeling and continuing to sell expired product ("Ukrainian Chemist" 2015). In 2016, the accused manufacturer was found guilty of serious commercial fraud and sentenced to three and a half years of imprisonment in Vienna, but he has appealed the verdict, and court proceedings are still ongoing (Der Standard 2016).

I learned this information on a Friday. That same weekend, I submitted a complaint to the Arizona Naturopathic Board of Examiners. Monday, I reported the events to the state's attorney. That afternoon, I confronted my boss and quit. Later that same evening, I received a call from a respected leader in the naturopathic profession. He urged me not to go to the authorities, pleading for me to reconcile with my boss because I was a "*naturo*-path, after all."

Was this code for dismissing my ethical duty to report a crime in honor of some kind of great, naturopathic philosophy? From this conversation, I lost my faith. This event forced me to confront the reality that patients whom I had come to deeply care about were being cheated out of money and duped into thinking that a so-called "natural" drug would help or even cure them.

I spent the next few months seeking out information that was scientifically critical of naturopathy and alternative medicine. I began scrutinizing my naturopathic education and training and, in parallel, learning more about evidence, standards of care, and medical ethics. After reading some great books and blogs, sending emails to leaders in the skeptical community—including Dr. Edzard Ernst and Jann Bellamy—and after a lot of tears, I concluded that naturopathic medicine is a scam. I was a fake doctor. How could I have been so wrong?

What Is Naturopathy?

Naturopathy is based on magic. Its origins trace back to a late nineteenth-century belief in vitalism (Whorton 2002; Brown 1988). In this worldview, all living beings embody a sort of supernatural energy force that is responsible for making you healthy or sick. Naturopaths refer to this power as the *vis* (Micozzi 2018). Other systems of pre-modern medicine have similar ideas about healing energies. In traditional Chinese medicine, it is called *qi*. In Ayurvedic medicine, it is called *prana* (Tabish 2008).

In naturopathy, the *vis* is manipulated using treatments that are based on food, water, plants, minerals, and physical contact. The mainstay naturopathic therapies are nutrition, hydrotherapy, herbal medicine, spinal manipulation, homeopathy, and traditional Chinese medicine. Some naturopaths utilize methods and substances that are more familiar to modern medicine, including intravenous injections and pharmaceutical drugs.

Today, naturopathy has been rebranded as naturopathic medicine. Naturopaths call themselves naturopathic doctors, or NDs for short. The profession has been somewhat successful in passing itself off as a credible medical specialty that claims to blend the best of both modern and traditional medicine. To get here, naturopaths aggressively and relentlessly lobby for laws to allow them to be licensed as medical practitioners. Almost half of the U.S. states license or regulate naturopaths, and five Canadian provinces regulate naturopathy. Current scopes vary wildly. In Arizona, an ND is considered a "physician" and can prescribe controlled substances and perform minor surgeries. In Alaska, an ND is restricted to providing nutritional advice, counseling, herbs, homeopathy, and physical therapies. Contrary to what naturopaths will tell you, regulating the naturopathic profession does not make the practice of naturopathy safe (Hermes 2017; Hermes 2016b; Weeks 2016).

A licensed ND needs to have graduated from a program accredited by the Council on Naturopathic Medical Education (CNME), which is granted programmatic accrediting status by the U.S. Department of Education. Many—especially NDs—often confuse CNME accreditation with government endorsement. In fact, CNME has accrediting power merely because it meets administrative criteria, *not* because the naturopathic curriculum is medically validated or scientifically sound. Tellingly, naturopaths accredit their own programs while shielding the process from outsiders.

Based on my experience, accredited naturopathic programs play fast and loose with biomedical science. Much of the curriculum is homeopathy, herbalism, hydrotherapy, craniosacral therapy, chiropractic manipulation, and naturopathic philosophy. In fact, when I started my master's of science program in Germany, which is ironically the birthplace of naturopathy, I was required to retake science courses, including anatomy, histology, and immunology.

Clinical training is mandated by the CNME to be no less than 850 hours of direct patient care (Council on Naturopathic Medical Education 2016). I reviewed my clinical training handbook and discovered this benchmark is attainable only through accounting tricks. For example, time counted when students reviewed a case with their peers or when we observed advanced ND students perform physical exams.

ND students are not trained in the same standards of care as medical students. It was common to hear instructors clinically discuss a patient's "vital force" as if this could somehow be detected and improved. Frequently, patients were diagnosed with dubious food allergies,[2] chronic Lyme disease,[3] or adrenal fatigue.[4] Patients would be prescribed an assortment of herbs and supplements—conveniently sold at the clinic's dispensary.

A residency is not required for naturopaths to practice any specialty of medicine, including oncology (Association of Accredited Naturopathic Medical Colleges 2019). After a naturopathic student graduates, he or she may move directly into practice after passing a two-part licensing exam that primarily focuses on naturopathic therapies, such as homeopathy and herbs. A small number of NDs voluntarily complete residencies, which are essentially encores of naturopathic clinical training. These positions are typically two years and are sponsored by naturopathic clinics that provide naturopathic treatments, including alternative cancer therapies, in an outpatient setting. It is not uncommon for naturopaths who call themselves "cancer specialists" to start practicing directly after passing their licensing exams.

During my time in naturopathic school, students took one lecture course (about twenty classroom hours) in cancer care. In my course at naturopathic school, I was taught that no natural substance has been proven to cure cancer. Despite this disclaimer, it is easy to find naturopaths advertising natural cancer cures.

Sued for Defamation by a Naturopathic Cancer "Specialist"

In early 2015, I started blogging at *Naturopathic Diaries* in an effort to help protect patients and other students from falling prey to naturopathy. I quickly became the naturopathic profession's number one enemy.

Sometime in 2016, a webpage misrepresenting my position on naturopathic medicine appeared. The site didn't post untrue statements about me or say anything negative. But it clearly made it seem as though I was an advocate for naturopathic medicine. I quickly learned that about half a dozen domains using versions of my name had been anonymously registered. Some of these domains forwarded to the websites of the American Association of Naturopathic Physicians and the Association of Accredited Naturopathic Medical Colleges. I called the company hosting the brittmariehermes.com page and asked for the email address associated with the account. Surprisingly, a forthcoming representative gave me an email address at natonco.org, the website for the Naturopathic Cancer Society—of which a naturopath named Colleen Huber is president and founder (Hermes 2016a).

I had never heard of Colleen Huber prior to that moment. When I started to look into Huber, I got really sucked down deep into what felt like a pit of despair. Here is what I learned: Huber's Naturopathic Cancer Society is a non-profit organization that previously claimed to help naturopathic patients raise money for alternative cancer therapies (Huber 2016b; Huber 2017a). Huber also owns a naturopathic clinic called Nature Works Best in Arizona, where she treats cancer patients with intravenous baking soda, megadoses of intravenous vitamin C, and a strict sugar-free diet (Huber 2015). Huber advocates against state-of-the-art oncology, especially chemotherapy and radiation, because she thinks these therapies strengthen cancer (Huber 2016a). Huber's clinic website and charity website previously linked to each other, which provided the impression that a patient of Huber's could use her charity funds to pay for treatments at Huber's clinic (Huber 2017b; Huber 2019a).

Huber is also the secretary and founding member of a naturopathic ethics review board that goes by multiple names. I later learned, according to a writ submitted by Huber's attorneys, this ethics board approved the cancer research that Huber conducts at her clinic. I should note here that Huber does not use a standardized cancer treatment protocol on her patients (Huber 2015). It all looked sketchy, and truthfully Huber seemed like someone I did not want to legally tangle with. But my views were being misrepresented, and I felt like I needed to act.

In June 2016, I hired an American lawyer to send Huber and one of her colleagues a cease and desist letter, asking Huber to relinquish the domain names to me. The letter went unanswered. About six months later, I decided to take the fight public.

In a blog post from December 2016, I laid out my theory for why I thought Huber, or someone in her close orbit, had purchased the domains of my name. I also wrote about her cancer therapies, charity, and so-called cancer research. I was highly skeptical of the relationship between her charity and clinic, and I also questioned the ethics of her research protocol. Nine months after this post, with the help of a German law firm, Huber sent me a cease and desist letter. Huber wanted me to remove the entire blog post about her and to pay her legal fees, because she claimed I had defamed her.

I hired a German lawyer and responded to this letter. We refused to comply with Huber's demands. I had not done anything wrong, and I found Huber's claims ridiculous. I felt strongly that Huber's treatment methods and research were questionable and that the public was entitled to this information. In September 2017, Huber filed a lawsuit against me in Germany for defamation.

Huber's Cancer Research

Colleen Huber conducts research out of her clinic and claims to have gathered data on the efficacy of her naturopathic cancer treatments. She characterizes her research as "groundbreaking" and states that "no other clinic has such a high documented success rate" (Huber 2016a; Huber 2019b). Her claimed success rate has varied, but on a recent version of her website, she writes, "85% of patients who completed our metabolic treatments and followed our food plan went into confirmed or assumed remission" (Huber 2019b). Regardless of what exact percentage you go by, these results, if true, would be remarkable and are certainly worth examining. In fact, Huber invites the public to do just that. Below these stats, she writes, "You can verify this data for yourself. . . . We invite you to take a simple calculator and 'crunch the numbers,' to see if you come up with the same percentages that we do" (Huber 2019b).

I did look at the data. And Thomas Mohr, a cancer researcher at the Medical University of Vienna, did crunch the numbers. Mohr calculated an odds ratio of 2:1 (95 percent confidence interval 1.014.40, p<0.05) in favor of state-of-the-art therapy (chemo, radiation, etc.). As Mohr puts it, "Patients under natural care have more than a two-fold higher risk to die" (Hermes 2018).

Mohr wanted to do a more robust analysis, because, frankly, we believed that the data Huber gathered is inconsistent and of questionable quality. When Mohr removed questionable data, he found that the odds ratio was 10:1 in favor of state-of-the-art therapy, meaning the risk of death with Huber's therapies appears to be ten times higher compared to those patients who pursue conventional cancer treatments (Hermes 2018).

In the lawsuit, Huber contested Mohr's analysis. She felt that it was defamatory for me to publish the results and wanted these statistics removed from my blog. Huber took issue with several other points as well, including me calling her a quack, my characterization of the relationship between her charity and clinic, and the cybersquatting (Hermes 2018).

On May 24, 2019, the District Court of Kiel, Germany, ruled against Huber. Reading through the judgment, I really felt like the judge understood me and the point of my blog. The judge wrote:

> The effectiveness of naturopathic treatment is a highly controversial issue among experts. This is exactly what is intended by the Defendant's article. It is a concern for information in connection with an issue that is discussed controversially and emotionally in the public. In medicine, in particular, a comprehensive and critical approach is required so even medical laymen can get an idea of different treatments and approaches. By making critical and partly exaggerated

statements the Defendant intends to initiate a discussion about treatments she regards as dubious.

The judge determined my statements were protected speech under Article 5(1) of the German constitution. The deadline has now passed for Huber to appeal, which means Huber cannot sue me again for these points in Germany, nor can she appeal at a later date.

Natural Cancer Therapies Kill

Even though I have won this battle, patients are still at risk of losing. A tragic example of this is Huber's former associate Rebecca Stephan. I found Rebecca's story through a GoFundMe page set-up by her sister-in-law (C. Stephan 2017). In 2016, Rebecca Stephan was diagnosed with mucinous ovarian cancer. She had surgery to remove the tumor and then was hospitalized due to complications. Her oncologist recommended she start chemotherapy right away. However, Stephan and her family didn't like this treatment plan. She was frail after surgery, and her family worried chemotherapy might kill her. So Stephan began to think about alternatives, writing in a blog post from August 2017: "My husband and I began to look elsewhere for natural treatment options and flew down to see Colleen Huber, NMD, Medical Director of Nature Works Best Cancer Clinic in Tempe, AZ. As soon as I met her I knew this is where I wanted to be treated. *Also, the percentages of doing treatment with Doctor Huber were better for me than with my oncologist*" (R. Stephan 2017; emphasis added).

Stephan moved her family, including her three kids, from Colorado to Arizona so that she could get weekly intravenous injections of baking soda and vitamins, a strict diet of no sugar, and supplements. Stephan claimed that after just three months of treatment, she was cancer free. She believed Huber's regimen cured her. Stephan then became the executive director for Huber's Naturopathic Cancer Society, which, in her own words, is a charitable organization "dedicated to raising funds for those who have cancer." She seemed keen to help other cancer patients afford the same naturopathic therapies as she did.

The latest update on Stephan's GoFundMe, from about a year and a half ago, reported that she was not doing well: "Friends it's with a heavy heart that I tell you Rebecca has cancer again! She is not doing well and the cancer has moved to other parts of her body! She has asked me to get as many people praying for her as we can. She is in a lot of unbearable pain and getting weaker."

It is important for cancer patients to understand the risks associated with using alternative therapies to treat cancer. The reality is that patients who use alternative therapies to treat cancer are more likely to die. There is clear scientific evidence to back this up. In 2018, Dr. Skyler Johnson and his colleagues at Yale University found that the use of alternative medicine to treat cancer is independently associated with a greater risk of death compared to conventional cancer care (Johnson et al. 2018a).

Later that same year, Johnson and colleagues published more disturbing results. They found that alternative medicine use by cancer patients was associated with refusing at least one component of conventional cancer care, which is not surprising in my opinion, and a two-fold greater risk of death compared with cancer patients who used conventional therapies alone (Johnson et al. 2018b). This may be due to delays in conventional care or refusal of specific therapies. The bottom line is this: If you mix conventional medicine with complementary care, there may be a higher risk of death.

This evidence is all here. I have to assume that Huber is ignoring it, perhaps because she is profiting off vulnerable cancer patients. She uses her own research and statistics to set up a bubble where she can convince patients to get potentially dangerous alternative therapies.

Thanking the International Skeptical Community

This defamation lawsuit has accomplished a few important things. First, it demonstrated the strength and the unity of the international skeptic community. Eran Segev of Australian Skeptics Inc. spearheaded a fundraiser to raise the money needed to cover my legal bills. Skeptics from around the world donated, and we raised over 100,000 euros. Over 50,000 euros was donated in just the first nine days of the campaign. I will not be out any money for this lawsuit. The legal fees that I was awarded from winning the lawsuit have been deposited back into this fund. Leftover money is being held for future cases in which skeptics are being legally bullied. We hope to establish an international permanent legal fund for cases like mine, in which the plaintiff would not be able to afford to defend themselves in court.

Without the financial and emotional support of this community, I might have given into Huber's demands. There was no way I could afford my legal fees, which were originally estimated by my lawyer to be about 50,000 euros. What I have learned from this experience is that skeptics are not lone wolves. We are a protective, and very clever, pack. I am very lucky and very proud to be part of this family.

Second, this lawsuit exemplified the Streisand effect. Huber chose to personally attack me. In doing so, she amplified the attention on her dubious cancer research, charity, and treatments. The international press picked up the story.

This lawsuit was written about, and Huber's actions were criticized all over the world.

This media attention also boosted my profile. The highlight of this attention was receiving the John Maddox Prize for my advocacy (see SI, March/April 2019). I want to especially thank Prof. Chris French and Dr. Chris Peters for nominating me for this award.

Finally, I think going through this lawsuit will help patients make informed and better choices regarding their cancer care. Hopefully Rebecca Stephan's story is a warning for other cancer patients. Since the lawsuit, Huber has edited the information presented on her charity and clinic websites. Huber now has a very stern disclosure statement on her clinic's homepage, in which she warns future patients that they may not have favorable results from her treatments.

Indeed, beware the naturopathic cancer quack.

Britt M. Hermes is a writer, scientist, and a former naturopathic doctor. She practiced as a licensed naturopath in the United States for three years and then left the profession after realizing naturopathy is pseudoscience. Her work focuses on the deceptions naturopathic practitioners employ to scam patients and contrive legitimacy in political arenas. Hermes's writings can be found at *Forbes, Science 2.0, KevinMD,* and *Science-Based Medicine.* She spoke on this topic at CSICon 2019. She hopes her stories will protect patients from the false beliefs and bogus treatments sold by alternative medicine practitioners.

Notes

1. Fellow of the American Board of Naturopathic Oncology (FABNO) is awarded by the American Board of Naturopathic Oncology (ABNO). The ABNO claims that the FABNO title is similar to board certification in oncology. To be eligible to sit for the FABNO exam, a naturopathic doctor must complete a two-year naturopathic residency in cancer care, submit medical cases for review by naturopathic examiners, and complete fifty continuing education credits approved by ABNO (Oncology Association of Naturopathic Physicians 2015). These qualifications are less stringent than those for medical doctors (American Board of Internal Medicine 2019).

2. Naturopaths commonly use IgG food allergy test panels to diagnose and treat symptoms that seem to be associated with foods. However, these tests are clinically invalid. Their use may lead to false diagnoses and unnecessary treatment (Moore 2019; Gavura 2018). To learn more, please visit the American Academy of Allergy, Asthma and Immunology webpage at www.aaaai.org.

3. *Chronic Lyme disease* (CLD) is a term commonly used to describe wide-ranging symptoms in people with various illnesses. These patients do not necessarily have clinical evidence of previous Lyme disease. The term is generally not accepted by the

conventional medical community. Naturopathic practitioners frequently diagnosis CLD and treat patients using intravenous injections, ozone, supplements, and long-term antibiotics. None of these therapies is recommended to treat CLD (Baker 2010; Feder et al. 2007). To learn more, please visit the National Institute of Allergy and Infectious Diseases webpage at www.niaid.nih.gov.

4. *Adrenal fatigue* is a term used by naturopathic and alternative medicine providers to describe a condition in which the adrenal glands are over-worked and run-down and, therefore, unable to work properly, resulting in excessive fatigue. This is a common alt-med diagnosis for highly stressed individuals. Scientific evidence to support the idea of adrenal fatigue is lacking, and the conventional medical community generally does not use this terminology (Felson 2019; Shah and Greenberger 2012). To learn more, please visit WebMD's webpage on adrenal fatigue at https://www.webmd.com/a-to-z-guides/adrenal-fatigue-is-it-real#1.

References

American Board of Internal Medicine. 2019. *Policies and Procedures* (October). Available online at https://www.abim.org/~/media/ABIM%20Public/Files/pdf/publications/certification-guides/policies-and-procedures.pdf.

Arizona Naturopathic Physicians Medical Board. 2015. Dr. Michael Uzick Board Disciplinary Actions. Available online at http://directorynd.az.gov/agency/pages/directorySearchDetail.asp?holderID=102.

Association of Accredited Naturopathic Medical Colleges. 2019. Post-Graduate Naturopathic Residencies. Available online at https://aanmc.org/naturopathic-residencies/.

Baker, Phillip J. 2010. Chronic Lyme disease: In defense of the scientific enterprise. *The FASEB Journal* 24(11): 4175–77. Available online at https://doi.org/10.1096/fj.10-167247.

Boehm, Katja, Edzard Ernst, and CAM-Cancer Consortium. 2015. Ukrain. Herbal Products. CAM-Cancer. CAM-Cancer (January 29). Available online at http://www.cam-cancer.org/The-Summaries/Herbal-products/Ukrain.

Brown, P. S. 1988. Nineteenth-century American health reformers and the early nature cure movement in Britain. *Medical History* 32(2):174–194. Available online at https://doi.org/10.1017/S0025727300047980.

Council on Naturopathic Medical Education. 2016. *2016 CNME Handbook of Accreditation for Naturopathic Medicine Programs* (April). Available online at http://www.cnme.org/resources/2016_cnme_handbook_of_accreditation.pdf.

Der Standard. 2016. Krebs-Heilmittel verkauft: 3,5 Jahre für Hersteller von 'Ukrain' (blog entry). *DER STANDARD* (May 23). Available online at https://www.derstandard.at/story/2000037504444/krebs-heilmittel-verkauft-3-5-jahre-fuer-hersteller-von-ukrain.

Ernst, Edzard, and K. Schmidt. 2005. Ukrain—a new cancer cure? A systematic review of randomised clinical trials. *BMC Cancer* 5(July): 69. Available online at https://doi.org/10.1186/1471-2407-5-69.

FDA. 2019. Actions & Enforcement. *FDA Office of Regulatory Affairs* (April 17). Available online at http://www.fda.gov/industry/import-program-food-and-drug-administration-fda/actions-enforcement.

Feder, Henry M., Jr., Barbara J. B. Johnson, Susan O'Connell, et al. 2007. A critical appraisal of 'Chronic Lyme Disease.' *New England Journal of Medicine* 357(14): 1422–30. Available online at https://doi.org/10.1056/NEJMra072023.

Felson, Sabrina. 2019. Adrenal fatigue: Is it real? (blog entry). *WebMD* (February 8). Available online at https://www.webmd.com/a-to-z-guides/adrenal-fatigue-is-it-real.

Friedman, Ronald. 2018. Importing prescription drugs remains risky business due to FDA and DEA regulation. *American Bar Association Health ESource* 14(7). Available online at https://www.americanbar.org/groups/health_law/publications/aba_health_esource/2017-2018/march2018/importing.html.

Gansauge, Frank, Marco Ramadani, Jochen Pressmar, et al. 2002. NSC-631570 (Ukrain) in the palliative treatment of pancreatic cancer: Results of a phase II trial. *Langenbeck's Archives of Surgery* 386(8): 570–74. Available online at https://doi.org/10.1007/s00423-001-0267-5.

Gavura, Scott. 2018. IgG food intolerance tests continue to mislead consumers into unnecessary dietary restrictions (blog entry). *Science-Based Medicine* (November 15). Available online at https://sciencebasedmedicine.org/igg-food-intolerance-tests-continue-to-mislead-consumers-into-unnecessary-dietary-restrictions/.

Hermes, Britt M. 2016a. Is dubious cancer 'doctor' Colleen Huber cybersquatting my name? (blog entry). *Naturopathic Diaries* (December 1). Available online at https://www.naturopathicdiaries.com/dubious-cancer-doctor-colleen-huber-cybersquatting/.

———. 2016b. Naturopathic medicine week and the problem of endemic quackery, like ozone therapy. *Forbes* (October 13). Available online at https://www.forbes.com/sites/brittmariehermes/2016/10/13/naturopathic-medicine-week-endemic-quackery-ozone-therapy/.

———. 2017. California has a deadly problem with regulating naturopathic doctors. *Forbes* (April 17). Available online at https://www.forbes.com/sites/brittmariehermes/2017/04/24/california-has-a-deadly-problem-with-regulating-naturopathic-doctors/.

———. 2018. I need your help: Naturopath Colleen Huber is suing me (blog entry). *Naturopathic Diaries* (January 13). Available online at https://www.naturopathicdiaries.com/need-help-naturopath-colleen-huber-suing/.

Huber, Colleen. 2015. Defeating cancer requires more than one treatment method: An 8-year retrospective case series using multiple nutritional and herbal agents, 2014 update. *Journal of Cancer Science & Therapy* 7(10). Available online at https://doi.org/10.4172/1948-5956.C1.059.

———. 2016a. Natural cancer treatments (blog entry). *Nature Works Best*. Available online at http://natureworksbest.com/.

———. 2016b. Naturopathic Cancer Society (webpage screenshot). *Nature Works Best*.

———. 2017a. About us (blog entry). *Naturopathic Cancer Society*.

———. 2017b. Donate—sponsor a patient, save a life (webpage screenshot). *Nature Works Best*.

————. 2019a. Doctors (blog entry). *Naturopathic Cancer Society*. Available online at https://www.natonco.org/doc.

————. 2019b. Nature works best (blog entry). *Nature Works Best*. Available online at https://natureworksbest.com/.

Johnson, Skyler B., Henry S. Park, Cary P. Gross, et al. 2018a. Use of alternative medicine for cancer and its impact on survival. *JNCI: Journal of the National Cancer Institute* 110(1). Available online at https://doi.org/10.1093/jnci/djx145.

————. 2018b. Complementary medicine, refusal of conventional cancer therapy, and survival among patients with curable cancers. *JAMA Oncology*. Available online at https://doi.org/10.1001/jamaoncol.2018.2487.

Micozzi, Marc S. 2018. *Fundamentals of Complementary, Alternative, and Integrative Medicine—E-Book*. Elsevier Health Sciences.

Moore, Andrew. 2019. The myth of IgG food panel testing (blog entry). *The American Academy of Allergy, Asthma & Immunology*. Available online at https://www.aaaai.org/conditions-and-treatments/library/allergy-library/IgG-food-test.

Moro, Paola A., Federica Cassetti, Gianni Giugliano, et al. 2009. Hepatitis from greater celandine (*Chelidonium majus* L.): Review of literature and report of a new case. *Journal of Ethnopharmacology* 124(2): 328–32. Available online at https://doi.org/10.1016/j.jep.2009.04.036.

Nowicky, Wassil. 2019. Ukrain (blog post). *Ukrainian Anti Cancer Institute*. Available online at https://ukrin.com/en.

Oncology Association of Naturopathic Physicians. 2015. FABNO Certification. Oncology Association of Naturopathic Physicians. Available online at https://oncanp.org/fabno-certification/.

Shah, Rachna, and Paul A. Greenberger. 2012. *Chapter 29: Unproved and Controversial Methods and Theories in Allergy-Immunology* (June). Available online at https://doi.org/info:doi/10.2500/aap.2012.33.3562.

Stephan, Charla. 2017. GoFundMe (blog entry). *Gofundme* (March 9) Available online at https://www.gofundme.com/f/3hk3tcw.

Stephan, Rebecca. 2017. Fighting cancer the natural way (blog entry). *Claire Preen's Healthy Blog* (October 27). Available online at https://web.archive.org/web/20171027121349/http:/clairepreen.com/2017/08/how-rebecca-fought-cancer-and-won-the-natural-way/.

Stickel, F., H. K. Seitz, E. G. Hahn, et al. 2001. [Liver toxicity of drugs of plant origin]. *Zeitschrift Fur Gastroenterologie* 39 (3):225–32, 234–37. Available online at https://doi.org/10.1055/s-2001-11772.

Tabish, Syed Amin. 2008. Complementary and alternative healthcare: Is it evidence-based? *International Journal of Health Sciences* 2(1): v–ix.

Thielking, Megan. 2016. "Essentially witchcraft": A former naturopath takes on the field (blog entry). *STAT News* (October 20). Available online at https://www.statnews.com/2016/10/20/naturopath-critic-britt-hermes/.

Ukrainian chemist lands in court for cancer "cure." 2015. *The Local* (January 1). Available online at http://www.thelocal.at/20150128/ukrainian-chemist-lands-in-court-for-cancer-cure.

Weeks, Carly. 2016. Are we being served by the regulation of naturopaths? Not if patients are still being misled. *The Globe and Mail* (April 28). Available online at https://www.theglobeandmail.com/life/health-and-fitness/health/canadian-naturopaths-need-to-follow-the-rules-if-they-want-regulation/article29785140/.

Whorton, James C. 2002. *Nature Cures: The History of Alternative Medicine in America.* Oxford University Press.

CHAPTER 40

Sources of Quantum Voodooism

Sadri Hassani

Vol. 44, No. 6
November/December 2020

In a series of episodes aired on her show in 2007, Oprah Winfrey talked about the then-new sensational New Age phenomenon known as *The Secret*, a movie by Australian film producer Rhonda Byrne, who later wrote a book of the same title that, due to Winfrey's enthusiastic endorsement, became an international bestseller. *The Secret* maintains that by merely *thinking* about losing weight, making more money, and falling in love, you can become thin, wealthy, and happily married. In one episode, Rhonda Byrne is joined by four "teachers"—well known self-help gurus who had chosen to disseminate the idea, much like the disciples of a prophet—in a speciously scientific discussion of the law of attraction, magnetic power, energy, frequency of mind vibration, and the vibration of the universe. All these buzzwords are the overture to the selling point of the conversation in which the author of *Chicken Soup for the Soul* proclaims, "If you go to quantum physics, we realize everything is energy" (see video at https://www.youtube.com/watch?v=9qwZMVe2WVY).

Marianne Williamson, former Democratic presidential candidate, designates "quantum realm of possibilities" as the source of "the good, the true, and the beautiful" and a solution to slavery, disenfranchisement of women, and segregation (Williamson 2019). She resorts to quantum physics to assert that "as our perception of an object changes, the object itself literally changes" (Williamson 1996) and that to change the world all we have to do is change our mind about the world. With this premise, would-be president Williamson's solution to all world problems is only a meditation away, and she has quantum physics to back her up! This is not a far-fetched, farcically concocted claim. When Hurricane Dorian was dashing toward Florida, Williamson advised her followers to stop it with their minds (Levin 2019).

The Indian guru Maharishi Mahesh Yogi instructs Deepak Chopra to "explain, clearly and scientifically, how [certain meditation techniques] work."

333

The result is a sophomoric concoction, which Chopra calls *Quantum Healing*, to explain its connection with Ayurveda (Hassani 2016). The publication of *Quantum Healing* was a milestone in the rampant trivialization of quantum physics by self-help gurus. Searching for the oxymoron "quantum spirituality" on Amazon .com yields several hundred titles. All have the word *quantum* in their titles or subtitles or on various pages inside. In these titles, one finds statements such as, "Quantum Angel Healing uses techniques rooted in the science of quantum physics, which proves that the thoughts and belief system of the observer influences the outcome of a situation" (Mora 2011); "[Quantum] physics suggested that the consciousness of the observer brought the observed object into being" (McTaggart 2002); and if that's too murky, "Quantum mechanics reveals that . . . your perception determines the shape of your reality" (Peirce 2009).

How did quantum physics become such a ludicrous gewgaw among modern gurus? The influence of *Quantum Healing* cannot be underestimated, but Chopra was a medical doctor with no background in physics. Why did he choose quantum physics to advance his absurd "scientific" theory of Ayurveda? One clue may be found in the 1960s, when the unpopularity of the Vietnam War opened the floodgates of the Western peace movement to the mystical beliefs of the Far East, where the atrocities of war were on display. To dissociate physics from the "military-industrial complex," books such as *The Tao of Physics* and *The Dancing Wu Li Masters* drew on quantum physics to paint physics as the embodiment of the "peaceful" Eastern mysticism. But why quantum physics? It turns out that the quantum physics–mysticism association goes back to the founders of quantum physics themselves, and the convergence of three factors facilitated that association: the infiltration of Eastern thought in Western philosophy, the rise of mysticism in the West, and the unique character of quantum physics.

Eastern Theosophy in Western Philosophy

As he was putting the finishing touches on his PhD thesis in 1813, Arthur Schopenhauer was introduced to a Latin translation of the *Upanishads*, ancient Sanskrit texts that contain some of the central philosophical concepts of Hinduism. To Schopenhauer, the *Upanishads* were the most elevating reading in the world, and he prophetically predicted that their philosophy would become the cherished faith of the West. He was so intrigued by the texts that he read passages of the book before going to sleep every night.

Schopenhauer's intellectual path to Eastern theosophy begins with his critique of Kant's *thing-in-itself*, a mind-independent entity that is beyond all human experience yet serves as the primary cause of our sensory perception. Schopenhauer maintains that our sensations cannot have an external cause, and that if we are to refer to the *thing-in-itself*, then we must come to an awareness of it, not by invoking the relationship of causality but by accepting that the world has a double-aspect, namely a "will" (a mindless, aimless, nonrational impulse at the foundational being of everything) and a "representation" (what we perceive around us). Will and representation are one and the same reality, regarded from different perspectives, like two sides of a coin, neither of which causes the other. The Hindu dualism of Brahman and Atman, with Brahman being "unlimited, unborn, not to be reasoned about, not to be conceived" (Müller 1884) and Atman being the true self, has a striking resemblance to Schopenhauer's will and representation. This should come as no surprise, because by the time his major work, *The World as Will and Representation*, came out in 1818, Schopenhauer had been perusing the *Upanishads* for five years.

According to Schopenhauer's philosophy, the great chain of being—the rocks, trees, animals, and human beings—is a complicated multitiered objectification of the meaningless will. The will's final tier of objectification appears when our minds introduce the forms of time, space, and causality, not to mention logic, mathematics, geometry, and moral reasoning. When the will is objectified at this level, the world of everyday life emerges. Thus, the laws of nature, along with the objects that we experience, *are our own creation* (Wicks [2003] 2017).

Rise of Mysticism in the West

One of the unintended consequences of the physics that began with Galileo and Newton was the eventual decline in the traditional western religions. Laplace's response, "Sire, I had no need of that hypothesis," to Napoleon's remark that there was no mention of God in Laplace's *Celestial Mechanics*, and Nietzsche's declaration—through Zarathustra's mouth—that "God is dead," created a moral vacuum by the end of the nineteenth century that could be filled only by a belief system that worshipped no supreme being. The filler turned out to be a salmagundi of spiritualism, esoteric Western philosophies, and Eastern thought.

Spiritualism was a movement rooted in the belief that the spirits of the dead existed and continued to evolve. Mediums were individuals gifted with the ability to communicate—in sessions known as séances—with the spirits and learn about the knowledge they had gained about God in the afterlife. Spiritualism gained enormous popularity among European intellectuals of that period. A prominent supporter of spiritualism was Sir Arthur Conan Doyle, the creator of the famous detective Sherlock Holmes. To assess the degree to which séances mesmerized the intellectuals of this period, suffice it to say that on one occasion, Harry Houdini, the American magician who became a leading opponent of the spiritualist movement, performed an impressive trick in the presence of Doyle. Houdini assured Doyle that the trick was pure illusion and that he was attempting to persuade Doyle not to endorse phenomena simply because he had no explanation for them. Doyle, nevertheless, refused to believe it was a trick and insisted that Houdini himself possessed supernatural powers ("Arthur Conan Doyle" 2019)!

The mediums' trickery was so impressive that even some well-known scientists ended up advocating spiritualism. In 1905, the well-known medium Eusapia Palladino came to Paris, where Nobel-laureate physicists Pierre and Marie Curie and some of their fellow scientists periodically investigated her. In 1906, five days before his accidental death, Pierre Curie wrote about his last séance with Palladino: "There is here, in my opinion, a whole domain of entirely new facts and physical states in space of which we have no conception" (Quinn 1995).

Another movement filling the moral void was theosophy, a potpourri of Western philosophies and Asian thought such as Hinduism and Buddhism. Because science had undermined the essence of traditional Western religions, theosophy proclaimed itself as an advocate of science. By 1902, Rudolf Steiner had transformed the teachings of the theosophy movement into anthroposophy, which he advertised was the science of an objective, intellectually comprehensible spiritual world accessible to human experience.

The flood of mysticism ravaging through Europe in the first decade of the last century eventually got a foothold in the mainstream science. In 1920, Arthur Eddington, the famous British astronomer, published a popular book in which he introduced the special and general theories of relativity to a nontechnical audience. But Eddington went beyond a mere exposition of the science. He arbitrarily subjected some of the mathematical symbols in the theory of relativity to his own philosophical interpretation and concluded, "All through the physical world runs that unknown content, which must surely be the stuff of our consciousness. . . . we have found that where science has progressed the farthest, the mind has but regained from nature that which the mind has put into nature" (Eddington 1920). Eddington's attribution of mysticism to relativity was too artificial to catch the attention of the public significantly, despite the public's appetite for the unification of science and the supernatural. For a "natural" unification, the public had to await the discovery of quantum physics.

Eastern Theosophy in Quantum Physics

Quantum physics is a highly mathematical theory that predicts probabilities of physical phenomena involving subatomic particles, and *probability defies explanation*. But the founders of quantum physics, unaccustomed to the new notion, could not swallow this fact. They looked for terrains of knowledge beyond mathematics and physics that could "make sense" of the strangeness of quantum physics, and philosophy was the only secular branch of knowledge that contained a huge repository of possible explanations.

As the probabilistic nature of quantum physics assented to the absence of causality, the founders' philosopher of choice became Arthur Schopenhauer, whose emphasis on the mindless, aimless, irrational will "made sense" of the randomness of quantum physics. Either through Schopenhauer or directly, the founders of quantum physics—Niels Bohr, Werner Heisenberg, Wolfgang Pauli, and Erwin Schrödinger—all developed a strong affinity for Eastern theosophy and, regrettably, *tied their science to that mystical viewpoint*. Here is Bohr talking about the parallel between Buddhism/Taoism and quantum physics:

> For a parallel to the lesson of atomic theory . . . we must in fact turn to quite other branches of science, such as psychology, or even to that kind of epistemological problems with which already thinkers like Buddha and Lao Tse have been confronted, when trying to harmonize our position as spectators and actors in the great drama of existence. (Bohr 2010)

Heisenberg was most influential in injecting Eastern thought in quantum physics. In his 1929 journey to the Far East, he had a long conversation with the Indian poet Rabindranath Tagore, subsequent to which he claimed to have realized that all fundamental aspects of physical reality, which had been so difficult for him and his fellow physicists to "make sense" of, "was the very basis of the Indian spiritual traditions. 'After these conversations with Tagore,' [Heisenberg] said, 'some of the ideas that had seemed so crazy suddenly made much more sense'" (Capra 1989).

Heisenberg recalls that during the famous 1927 Solvay Conference, some of the younger attendants gathered in the lounge of their hotel to converse about religion and science and the contrast between the religious beliefs of Planck and Einstein. While Planck firmly believed in a Christian personal god that was outside the realm of science, Einstein's god was the immutable laws of nature. To Pauli, who was present at the gathering, Einstein's perspective allowed the unity of object and subject (Einstein himself detested such unification and vehemently opposed any attribution of subjectivity to science). Pauli saw Planck's separation of object and subject as a threat to the ethics and values of society and found the solution in a spiritual framework where faith and knowledge, science and religion, object and subject are unified. He expressed hope in Bohr's complementarity because it implied that "the idea of material objects that are completely independent of the manner in which we observe them proved to be nothing but an abstract extrapolation. . . . In Asiatic philosophy and Eastern religions we find the complementary idea of a pure subject of knowledge, one that confronts no object" (Heisenberg 1971).

Pauli's belief in Eastern theosophy was tied to his great admiration of Schopenhauer, of whom he said, "Schopenhauer has exercised a lasting and fascinating effect on me, and he seemed to me to anticipate a future turn in the natural sciences" (Enz 2002). In fact, his veneration of Schopenhauer was so great that he defended the pseudoscientific notion of extrasensory perception because "Even so thorough critical a philosopher as Schopenhauer has regarded parapsychological effects as not only possible, but as supporting his philosophy" (Pauli 1994).

Schrödinger recalls how, after accepting a post as a lecturer in theoretical physics in Czernowitz, he had planned to spend all his free time acquiring a deeper knowledge of philosophy, having just discovered Schopenhauer, who introduced him to the *Unified Theory of Upanishads*. Schopenhauer's objectification by will is the essence of this Schrödinger statement: "Mind has erected the objective outside world of the natural philosopher out of its own stuff" (Schrödinger 2012). Schrödinger recognizes the paradox of individuals having different minds while there is only one world: "There is obviously only one alternative, namely the unification of minds or consciousnesses. Their multiplicity is only apparent, in truth, there is only one mind. This is the doctrine of the *Upanishads*" (Schrödinger 2012, 129).

Today's physicists, for the most part, are interested mainly in the theoretical and experimental ramifications of quantum physics, and in that pursuit we have been blessed with inventions such as transistors, lasers, and microchips, as well as the theoretical understanding of the tiniest constituents of matter and the largest galaxies. The majority of physicists brush aside the philosophical implications of quantum physics, because they have come to realize that if they try to understand quantum physics in terms of philosophy, they "will get down the drain" (Feynman 1967).

The founders' object-subject unification, which modern gurus have renamed "the experimenter effects," has played a prominent role in the academization of pseudoscience. Larry Dossey is the executive editor of *Explore: The Journal of Science and Healing*, a journal that, despite the scientific reputation of its publisher Elsevier, is devoted to pseudoscience. Dossey offers a list of suggestions on the future research in his field, prayer healing. His third suggestion is: "In view of the evidence for experimenter effects, the preexisting beliefs of prayer experimenters should be ascertained and recorded as part of the study" (Dossey 2008). In other words, the results of studies that negate the efficacy of prayer in healing could be attributed to the disbelief of the experimenters!

Messenger versus Message

Given a *printed* copy—to eliminate identification from handwriting—of a newly discovered piece of music, a musicologist specializing in Beethoven can not only identify it as the work of the master but also determine the period of the composer's creativity in which it belongs. That cannot happen in science! Einstein presented his general relativity field equation on November 25, 1915, to the Prussian Academy. Five days earlier, David Hilbert, the great German mathematician, had presented a talk containing the same equation to the Royal Academy of Sciences in Göttingen (Thorne 1995). A historian of physics, given the printed version of the two equations, could not tell which one is Einstein's and which one is Hilbert's. Such simultaneous creativity—so common in science that Nobel Prizes in a given field are often awarded to multiple scientists—is unheard of in other creative areas; the notion of two artists creating the same *Mona Lisa* is preposterous!

Scientific geniuses share many of the same kinds of strength and weakness that we possess. Outside their areas of expertise, they are quite ordinary characters who can be poor judges of politics, religion, morality, and philosophical outlook. (See "The Nobel Disease," Skeptical Inquirer, May/June 2020.) Einstein encouraged President Roosevelt to initiate the development of atomic weapons; Linus Pauling, winner of two Nobel Prizes (chemistry and peace), was the originator of orthomolecular therapy, a dangerous alternative medical procedure; and James Watson, the codiscoverer of the double helical nature of DNA, is an ardent racist. But these mistakes are not made right because of the science of their makers, just as the science is not made wrong because of the mistakes of its discoverers. However, the beneficiaries of the self-help industry proclaim that just as Eroica cannot be separated from Beethoven, the founders' mystical object-subject unification cannot be separated from quantum physics. They contend that ancient scriptures of the Far East are based on quantum physics. Nothing is further from the truth! The mystical views of quantum physics' founders were published in proceedings and trade books with no scientific editing or reviewing, because the publishers who printed their physics would have refused to publish their baseless viewpoints, and the founders themselves were fully aware of that. An important fact is buried under modern gurus' intense—and unfortunately successful—campaign of marrying their nonsense with quantum physics: *The messenger is not the message.*

Sadri Hassani is professor emeritus of physics at Illinois State University and author of several books at graduate, undergraduate, and introductory levels. His blog at https://skepticaleducator.org is devoted to exposing misconceptions and distortions of science by professional scientists. His previous SI

articles are "Why E-Cat Is a Hoax" (January/February 2019), "Does E=mc^2 Imply Mysticism?" (July/August 2016), "Deepak Chopra's 'Physics'" (January/February 2016), and "'Post-Materialist' Science? A Smokescreen for Woo" (September/October 2015).

References

Arthur Conan Doyle. 2019. Wikipedia. Available online at https://en.wikipedia.org/wiki/Arthur_Conan_Doyle#Spiritualism,_Freemasonry.
Bohr, N. 2010. *Atomic Physics and Human Knowledge.* Garden City, NY: Dover Publications, 20.
Capra, F. 1989. *Uncommon Wisdom.* New York, NY: Bantam Books, 43.
Dossey, L. 2008. Healing research: What we know and don't know. *Explore* 4(6): 341–352.
Eddington, A. 1920. *Space Time and Gravitation.* Cambridge, UK: Cambridge University Press, 182.
Enz, C. P. 2002. *No Time to Be Brief.* Oxford, UK: Oxford University Press, 426.
Feynman, R. 1967. *The Character of Physical Law.* Cambridge, MA: MIT Press, 129.
Hassani, S. 2016. Deepak Chopra's 'Physics.' *Skeptical Inquirer* 40(1): 55–58. Available online at https://skepticalinquirer.org/2016/05/deepak-chopras-physics/.
Heisenberg, W. 1971. *Physics and Beyond.* New York, NY: Harper and Row Publishers, 83–85.
Levin, B. 2019. Marianne Williamson insists she's not crazy after telling followers to stop Hurricane Dorian with their minds. *Vanity Fair* (September 4). Available online at https://www.vanityfair.com/news/2019/09/marianne-williamson-hurricane-dorian.
McTaggart, L. 2002. *The Field.* New York, NY: HarperCollins, 12.
Mora, E. 2011. *Quantum Angel Healing.* Amora Creations, ii.
Müller, M. 1884. *The Upanishads,* Part 2. Oxford, UK: Clarendon Press, 318. Also available online at https://www.sacred-texts.com/hin/sbe15/sbe15117.htm.
Pauli, W. 1994. *Writings on Physics and Philosophy.* Berlin, Germany: Springer, 163.
Peirce, P. 2009. *Frequency: The Power of Personal Vibration.* New York, NY: Atria Books, 28.
Quinn, S. 1995. *Marie Curie: A Life.* New York, NY: Simon and Schuster, 208, 226.
Schrödinger, E. 2012. *What Is Life?.* Cambridge, UK: Cambridge University Press, 121.
Thorne, K. 1995. *Black Holes and Time Warps.* New York, NY: W. W. Norton, 115–117.
Wicks, R. (2003) 2017. Arthur Schopenhauer. *The Stanford Encyclopedia of Philosophy* (Summer Edition), Edward N. Zalta (ed.). Available online at https://stanford.io/2T3CXIX.
Williamson, M. 1996. *A Return to Love.* New York, NY: HarperOne, 66.
———. 2019. *Politics of Love.* New York, NY: HarperOne, 224.

CHAPTER 41

Percival Lowell and the Canals of Mars

Matthew J. Sharps

Vol. 42, No. 3
May/June 2018

The planet Mars has always fascinated humanity. In fact, it seems to interest us considerably more than most things in the night sky.

This makes sense; Mars is often not only clearly visible but conspicuously red like blood. So many ancient societies associated Mars with war, always of considerable interest to the human species. Mars appeals to us both as a physical object for observation and as a lure for mythological speculation.

There is a duality here. On the one hand, there is the visible planet; the red coloration reflects its geology. On the other hand, there is the Mars of interpretation, whose red color reflects its attributional warlike nature; this says a lot more about human psychology than it does about the planet Mars itself.

The red planet causes us to observe and to speculate.

Speculation. That's where the problems come in. There is physical reality, and there is interpretation; and it is frequently the interpretation, rather than the reality, that seizes the attention of human beings. Our brains are remarkably predisposed to the interpretation of objective physical reality in psychological, self-referential terms. Unfortunately, these terms are frequently just plain wrong.

Examples of this are legion. In previous articles in SI, my coauthors and I have discussed ordinary objects that have metamorphosed, in the minds of their observers, into nonexistent phenomena ranging from UFOs to Bigfoot, and we have found specific patterns of mental processing that contribute directly to these misinterpretations (e.g., Sharps et al. 2016). In the more prosaic but more sinister worlds of eyewitness memory and officer-involved shootings, we have frequently found innocuous things such as power tools being transformed, psychologically, into far less innocuous firearms (e.g., Sharps 2017). It is very clear that our brains can lead us to see meaningful patterns where none actually exist and that we may extrapolate what we believe about a given perception to the

Percival Lowell in 1914, observing Venus in the daytime with the 24-inch (61 cm) Alvan Clark & Sons refracting telescope at Flagstaff, Arizona. *Public domain.*

perception itself. We tend to interpret what we see in terms of what we *believe*; this brings us back to the planet Mars.

Mars was the special focus of Percival Lowell, an important pioneer in planetary astronomy. Using his own considerable wealth, he created the great observatory at Flagstaff, Arizona. Lowell's computations there led ultimately to Clyde Tombaugh's discovery of Pluto, and Lowell's financial and intellectual support led to a literally stellar progression of Lowell Observatory discoveries to the present day. Many of his observations, and those of other scientists at the Observatory, have proven startlingly accurate (e.g., Schindler 2016).

Some of his other observations, however, present problems.

One of Lowell's most important discoveries, in his opinion, was finding canals on the surface of Mars. These long, straight, clearly artificial irrigation systems were ubiquitous. For Lowell, the dry landscape of Mars quite literally supported an intelligent race of beings with something like civil engineering degrees who were transporting water all over the place in their canals.

It wasn't just Lowell. Schiaparelli saw canals, or at least ditches (*canali*). Schiaparelli's ditches were long and straight and rectilinear, completely failing to obey the laws of perspective on the Martian planetary spheroid, but he saw and reported them anyway. Flammarion believed in canals, although he was also big on vegetation on the moon as well, so we might want to be a little careful there. Douglas, Lowell's assistant, also saw canals—until he decided they didn't really exist, was fired by Lowell as a result, and went on to invent dendrochronology at the University of Arizona. A lot of professional astronomers saw Martian canals, drew the things, and speculated on their nature.

But there aren't any Martian canals.

That's the problem: there just aren't any damn canals on Mars. Lowell, and many other expert observers, saw them.

But they're just not there.

The Mariner spacecraft thoroughly photographed Mars way back in 1964. Mariner found craters, rocks, flat bits and pointy bits, and bits with hills, but it didn't find a single canal. Anywhere.

Mariner did, of course, photograph many Martian surface structures of great interest to planetary astronomers. Lowell saw many of them, half a century earlier, through his excellent telescopes. The man was no fool; some of his drawings of the Martian surface are practically identical, in broad outline, to photographs of the planet taken from the Hubble space telescope. But his canals, drawn with equal clarity, simply don't exist.

You might assume that continuing progress in telescope technology would have reduced the observation of these canals in the early years of the twentieth century, but you'd be wrong. I had the honor of examining a number of globes and maps of Mars, held today in the excellent archives at the Lowell Observatory; these very clearly show an increase in the number and complexity of canals as new observations were made and new globes and maps created. Canals became more numerous and elegantly geometric as the observations poured in. Some canals even doubled in perfect parallel, an astonishing phenomenon termed *gemination*; all of this despite the fact that there were never any real canals to begin with.

These nonexistent canals had real staying power. As mentioned earlier, the Mariner orbiter, in 1964, proved that there were no canals on Mars, but I examined a 1969 map in the Lowell archives *that still showed the canals*, in all their impossible rectilinear glory. The ruler-straight lines of the canals were relatively faint, as if the planetary cartographers were somewhat ashamed of these nonexistent features, but these completely imaginary ditches were certainly there, in the imaginations and on the maps of scientific areographers. This was five years after Mariner had completely disproved their existence.

How do we explain this? Was Lowell, a fine observational astronomer, hallucinating? And were all the other astronomers who "saw" these bizarre ditches, straight and clear and marching over the Martian landscape, similarly afflicted with bizarre psychological disorders?

Of course not. Hallucinations derive from three sources: organic brain changes, psychosis, and extraordinary levels of stress. Lowell suffered from none of these. Granted, in the 1890s, Lowell left astronomy for four years due to a "nervous" condition, but nobody has ever suggested that he suffered from any of the conditions that produce hallucinations, and he kept seeing canals anyway. So did a lot of less nervous people; his predecessor Schiaparelli observed whole networks of Martian *canali*, as did a number of contemporaneous astronomers, none of whom were psychotic or brain damaged.

What on Mars Was Going On?

Well, that would be nothing. What was happening was not on Mars at all. The canal phenomenon was very clearly happening on Earth in the minds of the astronomers affected; for whatever reasons, a lot of them had canals on the brain.

Now, if anybody had a right to have canals on the brain, it was the aforementioned Giovanni Schiaparelli. Born a mere twenty-five miles from Canale, Italy, within thirty miles of several major transportation canals and living during a period in which the Suez and Panama Canals were the wonders of the world, it would be rather strange if Schiaparelli did not regard canals as the apotheosis of civilization, even though he himself only referred to the Martian canals as channels or ditches (*canali*). He may very well have had a *mental set* (e.g., Sharps 2017) about such things, a habitual way of looking at the world in canal-related or channel-based terms. This is of course speculation and can never really be anything more, but what we know for certain is that such habits of mind are intensely individualistic, based in our own idiosyncratic experience, and may form one of the first dynamics suggested to explain observation of the nonexistent canals of Mars: *Individual Differences*. Some people see canals. Others don't.

But why does anybody see them in the first place? As mentioned, research in my laboratory, published primarily in SI (e.g., Sharps et al. 2016), has elucidated some of the psychological dynamics of those who think they see Bigfoot, flying saucers, aliens, and ghosts. One of the things we found in that research was that people generally don't make something out of nothing. In other words, you don't see Bigfoot on a featureless plain; you see an ape-shaped tree stump or something similar, and your brain makes Bigfoot out of it for you. The same

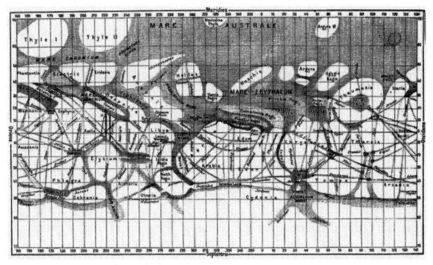

Map of Martian "Canals" by Giovanni Schiaparelli. *Public domain.*

brain-based phenomena can also create a Loch Ness monster out of a school of Scottish salmon, a Death Star out of a helicopter with a broken landing light, and so on. These *Gestalt reconfigurations* result from our mental misperception and misinterpretation of real things in the real world—or on the real Mars—and these phenomena are governed by specific psychological laws. These laws are suggested to be a major psychological source of the observation of the canals of Mars.

But how does an astronomer such as Lowell or Schiaparelli maintain his beliefs in these canals, to the point at which, in the face of mounting professional opposition, he sees more and more of them? Human beings are social creatures with the ability to develop strong investments in our ideas and beliefs. This is suggested to be another major source of the Canal phenomenon: *sociocognitive influences*, to be joined with *individual differences* and *Gestalt reconfiguration*.

Based on an intensive review of the relevant literature, and on the observations I was privileged to make in the Lowell Archives and also through Lowell's own 24-inch Clark telescope at the great Lowell Observatory, I submit that the erroneous observations of the canals of Mars can be better understood in terms of these three sets of dynamics.

Individual Differences

The precise influences on Lowell's thinking cannot now be ascertained. But it is clear that in 1901, when Lowell drew an "artificial planet," a mock-up disk designed to evaluate the accuracy of observational drawings, Lowell drew not one but two parallel canals, a "gemination," when in fact there had been "only a broad shading" in that portion of the model (Sheehan 1988). Part of Lowell's family wealth derived from investments in the great canal systems of the eastern United States. These were regarded at the time as among the modern wonders of the world and were used extensively to ship a tremendous variety of goods, including the textiles that were a major business interest of the Lowell clan (see Hoyt 1976 and Sheehan 1988); this was yet another source of his individual affiliation with canals and their builders. In the presence of this influence, he turned a "broad shading" into two very specific, but nonexistent, canals. It might readily be suggested that Lowell, perhaps like Schiaparelli, was something of a victim of a canal-based mental set. This speculation may or may not have merit, but we do know that when Lowell, as an individual, was offered the opportunity to draw a shadow, he drew a hydraulic engineering system.

These individual differences would of course have interacted with the conditions of any given observation—but in what way? In my own work at the Lowell Observatory, I observed both Mars and Jupiter through the great Clark telescope preserved there. Now, I am an aging researcher with very thick glasses, but what I can say is that the observations danced before me very swiftly, the result of atmospheric fluctuation. Sometimes I would seem to see a feature on Mars, and then it was gone, or obscured, within two or three seconds. This type of highly variable, atmospherically based visual fluctuation would certainly have been there for Lowell and his colleagues as well. Obviously their training and experience would have rendered them vastly superior observers to me. But expertise aside, the fact is that brevity of observation limits our precision, in astronomy as in criminal eyewitness identification. Brevity can completely change our interpretation of our observations (e.g., Sharps 2017), whether we think we see a criminal suspect with a gun or a canal on the planet Mars. In short, if we have strong individual psychological reasons to see canals, we will see them if the observational conditions permit them at all. Lowell saw them, to the degree that when his assistant A. E. Douglas questioned these interpretations, he was essentially fired. *Observations are subject to the psychology of the individual interpreting them*; this is a crucial principle that all scientists, in all fields, should consider.

Gestalt Reconfiguration

The astronomer E. M. Antoniadi was rather caustically critical of Lowell in most respects. Although he reported the odd Martian canal himself, he demonstrated, very enthusiastically, that many of the "canals" were in fact the result of observation of a series of surface features: craters, rocks, and so on, arranged by the forces of geology into linear patterns. Lowell, and the other "canal" observers, saw discrete surface features arranged by natural forces into relatively straight lines, and joined them, perceptually, into "canals" (e.g., Sheehan 1988).

How is this possible? Gestalt psychology, the venerable theoretical perspective that deals with perceptual and cognitive configuration, provides rather good answers (e.g., King and Wertheimer 2005; Kohler 1947). Consider two of the Gestalt laws of perception, the laws of *closure* and of *good continuation* (see Sharps 1993). When we see objects that are close together, we tend to see them as connected; and when they form contours, lines, or curves, we tend to see them as units. Lowell, and the other canal believers, saw craters and rocks very close together. These astronomers, with their human nervous systems, tended to see these things as contiguous. The contours thus created frequently formed lines, hence the canals. Contours of disconnected rocks were "closed," perceptually, into solid lines; under brief observation conditions, these lines appeared very solid, and they showed "good continuation" with other discrete features of the Martian surface. These factors would have created, perceptually, the "canals" (Sheehan 1988).

If an astronomer such as Lowell was individually predisposed to see canals and observed them with unavoidable fluctuating brevity, the Gestalt phenomena of closure and good continuation would practically ensure that he would see them, real or not (Sheehan 1988; Sharps 1993; 2017).

Sociocognitive Factors

In a letter to Lowell's brother, Lowell's assistant, A. E. Douglas, pointed out that the canals might have a psychological origin. Lowell discharged him.

Lowell regarded any psychological explanation for the canals as anathema. He may have seen the psychological idea as psycho*pathological*, rather than as rooted in the normal principles of cognition and perception; whatever the source, he fired Douglas. Lowell had invested enormously, in financial and in psychological terms, in the canals of Mars, and as has been demonstrated many times, strong investment leads to strong beliefs that are difficult to sway even in the presence of contrary evidence. The principle of *cognitive dissonance* (Festinger 1957; Sharps 2017) deals with this rather nicely. Even if a given idea proves to be completely wrong, we tend to hold to it, and even to defend it with relatively incoherent cognitive processing, *if we're sufficiently invested in it* (Festinger 1957).

Lowell had given the canals of Mars everything he had, in terms of a very long-term emotional and financial investment. The canals of Mars, in Lowell's mind, were the greatest discovery of his own observatory. To acknowledge error would have been virtually impossible for him, in view of this investment; he never gave up on his belief in the canals, even and especially in the face of mounting pressure and criticism from his colleagues and his detractors.

Conclusions

The Martian surface is densely covered with features derived from the geological processes of the planet and from astronomical impacts over an enormous span of time. These surface features create a variety of irregularities that are very clear in photographs from spacecraft and from modern telescopes. However, through the telescopes of the early twentieth century, these features would have been much less readily resolved. This relative lack of resolution would have resulted in perceptual and cognitive misinterpretation with reference to the Gestalt principles cited above. This is especially true when the fluctuating brevity of optical astronomical observation is considered and when we further consider the reinforcing factors derived from individual differences and from sociocognitive factors, cementing early interpretations of those observations into a form of cognitive concrete.

It's obviously impossible to perform experiments on the astronomers of the past. But within the realm of theoretical psychology, we can absolutely state that the observation of the canals of Mars demonstrates neither psychopathology nor incompetence on the part of pioneering scientists such as Lowell. Instead, we find an important lesson for our more modern inquiries. The scientist does not lie outside of the natural world. Rather, the scientist is entirely part of that world and is subject to scientific law; in the present case, to elements of the Gestalt laws of perception and cognition and to the laws of related areas of experimental psychology. It is important for all scientists, in all disciplines, to be aware of these essential facts and to use them to exert caution in the interpretation of what might otherwise be interpreted as purely objective observations.

Matthew J. Sharps is professor of psychology at California State University, Fresno, and serves on the adjunct faculty of Alliant International University in forensic clinical psychology. He specializes in eyewitness phenomena and related areas in forensic cognitive science. He is a Diplomate and Fellow of the American College of Forensic Examiners and the author of over 160 publications and professional papers, including the 2010 book *Processing under Pressure: Stress, Memory, and Decision-Making in Law Enforcement* (www. LooseleafLaw.com). He has consulted on eyewitness issues in numerous criminal cases.

References

Festinger, L. 1957. *A Theory of Cognitive Dissonance*. Evanston, IL: Row, Peterson.

Hoyt, W. G. 1976. *Lowell and Mars*. Tucson: University of Arizona Press.

King, D. B., and M. Wertheimer. 2005. *Max Wertheimer and Gestalt Theory*. New Brunswick: Transaction Publishers.

Kohler, W. 1947. *Gestalt Psychology*. New York: Mentor.

Schindler, K. 2016. *Images of America: Lowell Observatory*. Charleston, SC: Arcadia.

Sharps, M. J. 1993. Gestalt laws of perceptual and cognitive organization. In *Magill's Survey of Social Science: Psychology*. Pasadena, CA: Salem Press.

———. 2017. *Processing Under Pressure: Stress, Memory, and Decision-Making in Law Enforcement*. Flushing, NY: Looseleaf Law Publications.

Sharps, M. J., S. W. Liao, and M. R. Herrera. 2016. Dissociation and paranormal beliefs: Toward a taxonomy of belief in the unreal. *Skeptical Inquirer* 40(3) (May/June): 40–44.

Sheehan, W. 1988. *Planets and Perception*. Tucson, AZ: University of Arizona Press.

CHAPTER 42

The Roswell Incident at 70

FACTS, NOT MYTHS

Kendrick Frazier

Vol. 41, No. 6
November/December 2017

The seventieth anniversary of the so-called Roswell Incident came and went this past summer with a refreshing lack of fuss. One might even hope to think the passions it evokes among believers that a flying saucer crashed on a ranch in south-central New Mexico back in July 1947 have, over time, finally waned. But the rationalists in us realizes that is not likely. Maybe they are just tired and will be back again after a rest.

That's kind of what happened with Roswell. It was a big story back in early July 1947 for a few days, but then when the Air Force announced that what the rancher found was related to balloon flights and not to anything more mysterious, the story disappeared from public discourse until it was resurrected again by several factually unscrupulous writers in the early 1980s.

From your editor's vantage point as a Roswell-watcher from Albuquerque, only about a hundred air miles from the supposed crash site, the most noticeable recent blip on the radar was an anniversary story by the *Carlsbad Current-Argus* reprinted in the July 8 *Albuquerque Journal* and titled "Roswell Incident Lives on 70 Years Later."

The largest newspaper in the state, the *Albuquerque Journal* has been noticeably free of sensationalism about Roswell for a long time. This reprinted story was a bit of an anomaly. It basically recounted the myth and various claims believers have put forth about it since but unfortunately gave no information that explains the origin story.

That moved me to write a letter to the *Journal* that, to their credit, they published as a short op-ed piece in their Sunday, July 16, edition, "Roswell Myth Lives on Despite the Established Facts." (Available online at https://www .abqjournal.com/1033584/roswell-myth-lives-on-despite-the-established-facts .html.)

In it I simply pointed out some key facts the article failed to mention. It may be worthwhile reminding you, our readers, of those, and a few others as well.

What rancher W. W. (Mac) Brazel reported finding on his ranch, sixty miles northwest of Roswell, was simply this: Debris consisting of a large number of pieces of *paper* covered with a foil-like substance and pieced together with small *sticks*, much like a *kite*. And also some pieces of grey *rubber* (my emphases). All were small and hardly some high-tech alien flying saucer!

The reporter should have told readers what we now know (almost certainly) the debris to have been: remnants of a long vertical "train" of research balloons and equipment launched by New York University atmospheric researchers and not recovered—specifically, Flight No. 4. The research team launched NYU Flight #4 on June 4, 1947, from Alamogordo Army Air Field and tracked it flying east-northeast toward Corona. It was within seventeen miles of the Brazel ranch when the tracking batteries failed and contact was lost.

New York University's role in launching the "constant-level" research balloons was unclassified. In the 1990s, it was learned that the mission also had a classified purpose, called "Project Mogul," to learn whether such balloons could take highly sensitive microphones and keep them at a level in the atmosphere (the tropopause) where they might be able to detect acoustic signals channeled round the Earth from Soviet nuclear tests.

On the evening of February 8, 1995, I was present at a meeting in Albuquerque of New Mexicans for Science and Reason (NMSR) when the man who helped launch Flight 4, Professor Charles B. Moore, showed us some of what was on that flight. In 1947, Moore was an NYU graduate student, working on the balloon launches. He spent the rest of his career as a respected professor of atmospheric physics at New Mexico Tech in Socorro.

NMSR is the local science-oriented skeptics group in Albuquerque. I helped found it in 1990, based on CSICOP's inspiration, and it has been headed for years now by physicist/mathematician and Committee for Skeptical Inquiry Fellow Dave Thomas. Thomas works in Socorro and also teaches a course on pseudoscience there at New Mexico Tech.

In addition to Thomas and me (and many others), physicist and CSI Fellow Mark Boslough told me recently that he remembers being in the audience at that remarkable 1995 evening meeting.

Moore brought with him a radar reflector like the three that were attached to Flight 4. Specifically, they were Signal Corps ML 307B RAWIN targets. It looked much like a box kite but with some angular surfaces. The sticks and metallic paper are similar to what Brazel described. The rubber Brazel noted was similar to the neoprene balloons used to carry equipment aloft. The radar reflectors contained small metal eyelets, similar to those Brazel had described on the debris he found.

Moore also provided a new and very telling detail. The reinforcing tape used on the NYU targets had curious markings; UFO believers later described these markings on the debris Brazel discovered as "hieroglyphics," implying some form of alien writing. In fact, Moore told us the tape had been purchased from a New York City toy factory and the symbols on the tape were "abstract flower-like" designs made to appeal to kids.

These and other established facts of the Roswell incident will of course never catch up with the charming myth. It is understandable that UFO believers and Roswell city boosters will promote the myth as possible reality (wink, wink), but, as I wrote in my op-ed, "in this day of 'fake news,' let's not be a party to that."

All these facts, and many more supporting details, have been widely available since the mid-to-late 1990s in various scholarly publications. They include a book that Charles Moore himself coauthored with two anthropology professors, *UFO Crash at Roswell: The Genesis of a Modern Myth*, Benson Saler, Charles A. Ziegler, and Charles B. Moore (Smithsonian Institution Press, 1997); *The UFO Invasion*, a Skeptical Inquirer anthology I coedited with Barry Karr and Joe Nickell (Prometheus Books, 1996), which includes David E. Thomas's special report from the July/August 1995 Skeptical Inquirer "The Roswell Incident and Project Mogul" and many other Roswell-related articles; and two U.S. Air Force investigative reports, *Report of Air Force Research Regarding the 'Roswell Incident'* (1994) and *The Roswell Report: Case Closed*, Headquarters United States Air Force, written by Capt. James McAndrew, 1997.

The NYU balloon flight assemblages were huge. The diagram Moore supplied in his talk for flight 2, similar to flight 4, and published in the above-mentioned *Skeptical Inquirer* article, requires three vertical columns to display all the components. They include three radar reflectors, various measuring instruments, and twenty-four separate balloons. Charles Moore told us the whole interconnected array extended a vertical distance of 700 to 800 feet. So the common explanation of "weather balloon" is quite the understatement.

Some additional points: The director of research for the NYU balloon-launch experiments in 1947 was famous New York University geophysicist and meteorologist Athelstan Spilhaus. I knew Spilhaus when I was editor of *Science News* in the 1970s because he then was on the Board of Trustees of Science Service, *Science News*'s publisher. I knew nothing about Roswell then. Spilhaus died in 1998 at the age of eighty-six.

"Abstract flower-like designs" on toy-factory tape used on the NYU radar targets.

In 2009 when his son, Fred Spilhaus, retired from his longtime position as executive director of the American Geophysical Union in Washington, D.C., he wrote whimsically that his father was the man responsible for the Roswell Incident (*Physics Today*, February 2009, quoted in the May/June 2009 Skeptical Inquirer). Athelstan Spilhaus was quite a colorful character. He is the only scientist I have ever heard of who had his own Sunday newspaper comic strip. Titled "Our New Age," it ran in color in 110 newspapers all over the world from 1958 until 1975. When President Kennedy met Spilhaus in 1962, JFK told him, "The only science I ever learned was from your comic strip in the *Boston Globe*" (see http://www.smithsonianmag.com/history/sunday-funnies-blast-off-into-the -space-age-81559551/).

If what Brazel found was so mundane, why did someone think it had to do with a crashed "flying saucer"? The reason is that the term *flying saucer* had just hit the news media for the first time. On June 24, 1947, pilot Kenneth Arnold reported a series of what he described as boomerang-shaped objects flying up and down near Washington's Mt. Rainier. He said they "flew erratic, like a saucer if you skip it across the water," and from then on the term *flying saucer* took hold, even though he never said they looked like saucers (see Robert Sheaffer, *The UFO Verdict*, Prometheus Books, 1998, p. 15). This started a media frenzy and people began looking to the skies and seeing things they'd never seen before (including over New Mexico) and reporting more "flying discs" or "flying saucers." Many possible explanations for Arnold's sighting have been suggested. In their May/June 2014 *Skeptical Inquirer* cover article "Mount Rainier: Saucer Magnet," James McGaha and Joe Nickell describe McGaha's hypothesis that it was due to optical phenomena called "mountain-top mirages." (I wonder if at least some of the "flying disc" sightings in New Mexico were reflections of the huge NYU/Project Mogul balloon assemblages being launched fairly regularly.)

Brazel had been persuaded that the debris he had found might have something to do with the reports of "flying discs" that were then exciting everyone. His report was made public in Roswell July 8, 1947, at the height of the craze.

A public affairs officer at the local army air field was excited about the find and so made the now-famous announcement that it had something to do with saucer sightings, without further checking. That made front-page news. By then Brazel said he was amazed at the fuss and sorry he said anything about it.

Ironically, the report of what Brazel actually found, an explanation that it was a "weather balloon"(not quite right but kind of close), and the date he had found it, June 14, before the media frenzy of sightings started—all are reported in an Associated Press article published on page 2 in the July 9, 1947, *Carlsbad Current-Argus* ("'Flying Disc' turns Out to Be Weather Balloon"; I have a copy of it, see page 355). This is the same newspaper that unfortunately seemed to forget those facts in their July 2017 anniversary article.

In the subsequent mythmaking, one of the main sensationalist books (*The Roswell Incident*, Charles Berlitz and William L. Moore, 1980) claimed that the debris was from a flying saucer that passed over Roswell the evening of July 2, 1947. But in fact Brazel had found the debris much earlier, on June 14, just ten days after the NYU team had lost track of Flight #4, headed toward his ranch. ("This blows the whole yarn out of the water," wrote James Moseley in his *Saucer Smear* newsletter, v. 29, No. 4, May 15, 1982.)

Back then these "flying discs" didn't have the associations they have today. Nobody knew what they might be (and indeed some reports at the time did suggest that they were meteorological phenomena, or delusions, or mass hysteria, or visual misinterpretations of things seen in the skies). The idea of alien spacecraft hadn't gained hold yet. At best the concern was that if they were physical craft at all, they might be Soviet or even holdover Nazi aircraft.

In the 1980s and early 1990s, the mythmaking process really took off. More fantastic and wild stories emerged (or were concocted) in a process familiar to folklorists. Three out-and-out hoaxes were widely publicized, then exposed.

As for reports of sightings of alien bodies, the second (1997) U.S. Air Force report investigated and found there were no contemporary reports of alien bodies being found in 1947. These (unverified) reports came only in UFO books and articles published after 1978.

The Air Force report describes in detail a long series of Air Force experiments over decades in which instrumented lifelike anthropomorphic dummies were dropped out of high-altitude research balloons over New Mexico. This began in the 1950s and continued for many years. Most were launched over Holloman Air Force Base or the White Sands Missile Range, but the balloons soon floated beyond those boundaries. The idea was to measure the effects of extreme environments and situations deemed too hazardous for a human being (page 17).

Such instrumented crash-test dummies were not familiar at that time, and the report suggests that one "very likely could be mistaken for an alien." This conclusion was widely ridiculed by UFO believers at the time, but the report gives a large amount of supporting detail and shows dozens of photographs. That explanation, to most fair-minded observers, has stood the test of time. Among the kinds of eye-witness statements that support the crash-test dummies explanation were statements such as "his eyes were open, staring blankly," "their skin coloration . . . a bluish-tinted milky white." At other times, the report says, injured airmen, some seriously so, were brought to the Roswell base after accidents, and it suggests some reports are mixed-up remembrances of those.

One last piece of corroborating testimony: In 2001 journalist Guy P. Harrison interviewed Joe Kittinger (Colonel, U.S. Air Force, ret.), one of the great aviation pioneers of the twentieth century. In 1960 Kittinger had jumped out of a balloon over New Mexico from the very upper edge of the atmosphere (102,800 feet, or nineteen miles). In a free fall that lasted four minutes, he reached a speed of 600 miles per hour.

Harrison was mainly interested in those kinds of real adventures, but he hesitantly asked Kittinger about Roswell. "It never happened," Kittinger said, and went on to describe the events involving the NYU balloon experiments I have reported here. "The so-called alien spaceship was that balloon. . . . A lot of people want to believe it was aliens, and they want to believe there was a big cover-up. But I'll tell you, it never happened."

Associated Press article published July 9, 1947, reporting that the debris rancher Mac Brazel had found on June 14 (long before the July 2 "flying saucer" sighting over Roswell) consisted of "pieces of paper with a foil-like substance, and pieced together with small sticks, much like a kite." Plus "pieces of grey rubber. All were small." *From the collection of Kendrick Frazier.*

What *did* happen, he said, is that these high-altitude drops of humanlike dummies contributed to the Roswell myth. "Absolutely they did. These dummies we dropped from balloons were dressed in pressure suits, so they looked unusual." (These quotes are from Chapter 13 of Harrison's excellent 2012 book, *50 Popular Beliefs That People Think Are True* [Prometheus Books], and I relate them here with his permission.)

"One time we dropped one and it fell way up in the mountains," Kittinger said. "These dummies weighed more than two hundred fifty pounds. So how do you carry one out of the mountains? We put it on a stretcher and carried it in the back of an ambulance to take away. Now if somebody is back in the weeds watching this they are going to say, 'Wow, look at that alien they have there.' We think that a lot of the alien sightings were actually us doing our work with the test dummies."

Kendrick Frazier was longtime editor of the *Skeptical Inquirer* and a fellow of the American Association for the Advancement of Science.

CHAPTER 43

The "Roswell Slides" Fiasco

UFOLOGY'S BIGGEST BLACK EYE

Robert Sheaffer

Vol. 39, No. 5
September/October 2015

Sometime during 2012, video producer Adam Dew obtained a collection of Kodachrome slides reportedly taken during the 1940s. The slides were said to have been taken by the late Bernard and Hilda Ray, a well-to-do Texas couple who led an active life with much travel, and left behind no family. Two of the slides were of particular interest: they seemed to show the body of a small being laid out on a shelf. It looked like it might be an alien, Dew thought. So he contacted "Roswell experts" Donald Schmitt and Tom Carey, authors of *Witness to Roswell*. At that time, those authors were involved putting together something to be called the "Roswell dream team," intended to bring together expert investigators to do a fresh evaluation of the Roswell incident, and hopefully obtain long-elusive proof that the crashed saucer story was real.

UFO investigator Anthony Bragalia was a member of this team, as were Kevin Randle, a prominent Roswell proponent and author of many UFO books, and the Canadian investigator and author Chris Rutkowski. However, the harmonious Roswell dreaming was soon interrupted. Randle and Schmitt had been partners in earlier Roswell investigations during the 1990s. However, when it was discovered that Schmitt had falsified his credentials, among other things, Randle denounced him and severed all cooperation. After about twenty years, they were beginning to reconcile when the slides turned up. Randle stuck his toe into the water (or perhaps his whole foot), but didn't like what he saw, and withdrew from the effort. (In early 2015, as the slides fiasco gathered intensity and momentum, Randle sent me an email essentially saying, "I hope you realize that I have nothing at all to do with these Roswell slides!" I assured him that I did.) Rutkowski says that he was approached about being a member of the Dream Team, but when he expressed some reservations about the slides, his "membership" offer was withdrawn.

The remaining members of what should now be called the Slides Team apparently had no reservations whatsoever. The slides were supposedly being investigated by the best photographic and other experts, who said they appeared to be authentic. The cardboard mountings of the slides were said to prove that they must have been processed during the 1940s. The only problem was that nobody outside that group had actually seen the slides and the details of the supposed investigations were hazy. We were assured that when the time was right and the investigations were complete, this "smoking gun" evidence of the Roswell crash would be released to the world (a theme familiar to veteran skeptics).

For about two years, the existence of the slides was known mostly just to those who follow UFO-related blogs and such, and their content was only rumored. As might be expected, curiosity about them was building, along with a properly skeptical "wait and see" attitude. Then at a public forum in November 2014, Tom Carey announced, "We have come into possession of a couple of Kodachrome color slides of an alien being lying in a glass case. What's interesting is, the film is dated 1947. We took it to the official historian of Kodak up in Rochester, New York, and he did his due diligence on it, and he said yes, this filmstrip, the slides are from 1947. It's 1947 stock. And from the emulsions on the image, it's not something that's been Photoshopped like today. It's original 1947 images, and it shows an alien who's been partially dissected lying in a case." He described the being as "three and a half to four feet tall, the head is almost insect-like. The head has been severed, and there's been a partial autopsy; the innards have been removed, and we believe the cadaver has been embalmed, at least at the time this picture was taken. The owners of the slide—it's an amazing story. The woman was a high-powered Midland, Texas, lawyer with a pilot's license. We think she was involved in intelligence in World War II, and her husband was a field geologist for an oil company." This was widely reported in the press.

Dew released a professionally-produced teaser for his in-production documentary motion picture titled *Kodachrome* to hype the slides, from which blurry images of the two slides leaked, perhaps intentionally. A loosely-organized group of independent investigators came together, calling itself the Roswell Slides Research Group. But it is difficult to investigate something that one is not allowed to see except as a small, low-resolution icon. Anthony Bragalia wrote:

> Of the many scientists, PhDs, photography experts and other researchers who are among the very fortunate to have viewed the "alien slides"—not one has ever at any time mentioned that the 3 foot thing depicted in the 1947 photographs resembles a mummy. This includes KODAK experts, a NASA scientist of international standing who has left comments on his impressions of the creature on this blog and several UFOlogists. The creature depicted in the slides

(owned by an Oil Exploration Geologist in NM in the 1940s) in no way even remotely appears like any creature known on Earth. (See http://goo.gl/IfbmOa.)

Surprisingly, the "alien" body seems to have a placard on the shelf next to it. Why would a top-secret dead alien hidden away in Area 51 sport a placard, like a mummy in a museum exhibit? What would such a placard say? "Dead alien from Roswell. Top Secret—Don't Tell Anyone!" Was it possible to read the writing on the placard? Tom Carey said "there's a placard, very fuzzy, that can not be legibly read by the naked eye, yet we've had everyone from Dr. David Rudiak, to Studio MacBeth, even the Photo Interpretation Department of the Pentagon, as well as Adobe have all told us that it's beyond the pale, that it cannot be read, it is totally up to interpretation."

A date was finally set to reveal the slides: May 5, 2015, in Mexico City. A big extravaganza was being planned to reveal the slides on the holiday Cinco de Mayo, and it was organized by Mexico's best-known UFO huckster, Jaime Maussan. If the slide promoters were seeking credibility (as opposed to a quick buck), they could not have made a worse choice. A well-known sensationalist journalist, Maussan is Mexico's very own P.T. Barnum, having made a lucrative career peddling dodgy photos and videos of UFOs, alien beings, and the like. He previously promoted a skinned dead squirrel monkey as an alien creature, and even published a photo of what is supposed to be "un caballo en el cielo"—a horse flying across the sky. The Mexican website alcione.org lists "more than 40 frauds of a pseudo-journalist and charlatan," Maussan.

Maussan hyped the slides shamelessly, promising to reveal a Roswell "Smoking Gun" on May 5. This was your opportunity to witness an event that would change history! About 6,000 tickets were sold priced between US $20 and $100 (according to Ticketmaster in Mexico); some accounts claim that up to $350 per ticket was paid. Thousands paid $15 or $20 to watch the bilingual event on streaming internet video (which did not work well, angering many). The Twitter feed of those watching the streaming video (#RoswellSlides) was overwhelmingly negative, with most commenters mocking the presentation.

Promises to release high-resolution copies of the slides after the presentation did not materialize. However, at least one "high enough" resolution copy of the slide showing the placard did leak out. The French skeptic Nab Lator of the Roswell Slides Research Group quickly used the commercial software Smart Deblur to read the placard. The first line clearly read, "MUMMIFIED BODY OF TWO YEAR OLD BOY." Others quickly confirmed that finding. Researcher and satellite orbit guru Ted Molczan commented, "You folks solved in no more than 2–3 days what the promoters claimed not to have been able to solve in 3 years!" Perhaps the slide promoters had an incentive *not* to be able to read the placard.

When the unblurred copy of the placard was released on the Internet, Dew was furious. He called the Roswell Slides Research Group "a group of internet UFO trolls, claiming to be searching for truth but repeatedly spreading lies." He claimed that they created a "fake placard" using Photoshop, and hastily posted a copy of the "authentic" blurred placard on his own website, removing all doubt as to the provenance of the placard image. However, instructions were soon posted by slide debunkers showing how to take the copy of the placard *from Dew's own website*, and de-blur it to read at least the top line in less than two minutes using a trial copy of Smart Deblur. As for Bragalia, he quickly issued a somewhat tepid *mea culpa*, saying "I must be less trusting, more discerning and less accusatory of those with whom I disagree," and he blamed Dew for withholding the high-resolution scans from independent analysis. Considering that Bragalia had earlier called the detractors of the yet-unseen slides "rabid slide-skeptics" and even worse, some think that he has a lot more apologizing to do.

Amazingly, in the same posting in which Bragalia switched sides on this, he claims to have identified the specific mummy trumpeted as the "Roswell alien": it is a child mummy from Native American cliff dwellings in Colorado, at one time on display in a museum in Mesa Verde National Park. About a week later, Donald Schmitt threw in the towel saying, "I sacrificed my better judgement by being overly trusting when I should have known better." Tom Carey said that the matter "is still open to debate." As of this writing, there is no *mea culpa* from slide promoters Maussan, Dew, and Richard Dolan, with Maussan still insisting that the slides show an alien, joined by other UFO delusionists like Linda Moulton Howe and Whitley Strieber. The most interesting rationalization thus far comes from Dr. Richard O'Connor, M.D., who acknowledges that the placard does indeed read "MUMMIFIED BODY OF TWO YEAR OLD BOY," but it is a deliberate deception.

So while the fiasco of the Roswell Slides was a huge embarrassment to "Roswell research," and to UFOlogy in general, there were nonetheless some hopeful aspects of it. Many well-known pro-UFO researchers were very skeptical of claimed "smoking gun" photos of unknown origin and content, including Stanton Friedman, Kevin Randle, and Nick Pope. More encouraging still was the excellent cooperation between skeptics and UFO proponents, instead of the usual acrimony. Skeptics like Tim Printy, Nab Lator, Gilles Fernandez, and Lance Moody worked alongside open-minded UFO proponents like Paul Kimball, Curt Collins, Isaac Koi, and Chris Rutkowski to cooperatively solve the riddle of the Roswell Slides. Both skeptics and "skeptical believers" agree that the UFO field, as it now stands, is filled to the brim with rubbish. The latter group expects that, when the rubbish is cleared away, there will be a signal in the noise, while the former expects that nothing will be left. But both are natural allies in clearing away UFOlogical rubbish. It also allows us to identify those UFO researchers who are hopelessly mired in delusion, and still insist that the "Roswell Slides" do not show a mummy, even after deciphering the placard proclaiming that they do! The real fault line in UFOlogy lies between "UFO realists"—skeptics and skeptical proponents who are willing to look for weaknesses and prosaic explanations for UFO claims—and the "unrealists" who are ready to accept practically any exciting UFO claim on very little evidence.

CHAPTER 44

The Bermuda Triangle Mystery Delusion

LOOKING BACK AFTER FORTY YEARS

Larry Kusche

Vol. 39, No. 6
November/December 2015

Forty years have passed since my book *The Bermuda Triangle Mystery—Solved* was published in 1975. The most important chapter is "Flight 19," the account of the five Navy Avenger torpedo bombers and a Martin Mariner PBM that disappeared on December 5, 1945. Flight 19 was on an overwater navigation-training flight from Fort Lauderdale, Florida, to the Bahamas and back. The Mariner was searching for the Avengers after they got lost.

The disappearance of Flight 19 is the most famous, dramatic, complicated, and relevant incident in the part of the Atlantic Ocean off the southeastern coast of the United States that, twenty years later, would become known as the mysterious Bermuda Triangle. The original mystery story of Flight 19, as it was told for decades by those who did little or no research in authoritative sources, when compared to the true and accurate account—based on my research that included the official Navy report of the disaster, the personal records of flight leader Charles C. Taylor, the ninety-two personal interviews that I conducted, and my flight of the route—is a microcosm of how the mystery/delusion of the entire Bermuda Triangle story came about, and how I came to realize that the Bermuda Triangle is one of the biggest frauds/delusions that has ever been perpetrated. Flight 19 is such a significant part of the Triangle story that, if the planes had safely returned to base, the concept of the Bermuda Triangle would never have been created. We would never have heard of the Bermuda Triangle, and all the articles, books, documentaries, movies, and websites about it would never have been created.

The loss of the Avengers (see figure below) and the search plane was a legitimate, confusing, national front-page mystery at the time it occurred. Years later, magazines and newspapers began to publicize it and other supposed mysteries in

A Navy Avenger torpedo bomber. The pilot is Willard Stoll, one of the last men to talk with instructor-pilot Lt. Charles C. Taylor before Taylor left on ill-fated Flight 19 on December 5, 1945. Author Larry Kusche interviewed Stoll in his Michigan home. Stoll, the leader of Flight 18, heard Taylor on the radio when he was in the northern Bahamas. He told Kusche that Taylor "couldn't have been too far away at the time." *Courtesy of Willard Stoll.*

the area. UFOs were a new, popular, and exciting topic in the 1950s. Best-selling books such as *Flying Saucers on the Attack, The Case for the UFO, The Flying Saucer Conspiracy, Strange Mysteries of Time and Space, Stranger than Science,* and others speculated that Flight 19 had been captured by aliens from outer space.

Early in the popular 1977 Spielberg movie *Close Encounters of the Third Kind,* the Avengers suddenly appeared in the Mojave Desert. Near the end, the aviators, who had not aged, walked out of a huge UFO.

The part of the Atlantic Ocean where Flight 19 disappeared became popular when it was given a clever, catchy name by Vincent H. Gaddis in his February 1964 article "The Deadly Bermuda Triangle" in *Argosy,* a popular men's pulp adventure and modest girlie magazine. A shortened version appeared in the July/August 1964 *Flying Saucer Review.*

The Argosy article, with several additions, then became Chapter 13, "The Triangle of Death," in Gaddis's 1965 book *Invisible Horizons; True Mysteries of the Sea.* In what is a rarity among those who would later add to "the mystery," Gaddis listed the sources of information he used in the book:

Various newspaper articles. *The Books of Charles Fort. Sea's Puzzles Still Baffle Men in Pushbutton Age.* E. V. W. Jones, Associated Press, Sept. 16, 1950, was the first to mention the region as a place of mysterious disappearances. *Sea Mystery at Our Back Door,* George X. Sand, *Fate Magazine,* Oct. 1952, expanded on the Jones article.

The 1950s UFO bestsellers previously listed.

The Mystery of the Lost Patrol, Allan W. Eckert, *American Legion Magazine,* April 1962, which included fictionalized, dramatized messages that would be attributed to Flight 19 by many of the later mysteryans. [*Mysteryans* is my shortcut word for those who contributed to the creation of the mystery.]

Gaddis wrote:

Draw a line from Florida to Bermuda, another from Bermuda to Puerto Rico, and a third line back to Florida through the Bahamas. Within this roughly triangular area, known as the "Bermuda Triangle," most of the total vanishments have occurred. Others have happened in adjacent areas to the north and east in the Atlantic, south in the Caribbean, and west in the Gulf of Mexico.

Gaddis did not define how far the "adjacent areas" extended, while adding, "This relatively limited area is the scene of disappearances that total far beyond the laws of chance." He gave no information as to what were the laws of chance, but it sounded scientific, as if the matter had been thoroughly studied. Map 1 (based on my research) shows the intended routes or known locations of some of the mysteries in the "adjacent areas." Many of the other alleged "mysteries" did occur in or closer to the Triangle, but the details of their losses were often inaccurate.

Gaddis's writings started the tsunami of Bermuda Triangle magazine and tabloid articles, books, documentaries, movies, and the popular belief in unknown forces off the coast of the United States. His account of Flight 19 was largely based on the fictional quotations in the *American Legion* magazine article.

Many of the later mysteryans accepted Gaddis's version and then embellished it with their own speculations rather than doing original research. Some added another alleged mystery or two.

My source of information about Flight 19 in Chapter 22 of *The Bermuda Triangle Mystery—Solved* was the official Navy report: "Board of Investigation into five missing TBM airplanes and one PBM airplane, convened by Naval Air Advanced Training Command, NAS Jacksonville, Florida, 7 December 1945, and related correspondence." No one who had declared the loss of Flight 19 to be paranormal or UFO-related used the official Navy report.

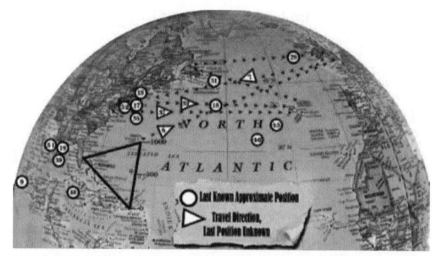

Map 1. Many of the "mysteries" attributed to the Triangle did not occur in or near it, as shown. Virtually all "mysteries" had solutions, or had no proof that they even occurred, when diligently researched and honestly reported. The numbers are chapters in the *Solved* book. *Map by the author.*

When I first heard of the Bermuda Triangle in the early 1970s, I realized it was quickly gaining in popularity. There were several reasons for my interest. First, it was an obviously unique and intriguing topic. Ships, planes, boats, and people were said to be disappearing off the coast of the United States! There were no survivors, no wreckage, no SOS, no clues. I had always wanted to write a book, but because I was entering a new profession and had a young family, there had never been a topic that so captured my interest that I was willing to embark on what would obviously be a huge research project. Second, I was an experienced pilot. By my early twenties I was a commercial pilot, flight instructor, instrument pilot, instrument flight instructor, advanced ground instructor, and flight engineer. I had logged several thousand flying hours, including more than four thousand takeoffs and landings, much of it as an instructor.

Third, my second master's degree was in library science. Back then I was working in the Reference Department at Arizona State University's Hayden Library, where we were barraged by student requests for information about the Bermuda Triangle for the term papers they had to write. Librarians know how to do research, which in the 1970s was far more difficult than it is with the technologies that exist today.

We could not find much about the Triangle, so I placed an ad in several library journals and soon received a large collection of magazine and newspaper articles, which I made available to the students. In retrospect, those students and many others across the country were a significant factor in the early growth of the Triangle story. Just as much false information today is spread by the Internet and social media, back then we knew that everything we found was not necessarily accurate. Today I still get letters and emails from students who are writing papers, only now they also have my information to use.

It might not seem likely, but pilots and librarians have an important common attribute. They absolutely hate to make a mistake, to be wrong. A librarian's primary job is to help people find the information they seek, which can often be complex and difficult. The inability to find accurate information is failure at the job and embarrassing.

If a pilot makes a mistake, he and others can end up dead. I had known a few people who died in flying accidents: A top-notch crop duster pilot for whom I was a loader while in high school. Two friends who learned to fly the same time as I did crashed on a small dirt runway on a sloping hillside that I often used. One of my high-school flying students died when his father crashed his plane. I remain adamant about not making "misteaks" (a small tweak of humor here).

After deciding in the early 1970s to write a book about the Triangle, my research consisted of locating the earliest accounts of the mystery incidents in legitimate sources rather than accepting the previous writers' versions and speculations. Searching for the mystery incidents was time-consuming, but many were listed in *The New York Times* and London *Times* indexes. I located the articles on microfilm and had them printed. I also received information from Lloyd's of London, the Navy, and the U.S. Coast Guard by U.S. mail. Most of the older incidents had no such thing as an official report.

My experience as a pilot was invaluable when analyzing many of the "mysteries." One case that was reported in a best-selling Triangle mystery book showed the writer's lack of flying knowledge. His story was that a plane flying in the Bahamas had lost the use of its hydraulic system, so the landing gear could not be lowered, but "the plane seemed to be landing as if buoyed up by a cushion of air." Every pilot I tell this story to laughs. Any student pilot with just a few flying hours knows what the "mysterious force" is. Ground effect is the increased lift and decreased aerodynamic drag that a wing generates when it is close to the ground or water. It can cause an aircraft to float far along the runway at a slowly decreasing airspeed before it finally settles in.

Another issue that confuses mysteryans is the difference between true north and magnetic north. When I was flying in Arizona in the 1960s, the difference was 12.5 degrees, and we had to adjust for it when navigating cross country. It changes minutely over the years as Earth's magnetic pole slowly moves. The difference is now 10.6 degrees. Mysteryans claim it is a problem in the Triangle because the difference there is near zero and that can confuse navigators, causing them to get disoriented and disappear. Actually, navigation is simplified because adjusting for zero is quite easy. Duh.

My research showed that the mysteryans rarely mentioned anything that conflicted with their "mystery," especially hurricanes and other severe weather. A well-known suicide that explained an abandoned boat found closer to the Azores than to Bermuda was ignored. See incident 46 on Map 1. Many "mystery" locations were thousands of miles away from reality. There was no credible evidence to support the oft repeated tales of only a canary being found on board, or of still-warm meals on the table of an abandoned ship.

In *The Disappearance of Flight 19*, published in 1980, I delved further into that case. After a lengthy search using the most advanced research methods of the time (library reference material, the U.S. mail, and long-distance phone calls), I located the (now deceased) sister and brother-in-law of Charles Carroll Taylor, the instructor pilot of Flight 19. I stayed at their home in Corpus Christi, Texas, several times in the late 1970s as they introduced me to local people who told me what they knew about Taylor. They visited my Arizona home, loved my swimming pool, and became friends with my family, especially my late father. We visited the Grand Canyon. They loaned me a large collection of reports, papers, letters, photos, and other correspondence that Taylor's late mother and aunt had gathered. An article in the *Corpus Christi Caller* included Taylor's picture and an interview: "Lt. Charles C. Taylor is visiting his mother after 28 month's service aboard a carrier in the South Pacific. . . . He will report to the East Coast for reassignment." I located and interviewed ninety-two people who had known Taylor. Childhood friends. Navy buddies and other personnel who were somehow involved in the events of December 5. The radioman and gunner who had flown with him off the carrier Hancock during the Pacific war. They told me he was an excellent pilot and the perfect southern gentleman who was proud to be a U.S. Navy pilot.

Another part of my research was to visit the Confederate Air Force in Harlingen, Texas, on April 3, 1978. (The name was changed to Commemorative Air Force in 2002.) I examined their huge single-engine Avenger torpedo bomber on the ground, then (as a passenger) headed out over the Gulf of Mexico for a loud, exciting, and informative ride.

Two days later I rented a Cessna in Fort Lauderdale (I was still an active pilot back then) and flew the intended route of Flight 19 to the Bahamas, landing for fuel at Walker's Cay, the chain's northernmost island. I then flew back across the Gulf Stream and the Everglades to Key West, Florida, refueled a second time, then flew along the Florida Keys on the way back to Fort Lauderdale.

The reason for my seven-hour flight was to see what Charles Taylor had seen on his flight and what he should have seen. Although the training group he was in charge of took off from Fort Lauderdale at 2:10 in the afternoon and headed *east* toward the Bahamas, an hour and a half later he radioed that he was sure he was in the Florida Keys, which are *southwest* of Fort Lauderdale, virtually in the opposite direction they had flown. Map 2 shows the Bahamas, east of Florida, and the Florida Keys, west of southern Florida.

"I don't know where we are," Taylor was heard to say on the radio. "We must have got lost after that last turn. I'm sure I'm in the [Florida] Keys and I don't know how to get to Fort Lauderdale."

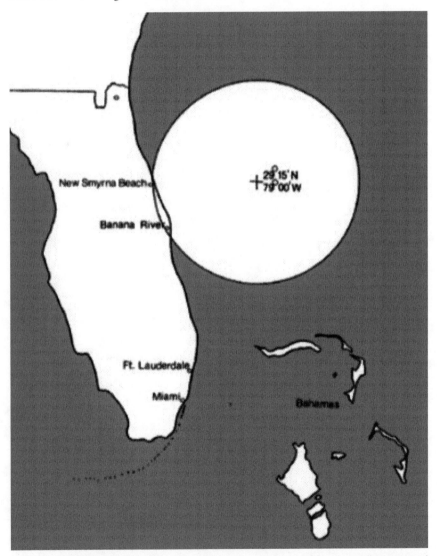

Map 2. Flight 19's 5:50 pm position fix was a circle of 100-mile radius, several hundred miles north of where Flight 19 should have turned west toward Fort Lauderdale. *Navy Department.*

He said "both my compasses are out." Taylor did not know where he was, which direction he was flying, or which way to go. He mistakenly identified himself as MT-28, which meant "Miami torpedo bomber." That revealed that, mentally, he was flying out of Miami; those flights were performed in the Florida Keys, not the Bahamas. His correct ID, FT-28, Fort Lauderdale torpedo bomber, was eventually learned.

Dead reckoning navigation, which Flight 19's students were practicing, does not use landmarks. It was used when flying over the ocean, out of sight of land. They fly in specific directions, according to their compass heading, for a specific length of time, based on the estimated wind direction and speed. Progress is marked on an erasable plotting board (see figure below). Taylor had extensive dead reckoning experience during his time in the Pacific war.

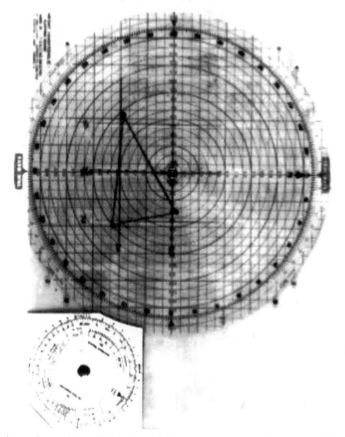

Plotting board used by pilots when flying over water in the 1940s. They mark their estimated course based on the estimated air speed and the estimated wind. A whole lot of estimating goes on when out of sight of land. *Courtesy of Larry Kusche.*

Prior to flying combat in the Pacific, Taylor had been based in Miami for a year, flying patrol over the Florida Keys and the Gulf of Mexico, watching for German U-boats. After his combat time in the Pacific, he was again based in Miami as an instructor. According to the information sent to me by the National Personnel Records Center, Taylor was "detached" to Fort Lauderdale on November 20, 1945, where the training flights went east to the Bahamas. Taylor reported for duty at Fort Lauderdale the next day, along with hundreds of other officers, student pilots, and enlisted airmen. He was flying again by December 1, but it is not known if he flew the Bahamas route before December 5. My conclusion, based on all my sources of information, is that this was his first time.

On December 5, when Taylor called, he was confused because there *are* parts of the smaller islands of the northern Bahamas (where he actually was) that do look like some of the islands in the Florida Keys. I saw both areas on my flight to the Bahamas and the Keys three decades after Taylor's fatal flight. The official Navy report states that Taylor "allowed himself to be led to believe he was in a position in which he could not possibly have been." Captain William O. Burch, commanding officer of the Naval Air Station, told Taylor's mother and aunt he thought Taylor had confused the string of cays north of the Bahamas with the keys south of Florida.

Some of the writers and websites report that Taylor was confused because his compasses failed, and that is what caused Flight 19 to get lost and eventually disappear. That story, which has largely been accepted, has only added to the mystery. What kind of mysterious force, the mysteryans ask, would cause compasses to fail? What other disappearances has this strange force caused? After all, compasses, whether mechanical or fluid–filled, are known to be extremely trustworthy. Distrusting a compass could be a response for a pilot when the landmarks he sees do not tally with his compasses, especially after he has made some turns. A former Avenger instructor pilot told me how easy it was for him to deliberately disorient a pilot just by doing a few turns to see how quickly he could reorient himself. When I was a flight instructor, I discussed with my students the issue of absolutely trusting the compass before I signed for them to take their solo cross-country flights. Shortly after takeoff from Fort Lauderdale, the planes of Flight 19 were to perform low-level bombing practice runs at Hen and Chickens Shoals, fifty-six miles east of the Naval Station. That would involve turns and altitude changes, which can also be disconcerting, especially at diving speeds and higher g forces.

Believing he was in the Florida Keys where he had previously spent many months flying, Taylor refused to head west, as pilots had been instructed to do if they were lost or confused in the Bahamas. If he had actually been in the Keys, as he mistakenly thought, flying west would have taken them into the huge Gulf of Mexico. He insisted on heading north because he was certain he was south of Florida. One of the stark truths of flying (and hiking) is that the longer a person stays lost, the more confused and the more lost he becomes, which is even worse if night is approaching. (At least a hiker can stop moving to try to get reoriented.) Flight 19 continued to fly north from the Bahamas, despite at least one of the pilots saying they should fly west. But Taylor's order had to be obeyed; he was the commanding officer.

Port Everglades Air Sea Rescue Unit 7 heard several messages between Taylor and the other pilots concerning their estimated position, their compasses, and which direction they should go. As best they could tell, no other plane ever assumed the lead.

Navy personnel had no idea where Flight 19 was until 5:50 pm, a half hour after sunset, nearly four hours after takeoff. An approximate position fix based on some of their radio calls was calculated. Flight 19 was somewhere in a huge area, 200 to 400 miles north of the Bahamas and sixty to 260 miles northeast of Cape Canaveral! (See Map 2.)

By that time, it had turned dark and stormier over the Atlantic. Fuel was running down and a strong wind was blowing out to sea. Turbulence, storm clouds, and the setting sun made it even more desperate. The only chance they had to survive was to immediately turn west.

But none of the Navy stations could contact the flight to let them know where they were and what to do. Radio contact had faded as the planes moved farther away and the sun had set. The atmosphere changes at night, drastically affecting radio reception: static and background noise increases, as does interference from commercial radio stations, especially the powerful music coming from Cuba. No one sent "blind" broadcasts, hoping that someone in the flight would hear it. The storm worsened after darkness fell; they surely were flying on instruments in turbulent air.

An old cliché is that flying is 99 percent boredom and 1 percent stark raving terror. I never did feel the boredom part during my flying career, but on a flight in northern Arizona in 1964 I had a scary hour-long experience in a Piper Comanche, concerned about if I would manage to survive. I was over a mountainous area at 17,500 feet (the plane carried no oxygen), trying at full throttle and full cabin heat to outclimb or outrace the dark thunderhead clouds that were rapidly developing below me as far as I could see. I was in serious trouble, but my anxiety and fear surely were nothing compared to what the men of Flight 19 must have felt, knowing they were doomed to go down in the cold, raging ocean in the darkness, with no chance of survival. I survived by spotting a small break between the huge clouds, diving between them, hoping the break would not close, hoping I would not crash into a peak, and then, miraculously, seeing an abandoned strip on a hillside and making a safe landing. I spent the night at a nearby ranch. I played a lot of pool.

On December 6, 1945, the day after Flight 19's disastrous flight, more than 200 planes and seventeen ships from Florida and the Bahamas went out by dawn, braving high winds and heavy seas. Some small boats and their crews were rescued, but no trace, no one from Flight 19 could be found.

The Martin Mariner search plane that became a part of the mystery was seen (from a search ship) to explode in the air on the night of December 5 at 7:50. Mariners were called "flying gas tanks" because they carried almost 2,000 gallons of fuel. The next morning the ocean was so rough that not a trace of the plane or its crew could be found, even though they knew where it had come down. One officer I interviewed said that fumes were occasionally present inside the plane, and that he, while on duty in Greece, had seen a Mariner explode in the air.

Taylor's flying skills and combat experience surely were extensive and heroic during the war, but I did interview two men who were with Taylor other times when he ditched. The first was his gunner, who told me "Taylor lost his bearings" on June 14, 1944, near Trinidad. They ran out of gas, could not get the raft out before the plane sank, and their depth charges blew up beneath them, but they were quickly rescued. The other ditch was January 30, 1945. After losing radio contact, Taylor couldn't find Guam. He climbed so they could find him on radar. Almost out of gas, he ditched in a rough sea. He and his passenger spent a wet, cold, turbulent night in a raft and were saved late the next day. Photo 19 in the Flight 19 book shows Taylor in the raft, being rescued by the USS Bailey. December 5, 1945, was the third time he got lost. All these years after the publication of both books, I still get the question: "How did you solve the Bermuda Triangle mysteries?" The question I rarely get, which shows a more analytical mind, is: "Did the alleged disappearances actually occur the way the mystery writers said they did?" The answer is no, they did not.

Map 1 shows the Triangle, a thousand miles on a side. While the mysteryans claim it is small and limited, roughly half of the so-called unsolved mysteries occurred far from the Triangle; thousands of miles away in some cases. The numbers on Map 1 are the chapters in *The Bermuda Triangle Mystery—Solved*. "Research" by the mysteryans was so poor that one ship that had been dismasted and lying on its side (obviously a storm victim) in the Pacific Ocean was included as a Triangle mystery.

Some occurred closer to Newfoundland and the route of the *Titanic* than to Bermuda. Many of the "mysterious losses" occurred because of storms, even hurricanes that the mysteryans did not mention. Today, the Bermuda Triangle is not the hot topic it was forty years ago, but it still gets plenty of interest. The numbers vary considerably, but as I write this, a Google search returns 3,650,000 entries for it; Flight 19 shows 285,000,000; my name shows 11,000. Some of the websites I perused show much of the same old stuff: accepting the mystery stories, then listing various "theories" such as death rays from Atlantis, a drawing of a UFO over Columbus's ship, and so on. Some of the sites do present good information; many refer to my two books.

The concept of mysterious disappearances has become part of the language. In baseball, a batted ball that cannot be caught is said to have gone into the Bermuda Triangle. It is where money that disappeared in the stock market went, and where anyone who is lost is said to be. The disappearance of Malaysia Airlines Flight 370 in March 2014 in the Indian Ocean has been mentioned as a Bermuda Triangle–type situation, as expected.

The word *solved* in my title elicited comments from some who apparently believe that a solution had to involve UFO captures, death rays, or some other mystery solution. My solution is that the Bermuda Triangle is a fraud, a delusion, and that reality is far more interesting than the phony stories.

Someone wrote that I was a debunker, that my purpose was to snitch on those who I claim got it wrong. That comment is also devoid of reality. When I began my research, I had no idea how true or untrue the mystery stories were. Early on, I hoped to find that the mysteries were true, because I knew the first hardcover book on the market would be a huge bestseller. I wanted mine to be the first! Then reality reared its head. Good, honest research, which takes much more time than cobbling together "mysteries" that previous writers have created, revealed that virtually every incident had been distorted to make it look mysterious. Thus, I did not get the pleasure of having the first hardcover book to be published.

At the end of *The Bermuda Triangle Mystery—Solved*, I stated that the Triangle was a "manufactured" mystery. That was a polite way to say it was a fraud. The "mystery" of the Bermuda Triangle is one of the most widespread frauds that has ever been perpetrated. It was based on poor research and distorted, untrue, inaccurate information that was uncritically copied, embellished, and sensationalized. Does it matter if people believe that forces "beyond the laws of science as we now know them" are capturing ships, boats, planes, and people, and that scientists, the Coast Guard, Navy, Lloyd's of London, and other experts are said to be baffled by it all? If those alleged unknown forces in the Triangle are only light entertainment for the masses, does it matter that some people believe in them? So what if they believe in UFO captures, psychic powers, astrology, ancient astronauts, and ads to lose thirty pounds by the end of the month? What is the harm in it? Few of us will ever be affected by whether there are unknown forces in the Triangle or anywhere else.

Actually, there *is* an issue of greater importance than whether "paranormal forces" are at work anywhere. In this age of information explosion and social media, it is worrisome that so many people believe so many things without requiring any supporting evidence, that they employ little skepticism, have such a lack of curiosity, and such a bias toward what they want to be true, that they ignore what is true. Once false information becomes "common knowledge," no matter how thoroughly it might be shown to be false, the false version will continue to be believed by some, either because they remain uninformed about the correct information or because they refuse to accept any information that is contrary to the beliefs they hold.

The need for skepticism, for paying close attention to detail, is of critical importance in everyday life. A healthy dose of skepticism might have saved billions of dollars from disappearing during the dot com debacle in the early 2000s and the financial meltdown in 2008. It would have kept millions of dollars from vanishing during the more recent Ponzi schemes.

Skepticism and critical thinking are important in politics when voters let their emotions rule rather than becoming informed on the positions of the candidates. It is important in issues of health, such as the vaccination/autism controversy, which is resulting in diseases that were virtually wiped out to start coming back. It is important in the discussion of the use of genetically modified food and the global warming situation.

Skepticism. Critical thinking. Honesty.

Endorsements of Kusche's Decisive Investigation

Editor's note: Prominent scientists, skeptics, naval experts, and historians endorsed Larry Kusche's 1975 book *The Bermuda Triangle Mystery—Solved*. Here are some of those comments:

- Carl Sagan, Astronomer: "Your book is a welcome alternative to the standard credulous and uncritical works on the subject."
- Isaac Asimov, Scientist, Author: "Reading it was like drinking a glass of sparkling spring water after being offered nothing but mud."
- T. M. Dinan, Lloyd's of London: "This book is essential reading, for here the mystery is solved."
- Admiral O. W. Siler, Commanding Officer, U.S. Coast Guard: "Your position is wholeheartedly endorsed by the Coast Guard."
- Walter Sullivan, Science Editor, *New York Times*: "Thank goodness there are a few rational human beings dealing with this subject."
- Martin Gardner, *Scientific American*: "Kusche's research is impeccable, his arguments unanswerable. His solution will convince everybody except those simple-minded believers who refuse to listen to anyone except the hucksters of irrationality. . . . I agree with him totally."

- James Randi: "Congratulations on the quite excellent job you've done to bring some rationality to the very aggravating Bermuda Triangle 'mystery.'"
- Philip J. Klass, Author of *UFOs Explained*: "The book should resolve all doubts."
- J. Allen Hynek, Director of the Center for UFO Studies: "Head and shoulders above what else has been written. It is bound to clarify matters for all people who think."
- Samuel Eliot Morison, Naval Historian: "Anyone who has been bemused by the Bermuda Triangle literature, which mostly belongs in the categories of myth and fable, should read Mr. Kusche's book. You will find there an objective estimate of all the alleged mysteries of the alleged triangle, and learn that most of them can easily be explained by natural causes."

Larry Kusche is the author of *The Bermuda Triangle Mystery—Solved* (1975) and *The Disappearance of Flight 19* (1980). The books set high standards for investigative research and reporting on popular subjects. They demonstrate the need for critical thinking and being a skeptic. He has been a commercial pilot/flight instructor and a technical writer, and he was a founding CSICOP fellow. He has lived in the Phoenix area since the second grade (1947).

CHAPTER 45

Magic Waters

Joe Nickell

Vol. 43, No. 5
September/October 2019

Water's ability to cleanse obviously made it a natural choice for ritual washing (such as baptism in Christianity and *mikvah* in Judaism). And its power to soothe a minor burn, quench a thirst, or provide other relief naturally inspired further medicinal uses. Belief in the supernatural extended the supposed powers of water in various ways—hence this overview of what we may simply call "magic waters." It is divided into two parts, Natural Water and Imbued Water.

Natural Waters

The "natural waters" category of magic waters consists of those directly produced by nature—no matter how people may regard them. Examples are those that are "miraculous" (such as the River Jordan or the spring at Lourdes), talismanic (say, kept in a sealed vial as a charm), mineral-enriched (such as those of secular "healing" springs), and fakeloric (such as the pseudo-legendary Fountain of Youth in St. Augustine, Florida).

MIRACULOUS

So important was water to the ancients that temples—and even cities and empires—were built around it. Often considered sacred, these rivers, lakes, and springs played central roles in the lives of worshippers.

Such was the status of the Jordan River, which connects the Dead Sea and the Mediterranean and is sacred to two faiths. For Jews, it miraculously parted for Joshua to lead his people on dry ground across the Jordan and into the

Promised Land (Joshua 3–4). For Christians, it was the water in which John the Baptist baptized Jesus (Matthew 3:13–16). It was also, for both, a source of healing, as told in 2 Kings 5:14, where Naaman bathed in Jordan's waters and was cured of leprosy.

In ancient Greece, springs were believed to have supernatural powers because they were the dwelling places of gods. (Sacred waters can be differentiated from secular "healing" springs, discussed in a subsequent section.)

Around the world are alleged "miracle" springs, many promoted by Roman Catholics. The most famous of these is Lourdes in southern France. There, in 1858, fourteen-year-old Bernadette Soubirous (1844–1879) claimed to see the Virgin Mary, who directed her to the spring at the back of a grotto. Soon tales of miraculous healings surfaced, attributed to Bernadette. Ironically, the healing powers of the water were not for her, who died young. That fact did not provoke much skepticism, and she was nevertheless canonized as a saint in 1933.

Skeptics observe that the criterion for proclaiming cures to be miraculous is that they are "medically inexplicable," but that is only engaging in a logical fallacy known as an argument from ignorance. (That is, one cannot draw a conclusion from a lack of knowledge.) Moreover, there are other explanations for apparent cures: misdiagnosis, psychosomatic conditions, prior medical treatment, the body's own healing power, and other effects.

Supposed cures were once certified by the medical bureau of Lourdes (beginning in 1884). However, in 2008 there was something of a rebellion by the doctors, who objected to being in the "miracle" business. They determined from then on only to indicate whether cases are "remarkable"—since remarkable healings can happen to anyone, independent of religious shrines and supposed magical water (Nickell 2008).

TALISMANIC

It is only a short step from conviction that the application of water from a miraculous source can be curative to the belief that a small amount contained as a talisman (for example, in a little vial or bottle) can also effect cures. (A talisman is an object believed to have magical properties, such as bringing good luck or offering protection.) For talismanic power the water does not actually need to be *used*; supposedly it works its magic even from the closed container.

It should not be surprising that pilgrims visiting the Holy Land during the Byzantine era could purchase small *ampulla*—bottles now known as "pilgrim jars"—suitable for obtaining some sacred water, as from the Jordan River, to take home as a talisman or souvenir. For Christians, the bottles were embossed with crosses—and so might alternately hold sacred oil from a lamp at the Holy

Sepulcher, a splinter from the (alleged) True Cross, or other holy item. There were similar pilgrim jars for Jews as well, embossed with menorahs (Goudeau et al. 2014).

Pilgrim jars were especially common during the period 600–900 CE, like the one shown here (see figure below) from the author's collection (measuring about 3.5" high). Of course, we can only say it *might* have held magical water from the Jordan River itself, or from a sacred spring, or the like.

Shown with more certainty is the tiny 1.75" high bottle beside the pilgrim jar—it actually being embossed "JORDAN/WATER//PAN AM/1901." As indicated, such vessels were souvenirs of the Pan American Exposition, a world's fair held in Buffalo, New York, from May 1 through November 2, 1901. More than one eBay listing has opined that the bottles held "Holy Water," but I think these sources have jumped to a conclusion. I suspect the bottles held just what they say they did, no more or less, and were not exclusively for Catholics (like those bottles we will discuss presently). These bottles—whether thought talismans or not—show the popularity of Jordan water a century ago, and bottles of the venerated liquid are still sold.

Natural waters. From left to right are containers for "miraculous" water—as from the River Jordan (Byzantine era pilgrim jar and "Jordan Water" bottle from 1901 Pan Am Exposition) and the spring at Lourdes (plastic figurine of Mary containing Lourdes Water and in front of it a boxed cross with drops of the water intended to be used talismanically); ca. 1930 amber bottle for White Rock Mineral Springs water; and a mid-twentieth century souvenir bottle of water from the pseudo Fountain of Youth in St. Augustine. *Author's photo: items from his collection.*

Also shown in the above figure is a souvenir "Lourdes Cross" containing drops of water from the famous "miracle" spring. For the superstitious, this could represent combined talismanic powers—that of the cross itself and that also emanating from the water.

MINERAL-ENRICHED

Among the earliest "healing" practices is hydrotherapy—the internal or external use of water for treating disease. Indeed, drinking from, or bathing in, springs, pools, or streams for therapeutic purposes predates recorded history. Ancient Greece and Rome placed therapeutic centers at mineral springs across their realms. In the Americas, too, native peoples also believed in the restorative powers of mineral waters; Aztec emperor Montezuma, for instance, used a spa called Agua Hedionda (Nickell 2005).

Imbued waters. Left to right, front row: patented cross-shaped bottle for holy water; a homeopathic tincture with booklet; a magnet and vial for creating "magnetic" water. Back row: New Age pyramid for making pyramid-"energized" water; and a goblet with a mineral and set of "Healing Stones" for making "gem-informed" water. *Author's photo: items from his collection; healing stones gift of Barry Karr.*

In my travels I have visited many famous spas around the world. A historic one at Pozzuoli, Italy (which I visited with investigator Luigi Garlaschelli), is in a volcano that formed 4,000 years ago and last erupted in 1198. It yields sulfurous steam that was considered beneficial for respiratory ailments, hot mud used to treat rheumatism, and thermo-mineral waters with widespread curative claims (Nickell 2005).

The emphasis at a great number of European and American springs was on the natural inclusions in the water: iodine to treat goiter, lithium for manic depression, etc., as well as assorted minerals in general, the springs often being billed as "health-giving mineral waters." Actually, most such springs (such as New York's celebrated Saratoga Springs) contain dissolved salts, and it became usual to set the level at fifty grains per gallon to justify the designation "mineral water." The beneficial effects were commonly debated and probably often negligible. With exceptions, most mineral springs simply offered a drink of cool water or a soothing bath along with the placebo effect.

The tallest bottle pictured in the first figure (measuring about 8" tall) is a rather typical one for mineral-springs water. Its label bills it as "Delicious/Sparkling/Healthful." It was "Lithiated and Carbonated" (i.e., contained additives), sold by the "White Rock Mi[nera]l Springs Co." of Waukesha, Wisconsin. Interestingly, the upper label instructs, "Lay this bottle on its side / Always cool before opening." This shows us the early bottles were corked (and so predate the modern crimped-metal cap known as a "crown cork"). When a bottle was laid on its side, the cork would be kept wet; otherwise, it would dry and shrink, and the pressure of the gas would cause it to "pop" out (hence the name "pop bottle") (Nickell 2011). This White Rock bottle's label continued to bear the words quoted, "Lay this bottle on its side," for decades after the crown cork was in use—apparently for historical continuity. The earliest cork-stopped soda bottles often had rounded bottoms so the bottles could not be stood upright.

FAKELORIC

The legend of a "Fountain of Youth" may have originated in India. By the seventh century, the tale had arrived in Europe where it was widely discussed through the Middle Ages. In 1546, German Renaissance artist Lucas Cranach produced a popular painting of the miraculous spring. It depicts wrinkled, frail women entering the pool and exiting as young beauties on the other side. (But since the fountain spouting the water is graced with the statues of Venus and Cupid, it is really a metaphoric fountain of love—the true source of immortality.)

Earlier, reportedly, Spanish explorer Ponce de León (ca. 1460–1521) had searched for the fabled spring. Having accompanied Columbus on his second

The Fountain of Youth, 1546 painting by Lucas Cranach the Elder. *Public domain; Lucas Cranach the Elder, 1546.*

voyage to America (1493) and conquered Puerto Rico in 1509, he was rewarded with permission to search for a land called "Bimini." Supposedly, according to a Native American legend, that region had a fountain with marvelously curative water, and anyone who drank from it would never age. Ponce de León did land at St. Augustine and sailed through the Keys and on to Cuba but, finding nothing, abandoned his quest. Returning in 1521 intending to conquer the Native Americans and colonize Florida, he was mortally wounded by an arrow (Nickell 2005).

The spring that is now advertised as the one sought by the conquistador lacks any historical or archaeological evidence of such a connection. It was a tourist attraction as early as the 1860s, but its present form awaited purchase by Luella Day McConnell in 1904. She had returned from the Klondike gold rush of the late 1890s with money and the nickname "Diamond Lil," and by 1909 she was selling promotional postcards and well water. Until her death in an auto accident in 1927, she regaled tourists with tall tales about the site. She was succeeded by Walter B. Frazier, who made it a great attraction ("Fountain" 2018).

The souvenir bottle shown in the first figure (measuring about 1.5" in diameter by 3.75" tall) is machine made and dates from about the middle of the twentieth century. I recall drinking from the Fountain of Youth myself as a boy, and I can only say, look at me now.

Imbued Waters

This second category of magic waters comprises those that have supposedly been imbued with alleged "energy" or "power." They may be made with water that contains natural inclusions such as minerals or additives such as chlorine, etc., but they could also be made with distilled water. The important thing is that whatever may be added during the transformational process does not substantially remain; only the imbued power does. These waters are called Holy, homeopathic, magnetized, pyramid-"energized," and gem-"informed."

HOLY

Holy water is blessed by a religious figure. In Catholicism, its use is nonscriptural, dating back to ca. 400 CE. It is used as a sacramental (recalling baptism), as well as a means of purification, blessing places and people, healing, and repelling evil. During the Middle Ages, holy water was kept under lock and key to prevent its theft for unauthorized magical practices ("Holy" 2018).

Some churches offer take-home holy water at external spigots. Otherwise it is found in a font, usually at the entrance of a church. (Unfortunately and ironically, these may become sources of viral and bacterial infection [Fields 2013].)

That holy water can repel evil has long been believed by Catholics. St. Theresa wrote, "I know by frequent experience that there is nothing which puts the devils to flight like Holy water" ("Holy" 2018). However, in my investigations of supposed demon-possessed houses—like that on which the horror movie *The Conjuring* (2013) is based—I have discovered that "demons" are best routed simply by learning the true facts behind an "outbreak" (Nickell 2016a).

In addition to Roman Catholics, Anglicans and Eastern Christians also use holy water, as do some other religious sects, including Buddhist and Hindu traditions. In modern Wicca practice, "holy water" may be any water held sacred to the user (e.g., that from a shrine or sacred spring), rainwater collected on a special day, or water "charged" by a full moon or eclipse—among many possibilities. (I am omitting here those with added herbs or other substances) ("Magic" 2018).

HOMEOPATHIC

A type of "alternative" medicine, homeopathy takes the idea of magic water in an incredible (i.e., not credible) direction.

Homeopathy was created by Samuel Hahnemann in Germany in 1796 based on his belief that "like cures like"—that a substance capable of causing disease symptoms in a healthy person can cure symptoms in a sick person. Moreover, Hahnemann advocated "homeopathic dilution" by which the supposedly curative substance was diluted and re-diluted—in water or alcohol—to an extreme degree. In fact, scientists observe that the remaining amount of a supposedly curative substance is so infinitesimal that it cannot have any effect because not a single molecule of the original remains!

In other words, one is left with water that is just water, although homoeopathists insist the water has a "memory"—a claim that is scientific nonsense. So homeopathic remedies, while worthless, are not harmful—right? Unfortunately, no. While homeopathic products may contain only water, there are exceptions of which buyers should beware. For instance, several brands of homeopathic teething products were discovered to additionally contain belladonna (a.k.a. "deadly nightshade"). While that is allegedly used in trace dosages, supposedly to ease inflammation, laboratory analysis showed some of the products contained varying amounts of the potentially toxic ingredient. Relatively unregulated, the $6.4 billion homeopathic industry is facing increasing scrutiny by federal drug overseers (Nickell 2016b). In 2018, our international organization, the Center for Inquiry (CFI), initiated a lawsuit against CVS Pharmacy for fraud for advertising and selling useless homeopathic "remedies" (Brayton 2018). In May 2019, CFI served a similar lawsuit against Walmart (Center for Inquiry 2019).

MAGNETIZED

German physician Franz Anton Mesmer (1734–1815) postulated an invisible natural force in all living beings called "animal magnetism." It was later termed *mesmerism,* and it is now known that Mesmer was inadvertently using hypnotic suggestion to successfully relieve "the hysterical symptoms of susceptible young females" (Lyons and Petrucelli 1978, 489).

In his experiments, Mesmer used iron magnets and water "magnetized" by them for healing, and patients reported feeling "unusual currents" flowing through them. Later, Mesmer discovered he could achieve the same effect by merely moving his hands above his patients—the result of what is now understood to be the power of suggestion ("Franz Mesmer" 2018).

Today, various questionable therapies involve magnets and "magnetized" water. In fact, only iron and some other metals can be permanently magnetized. An external magnetic field applied to water will exert only a miniscule effect, and that goes away upon removal of the external field. Quackwatch's Dr. Stephen

Barrett (1998) says a marketed "Magnetic Mug" would cease to have even a very tiny effect "as soon as the liquid leaves the mug." Barrett goes on to call the mug claims "imaginative nonsense." Any pain relief or other alleged beneficial effect would be attributable to suggestion (i.e., the placebo effect).

PYRAMID "ENERGIZED"

Launched by the pseudoscientific book *Psychic Discoveries behind the Iron Curtain* (Ostrander and Schroeder 1970, 366–376), the supposedly mystical power of the pyramids became an American craze of the 1970s and beyond. Supposedly, small models of the Great Pyramid of Cheops generated some mysterious power that the authors, in New Age fashion, referred to as "energy."

Initially based on a Czechoslovakian patent, models of cardboard, sticks, wire, and other materials were alleged to sharpen razor blades(!), preserve food (especially "mummify meat"), impart a mellower taste to wine, perk up houseplants, and perform other marvels, including (when worn like a dunce cap) relieving headaches and restoring vitality. Not surprising, scientific evidence was lacking (Nickell 2004, 200–203).

On an investigative tour of Russia in 2001 (Nickell 2004, 200–206), I found pyramid power on the rise—almost literally: one pyramid I visited (the tallest of about twenty) towered forty-four meters (about 144.4 feet, or some twelve stories). It was constructed of translucent Plexiglas panels over a wood framework. It was closed when I arrived with Russian friends, Valerii Kuvakin and his wife, Uliya, but a custodian consented (following a small bribe) to show us around. The pyramid was largely empty but held cases of bottled water that were supposedly being energized for curative purposes.

A booklet I bought titled (in translation) *Pyramids of the Third Millennium*, foresaw a new physics, a new biology, and so on, even claiming that the new pyramids had reduced the incidence of cancer, AIDS, and other diseases in the areas surrounding them. However, Valerii noted the claims were published without any supporting data, and physicist Edward Kruglyakov, who had previously visited the site and suggested it to me, also found them without any scientific merit. A cordoned-off central area of our pyramid was presently "energizing" some crystal spheres, and we were cautioned that the energy there was so intense that one could lose consciousness. On impulse, I ducked under the rope and stood in that area for a time, while Valerii photographed me, barely containing his amusement. But I felt no effect whatever. No effect is what we also can suspect of pyramid power applied to water.

GEM-"INFORMED"

Supposed "information" contained in rocks, minerals, and crystals can—according to some New Age sources (e.g., Gienger and Goebel 2007)—be transferred (in some mystical way) to water. Thus, proponents say, the water has become "infused" with so-called "crystalline energy"—which is not a scientific concept. (Also known as "crystal water," it is distinguished from mineral water, which, as discussed earlier, simply contains traces of minerals.) The "infusions" can be made in various ways—as by soaking, steaming, or boiling, although one is cautioned to beware of the multitude of toxic stones! Alternately, one can insert a test tube with the stone into a container of water or stand a glass of water on a slice of the stone; thus, the stone and water never touch! It is not surprising, therefore, that proponents compare the liquid remedies to homeopathic ones (Gienger and Goebel 2007, 6).

The practice derives from the therapeutic effects that allegedly come from applying a "healing" stone to the body or wearing or carrying it as a talisman. For instance, amethyst is said to have (among other things) "a blood-pressure lowering effect." By supposedly transferring the "information" and/or "energy" to water (the stone is not intended to *dissolve* in it), the "remedy" may be utilized more effectively. For example, it can be drunk, used in a compress, generously applied over a larger area, or even bathed in (Gienger and Goebel 2007, 6–10, 18–33, 66).

Claims to the contrary, however, there is no credible scientific evidence for special therapeutic effects of gem water per se—other than as a placebo.

* * *

As these many examples show, water deemed to be in some way "magic" has existed in some form since the most ancient times, and it continues today along with newer forms, including that allegedly imbued with various so-called "energies." As a practical matter, however, we see that—except for certain natural additives such as lithium or various artificially introduced substances—water is basically water. Claims that it is magical are pseudoscientific—indeed, examples of what is called "magical thinking"—and such belief is perpetually trumped in the real world by science.

Joe Nickell, PhD, was a senior research fellow of the Committee for Skeptical Inquiry (CSI) and "Investigative Files" columnist for *Skeptical Inquirer*. A former stage magician, private investigator, and teacher, he is author of numerous books, including *Inquest on the Shroud of Turin* (1983), *Pen, Ink and Evidence* (1990), *Unsolved History* (1992), and *Adventures in Paranormal Investigation*

(2007). He has appeared in many television documentaries and has been profiled in *The New Yorker* and on NBC's *Today Show*. His personal website is at https://joenickell.com.

References

Barrett, Stephen, MD. 1998. Magnetize Your Beverages? Available online at https://www.mwa.co.th/downloadprd01/article/eng/magnets.pdf; accessed August 1, 2018.

Brayton, Ed. 2018. CFI Sues CVS for Homeopathy Fraud. Available online at www.patheos.com/blogs/dispatches/2018/07/10/cfi-sues-cvs-for-homeopathy-fraud/; accessed August 16, 2018.

Center for Inquiry. 2019. Walmart Sued for Fraud. May 20. The full text of the complaint against Walmart is on the CFI website at https://centerforinquiry.org/wp-content/uploads/2019/05/CFI-v-Walmart.pdf.

Fields, Liz. 2013. Holy Water May Be Harmful to Your Health, Study Finds. Available online at https://abcnews.go.com/Health/study-holy-water-harmful-health/story?id=20257722; accessed August 16, 2018.

Fountain of Youth. 2018. Available online at https://en.wikipedia.org/wiki/Fountain_of_Youth; accessed August 15, 2018.

Franz Mesmer. 2018. Available online at https://www.sciencedirect.com/topics/nursing-and-health-professions/franz-mesmer; accessed August 1, 2018.

Gienger, Michael, and Joachim Goebel. 2007. *Gem Water: How to Prepare and Use More than 130 Crystal Waters for Therapeutic Treatments*. Forres, Great Britain: Earthdancer.

Goudeau, Radbond, et al. 2014. *The Imagined and Real Jerusalem in Art and Architecture*. Boston: Brill, 170–183.

Holy water. 2018. Available online at https://en.wikipedia.org/wiki/Holy_water#History; accessed August 16, 2018.

Lyons, Albert S., and R. Joseph Petrucelli. 1978. *Medicine: An Illustrated History*. New York: Harry N. Abrams.

Magic formulary. 2018. Available online at www.thesmartwitch.com/The_Smart_Witch_Formulary.html; accessed August 22, 2018.

Nickell, Joe. 2004. *The Mystery Chronicles*. Lexington: The University Press of Kentucky.

———. 2005. Healing waters. *Skeptical Briefs*. Part I: Spas (September): 5–7; Part II: Miraculous springs (December): 6–7.

———. 2008. Lourdes Medical Bureau rebels. Available online at https://centerforinquiry.org/blog/lourdes_medical_bureau_rebels/; accessed August 14, 2018.

———. 2011. 'Pop' culture: Patent medicines become soda drinks. *Skeptical Inquirer* 35(1) (January/February).

———. 2016a. Dispelling demons: Detective work at the Conjuring house. *Skeptical Inquirer* 40(6) (November/December).

———. 2016b. Harmless homeopathy horror? Available online at https://centerforinquiry.org/blog/harmless_homeopathy_horror; accessed August 16, 2018.

Ostrander, Sheila, and Lynn Schroeder. 1970. *Psychic Discoveries behind the Iron Curtain*. New York: Bantam Books.

CHAPTER 46

Reasons to Believe's Continuing Assault on Science

Brian Bolton

Vol. 44, No. 5
September/October 2020

On a Sunday morning in early February 2020, I attended a presentation at a local Baptist church titled "Science and Faith: Finding Truth in All Things." The speaker was Anjeanette (A. J.) Roberts, a staff member with Reasons to Believe (RTB), a Christian ministry headquartered in Southern California that promotes old-earth biblical creationism.

Roberts is a pleasant person with an academic background in molecular biology, specializing in virology. I had anticipated that she would address the topic of "theistic science," the pseudo-discipline that incorporates Christian beliefs into scientific claims. Instead, the presentation was mostly her Christian testimony, with some scripture reading and a variety of anecdotes. It concluded with a prayer.

When leaving, I received a complimentary copy of *Building Bridges: Presentations on RTB's Testable Creation Model*, Reasons to Believe's latest overview of the ministry's version of biblical creationism. Published in 2018, it consists of six brief chapters by Fazale Rana, A. J. Roberts, and Jeff Zweerink.

It's a remarkable publication because the essays demonstrate beyond any doubt that Reasons to Believe's creationism model is not science. It is in fact the very antithesis of scientific reasoning, because the authors start by assuming they know the truth—biblical Christianity—and then search selectively for evidence that is compatible with their faith.

Christian Faith, Science, and Creationism

Jeff Zweerink is an astrophysicist who specializes in intentional design from multiverse theory, dark energy and dark matter, and exoplanets. He speaks and writes on the compatibility of faith and science. The following quotes summarize his perspective.

- "The universe that God revealed to us through the Bible matches the universe that we see when we study creation."
- "It seems natural to conclude that when we see design it is because a designer exists and created the universe to support humanity."
- "Not only does the scientific evidence demonstrate the rationality of belief in God, I would argue that God's existence provides the best explanation of our scientific understanding of the universe."

A. J. Roberts became a Christian at age twelve when she became convinced of the gospel message of Jesus's atoning death and resurrection. She has expertise in microbiology, immunology, and infectious diseases. Her special interest is integrating science and Christianity, as illustrated by the following quotes.

- "As a Christian who is also a scientist I often use scientific discoveries to help others see how God reveals himself to us in creation."
- "I believe that this evidence strongly supports a view of progressive creationism and a common design model. God created life, over long epochs of time, according to specific kinds."
- "In contrast to the restrictive commitment to naturalistic explanations, a Christian or theistic paradigm is actually better for science. If we seek we will find when we seek with all of our heart. This is a promise of Jesus, recorded in the gospel. Thanks be to God!"

Fazale Rana is a biochemist who has written about the origins of life, the cell's design, and the creation of life. He is dedicated to communicating to skeptics as well as believers the evidence for God's existence and scripture's reliability. The following quotes express his Christian viewpoint.

- "There is scientific merit to a view that regards human beings as the product of a creator's handiwork, specially endowed with intelligence and the capacity for speech—qualities necessary for beings assigned as God's vice regents on Earth and that align with the image of God."
- "The human genome appears to be far more elegant and sophisticated than we could have ever imagined. It displays features that bespeak a creator's handiwork."
- "If we allow ourselves to relax the philosophical (nonscientific) restrictions on science, we are free to admit that the data from anthropology supports the notion that human beings are God's exceptional creations."

Christian Anxiety and Science Denialism

These testimonial statements by the authors of *Building Bridges* establish unequivocally that their Christian faith constitutes the framework and filter through which they interpret scientific findings. The universal causal explanation for everything that exists is the god of the Holy Bible. All phenomena must be integrated into the Christian worldview, and nothing makes sense otherwise.

As Rana emphasizes, "If human evolution is truly a fact, then there can be no ultimate meaning or purpose to human existence, then human life has no inherent value and people lack any dignity, then we are not accountable to a creator." In other words, life is completely meaningless without the assurance and direction of Christian faith.

Rana's declaration is obviously not a scientific statement but instead an expression of utter dependence on belief in a cosmic father figure, referred to as the "divine mind" and "divine creator." Based on their testimonials, it can be concluded that Roberts and Zweerink agree wholeheartedly with Rana's avowal.

Because the three authors have backgrounds in traditional scientific disciplines, it is reasonable to expect them to be versed in the philosophy and methodology of science. This expectation is not realized. In their efforts to justify the distortion of science by making it subservient to Christian theology, the Reasons to Believe creationism advocates commit four major violations of the principles of logic and scientific investigation.

1. SELECTIVE ACCEPTANCE OF EVIDENCE

In his chapter on human origins, which concludes with "The Scientific Case for Adam and Eve," Rana mentions only the evidence that can be viewed as compatible with, or at least not contradicting, Reason to Believe's divine creation story. In a previous discussion of this serious violation, I gave twenty examples where Rana and colleague Hugh Ross rejected the scientific consensus, because they said accepting the evidence would compel them to deny the biblical foundation of their human origins model (see "Sorry, 'Theistic Science' Is Not Science," *Skeptical Inquirer*, May/June 2018).

The basic principle of scientific hypothesis testing is that if the empirical evidence does not support the claim being evaluated, then it is the hypothesis that must be rejected or modified, *not* the evidence. Rana and Ross have the logic of hypothesis testing backward. They assume the truth of their Bible-based hypothesis and dismiss all evidence that fails to support the Reasons to Believe theological conception of human origins.

2. INVOKING A DESIGNER

The Reasons to Believe creationism model considers shared biological features of organisms evidence of "common design, not common descent." To justify this explanation, the pre-Darwinism theological conception of archetypal designs that existed in the mind of the first cause is invoked. The originator of this idea was nineteenth-century British biologist Sir Richard Owen, who defined the archetype as "an exemplar on which it pleased the Creator to frame certain of his living creatures."

Rana summarizes by saying, "In Owen's view, the archetype existed only in God's mind and was manifested in the created order in the form of shared biological features, thus representing a teleology of a higher order." This religious explanation is the basis for Rana and Roberts's numerous references to the appearance of design in the basic expressions of life on earth. Based on an entirely different rationale, Zweerink states repeatedly that the universe "looks like" it was designed for life.

Rana and Ross have previously described Owen's proposal as a respectable scientific explanation, which it obviously is not. It clearly involves a theological explanation for observed regularities in nature, a claim that cannot be subjected to objective evaluation. Specifically, the mechanism is impossible to verify or refute, because the archetypes are alleged to exist only in God's mind, a hypothetical entity that is beyond examination by science. It is more parsimonious to assert that the causal agent (God) exists only in the human imagination.

3. SUPERNATURAL CAUSATION DEFENDED

Rana complains endlessly that methodological naturalism is unfair to theists, because supernatural beliefs are not accepted as legitimate explanatory concepts in the scientific endeavor. He then alleges that this restriction renders science an "inherently atheistic enterprise."

The reason for this exclusion is incontestable: there has never been one single demonstration of a supernaturally caused event. The best modern example of the failure to document a common supernatural claim is the case of Christian prayer. Five major replicated clinical trials carried out by devout Christian investigators demonstrated conclusively that God doesn't answer prayer (see "Have Christians Accepted the Scientific Conclusion That God Does Not Answer Intercessory Prayer?," *Free Inquiry*, December 2018/January 2019).

Yet Rana repeatedly pleads for a "relaxation on the restrictions of methodological naturalism," so that theological explanations like those invoked by the book's authors can be treated as scientifically verified causal factors. Rana's argument is really pointless for two reasons: there is no scientific support for claims of supernatural causation, and Rana and his colleagues disregard this undisputed canon of science anyway. Instead, they view the natural world exclusively through the lens of their Christian faith.

There is a straightforward distinction between science and faith. Science has proven to be the best strategy for ascertaining knowledge about the natural world. In contrast, faith is the exclusive "mode of knowing" about the postulated supernatural realm.

No assumed supernatural agents or processes—not gods or devils, not angels or demons, not prayers or curses—can cause or influence events in the real world. Purported knowledge of the supernatural is accessed through faith, whereas reliable information about the natural world is discovered through science.

4. CREATIONIST ILLOGIC

Two of the authors, Rana and Roberts, commit the logical error of arguing that alleged problems with the overwhelming evidence for human evolution actually constitute support for their old-earth biblical creationism beliefs.

This false-choice fallacy would be valid only if there were only two possible explanations for human origins and only one could be correct. In this hypothetical circumstance, evidence that discredited one claim would constitute support for the other. But this is not the case here.

In fact, there are numerous competing creationism scenarios, including three very different biblical creation formulations (young-earth, old-earth, and intelligent design), as well as many other religious creation narratives, such as Hindu, Confucian, and Islamic, and the creation stories of native peoples around the world, including Australian Aborigines and American Indians. All these stories are just as reasonable as Reason to Believe's Bible-based version.

I suggest that an initial step in resolving this creationist confusion would be for the Reasons to Believe staff to hold a summit meeting with representatives from Answers in Genesis (young-earth) and the Discovery Institute (intelligent design) for the purpose of developing a unified Christian view of the origin and history of humankind on earth that all Christians could endorse.

Of course this would not be easy, given the long history of disrespect, acrimony, and disparagement among the participants. For example, Answers in Genesis leaders have charged that the Reason to Believe's creation model "contradicts Scripture, assaults God, and undermines the gospel."

On the other hand, in his introduction to *Building Bridges*, Rana stresses that nothing is more effective when reaching out to adversaries, whether they be skeptics, nonbelievers, secularists, or people of other faiths, than establishing friendships first. Surely this advice applies to fellow Christians too.

For all biblical creationists, the central complaint about evolution is that humans are reduced to just another species of animal. This is unacceptable to many Christians, because the Genesis account of separate creation supports the fundamentalist dogma of "human exceptionalism."

One point on which all Bible-based creationist approaches must agree is the exact process by which God created humankind, because scripture is unambiguous on this assertion. God created humans in his image (Genesis 1:27), forming the man from dust (Genesis 2:7) and the woman from the man's rib (Genesis 2:22).

Doesn't this unequivocal specification provide the biblical platform for an affirmative program of exploratory investigation that would put a unified Christian creationism model on a solid foundation of scientific research?

Conclusion

It's clear from the essays in this slim volume that Reasons to Believe is no longer attempting to achieve legitimacy in the mainstream scientific community but rather has decided to direct its efforts to indoctrinating Christian believers in its idiosyncratic and incorrect view of science. Instead, its members should publicly acknowledge that their work is entirely an expression of their sincere Christian faith and end their embarrassing misrepresentation of the philosophy and methodology of science.

Brian Bolton is a retired research psychologist with a background in mathematics, statistics, and psychometrics. His contributions in psychological measurement, personality assessment, and the psychology of disability have been recognized by universities and psychological societies. He lives in Georgetown, Texas.

CHAPTER 47

Twenty-One Reasons Noah's Worldwide Flood Never Happened

Dr. Lorence G. Collins

Vol. 42, No. 2
March/April 2018

Young-Earth creationists claim that the Paleozoic sedimentary rocks in the Grand Canyon and the Mesozoic sedimentary rocks of the Grand Staircase north of the canyon, in which Zion and Bryce Canyon National Parks occur, were deposited during Noah's worldwide flood about 4,500 years ago (Hill 2002; Hill and Moshier 2009). I realize that readers of *Skeptical Inquirer* accept modern scientific views on this subject, but this examination of the creationist claims might be useful when communicating with others less imbued with scientific thinking.

There are at least twenty-one scientific reasons a worldwide flood recounted in the Bible cannot have happened.

1. The stair-stepped appearance of erosion of sedimentary rocks in the Grand Canyon with sandstones and limestones forming cliffs and shales forming gentle slopes cannot happen if all these rocks were deposited in less than one year. If the Grand Canyon had been carved soon after these rocks were deposited by a worldwide flood, they would not have had time to harden into solid rock and would have been saturated with water. Therefore, the sandstones and limestones would have slumped during the carving of the canyon and would not have formed cliffs (Hill et al. 2016).

2. Salt and gypsum deposits, more than 200 feet thick, occur in the Paradox Formation in Utah just 200 miles north of the Grand Canyon, and these deposits are the same age as the Supai rocks in the Grand Canyon that were supposedly also deposited by Noah's flood. Similar salt deposits, up to 3,000 feet thick, exist in various places on all continents and in layers of all geologic ages, and these deposits can only be produced by evaporation of

sea water. Such evaporation could not have happened in repeated intervals in the midst of the forty days and forty nights of raining and during the supposed continuous deposition of sedimentary rocks by a worldwide flood and in which the only drying and evaporation is said to have occurred at the end of the flood (Collins 2006; 2009; 2012; Hill et al. 2016).

3. Sand dunes with giant cross bedding occur in the Mesozoic rocks in Zion National Park and are further evidence that desert conditions occurred at the time of the supposed flood (Senter 2011; Collins 2017).

4. Fossilized mud cracks occur in the Cambrian Tapeats Formation on top of the Precambrian Vishnu schist at the bottom of the canyon and indicate that drying conditions existed during the alleged worldwide Noah's flood, and these drying conditions occurred at the very beginning of this supposed flood. Although mud cracks can also form in mud under water by compression that squeezes out water from the mud, such compression is not likely to occur during a flood. Moreover, fossilized mud cracks are found in other formations that were supposedly deposited during Noah's flood, and these mud cracks occur in red shales that coexist with salt and gypsum layers. Therefore, these mud cracks were likely formed in deltaic mud flats that were exposed to the atmosphere where their iron-bearing minerals reacted with oxygen in the air to form red hematite (Collins 2006; Senter 2011; Hill et al. 2016).

5. Raindrop prints occur in many places around the world, which could not have been formed or preserved if the muds (now in shales) containing these prints were deposited under water during Noah's flood (Senter 2011; Hill et al. 2016).

6. Nests of dinosaur eggs are found in several places around the world, and it is illogical that dinosaurs could have had enough time to create these nests and lay their eggs while they were fleeing from rising waters to reach higher ground (Senter 2011; Hill et al. 2016).

7. The White Cliffs of Dover on the eastern coast of England consist of chalk layers, up to 350 feet thick, that are composed of fossilized coccoliths (a kind of algae), and these layers are the same age as the sedimentary rocks that overlie giant cross-bedded sandstones in Zion National Park. Therefore, they were supposedly also deposited by Noah's flood. But coccoliths are very tiny and have chloroplasts that require sunlight and must float close to the ocean surface to get energy from the sun. Because of this, all of them could not have been living at the same time to depths of 350 feet in the one year in which the flood is said to have occurred because that many organisms in the water at the same time would have blocked out the sun from organisms below the near-surface (Collins 2015a).

8. Up to 4,590-foot thicknesses of radiolarians occur on the Pacific Ocean floor. Radiolarians are tiny marine organisms with silica skeletons; they contain chloroplasts and must float near the ocean surface to obtain sunlight. The rate at which dead radiolarians settle to the bottom of the ocean is too slow for that thickness and number of radiolarians to accumulate in the 4,500 years since Noah's flood. Although radiolarians are not found in the sedimentary rocks of the Grand Canyon, fossilized radiolarians are common in sedimentary layers in other parts of the world of the same geologic age, and each of these layers contains distinctly different radiolarian species that are among more than 4,000 different species that have been identified. Chaotic rushing waters of a tsunami in Noah's flood would have been unable to sort out these different species in different geologic ages from those living early in the flood to those created late in the flood (Collins 2015a).

9. None of the sedimentary rocks in the Grand Canyon contains fossilized pollen grains that are produced by grasses, pines, and flowering trees and plants, whereas these same rocks in the canyon contain only spores of algae, ferns, moss, and fungi. A worldwide flood would be expected to mix these tiny structures if all were alive at the same time of Noah's flood, and this mixing did not occur. How can a rush of water in a tsunami sort out and separate such tiny reproductive structures from each other? (Hill et al. 2016).

10. The Redwall limestone in the Grand Canyon contains billions and billions of jumbled sea lily (crinoid) fossils in multiple layers, and such marine animals would have had to grow on stalks on the ocean floor and cover the whole Earth at space intervals of one foot apart if all were alive at the same time during Noah's flood. That distribution and abundance is extremely unlikely in less than one year's time. Moreover, there would have been the need for already available, precipitated, calcium carbonate crystals somewhere to be carried into the Grand Canyon area to be deposited as limestone to host these fossils. This seems highly unlikely because the source of the calcium requires long periods (tens of thousands of years) of chemical weathering of calcium-bearing rocks, such as basalt lava flows, to produce such a large volume of limestone that extends, not only in the Grand Canyon, but also in the Redwall limestone across most of western and central United States in buried sedimentary layers (Collins 2015a; Hill et al. 2016).

11. If all limestones were deposited by Noah's flood during a giant tsunami, then all limestone layers should show evidence of fossils being jumbled by rushing water. This is not the case. The presence of Silurian limestone layers that are older than the Redwall limestone occur with consistent sequences at constant thicknesses over hundreds of square miles in Illinois, Iowa, and Wisconsin, and they lack any fossils in a jumbled array. These consistent thicknesses indicate that these limestone layers could not have been

deposited by a tsunami, and these layers could only have been formed in quiet water by slow chemical precipitation of the calcium carbonate during tens of thousands of years. Thus, the limestones around the world, alleged to be deposited by Noah's flood, were not deposited by a rush of flood water in a tsunami in less than one year. Many other examples occur in sedimentary rocks around the world where fossils of communities of many different marine animals are totally undisturbed (Senter 2011; Collins 2015a; 2017).

12. Abundant fusain (charcoal) is found in several different sedimentary rocks around the world that were supposedly deposited by Noah's flood, which is good evidence that a worldwide flood never happened. Fires that burn forests are not likely to occur in the midst of forty days and nights of rain (Senter 2011).

13. River terraces exist on the sides of Colorado River canyon walls that give ages of deposition at the top of 350,000 years and at the bottom of 38,000 years, and these ages were determined by two entirely different methods and are much beyond the 4,500 years that young-Earth creationists claim is the age of the sedimentary rocks deposited during Noah's flood (Collins 2015b; 2017).

14. Scientific observations and measurements show that the canyon of the Colorado River was eroded by rates of 80 to 458 meters per million years at different places along the canyon (Collins 2015b; 2017).

15. The coarsely crystalline Zoraster granite occurs in the Vishnu schist in the bottom of the Grand Canyon. Experimental work shows that coarse crystals in granite are formed only at depths of five miles or more in the Earth's crust and that at these depths, the rate of cooling of rock melts (magma) at temperatures of more than 800 degrees C require at least 5 million years before temperatures are low enough that crystals can start to form and slowly increase in size. Therefore, the age of the Earth must be more than 6,000 years and much older than when Noah's flood supposedly occurred (Collins 2017).

16. The rate of erosion of the Zoraster granite in the floor of the Colorado River in the Grand Canyon, as measured by how fast the erosion occurs on a yearly basis from year to year, is about a thousandth of an inch per year. This means that the carving of the Grand Canyon took millions of years—not less than a year in a sudden rush of water draining from three lakes at the end of Noah's flood (Collins 2017).

17. The average thickness of sedimentary rocks around the world in the continents that were supposedly deposited by Noah's flood is about 1,800 meters (5,905 feet) (Nelson 2012). If just 1 percent of this thickness represents fossil remains of marine animals that were alive at the same time during Noah's flood, then the whole world would have been covered with 59 feet of

living marine animals, such as clams, snails, corals, trilobites, and sponges. That many animals living at the same time during that single year would have been impossible. The value of 1 percent is not unreasonable when some limestone layers are composed of nearly 100 percent fossils. Even if 0.1 percent of the sedimentary rock thickness contained all marine animals that were alive at the same time in the year of Noah's flood, that means that the whole world would have been covered with 5.9 feet of animals, and that still is too many animals. The impossible numbers of fossils of coccoliths (Reason 7), radiolarians (Reason 8), and sea lilies (Reason 10), are illustrative of this fact. Moreover, if only one out of 10,000 animals living on the land (amphibians, reptiles, birds, and mammals) are preserved as fossils in the sedimentary rocks deposited by Noah's flood, then Noah would have had no space in which he, his family, and cattle could have existed with all these creatures living at the same time, particularly if tens of thousands of huge dinosaurs were alive when Noah lived prior to the flood.

18. An experiment done by Charles Munroe III shows that the submergence of an olive tree under water for more than three months kills the tree. On that basis, when a dove brought an olive twig with fresh leaves to Noah on the ark (Genesis 8:11), the whole world could not have been submerged under water during the flood. Otherwise, all olive trees would have been killed in six months of their submergence under the flood waters. Therefore, the flood must have been local in southern Mesopotamia with some land (say, 100 feet above water) present on which olive trees must have been growing and from which an olive tree with such a twig with fresh leaves could have been obtained by the dove (Collins 2017).

19. Wave action at high tide from a powerful category 5 hurricane with sustained winds of more than 156 mph can move an offshore barrier sandbar as much as 50 to 100 feet inland toward the continent, but such a major storm never moves sand for distances of thousands of miles across the United States, as creationists claim for the Tapeats, Coconino, and other sandstone deposits in the Grand Canyon or the calcite grains and fossils in the Redwall limestone. Therefore, producing such deposits by such winds and waves in Noah's Flood has no scientific support. Noah's ark could not have survived such wind power and large waves. Moreover, moving water cannot carry such large, suspended sediment loads for that distance (Collins 2015a; 2017).

20. The claim that the erosion surface that underlies the Cambrian Tapeats sandstone on top of the Precambrian basement rocks, which forms a major unconformity at the bottom of the Grand Canyon, was caused by a tsunami makes no scientific sense. A tsunami-created wave is caused by either a) a large earthquake following a sudden fault movement that shifts the position

of the ocean floor; or b) a giant explosion of a volcano in an oceanic region, such as the explosion of Krakatoa—but never by flood waters of a large wave washing across the continent for thousands of miles. No such huge earthquake or explosion of a volcano is mentioned in the Bible, so postulating a tsunami to cause widespread rapid erosion across the world and at the bottom of the Grand Canyon at the beginning of Noah's Flood is without biblical or scientific support (Collins 2017).

21. If the sedimentary rocks in the Grand Canyon were all deposited by Noah's worldwide flood, then these rocks should all be deposited in less than one year *under water* and show continuous deposition from one formation to the next up the canyon walls without disruption. But that is not the case. For example, the Redwall limestone formation has ancient karst topography with caves and sinkholes, indicating that this formation was once lifted out of the water so that percolating rain water dissolved out cavities and tunnels in the limestone; erosion channels of the Temple Butte Formation, as much as 100 feet deep, cut the top of the Muav limestone, and networks of channels of the Surprise Canyon Formation, up to a half mile wide and 400 feet deep, cut the top of the Redwall limestone, indicating that these rocks were exposed at the Earth's surface to river erosion (Hill et al. 2016). These structural land *surface* features would take much more than one year to be formed.

A Reasonable Possibility

Thomas H. Huxley (1825–1895) said the following: "The great tragedy of science—the slaying of a beautiful hypothesis by an ugly fact." Any one of the above twenty-one facts destroys the idea that Noah's flood was a worldwide occurrence. Nevertheless, a local large flood in Mesopotamia in biblical times could have been possible (Collins 2009). Young-Earth creationists commonly point out that Jesus supported the existence of Noah's flood (Luke 17:27; Matthew 24:38–39), but in the context of the time in which Jesus lived, the translation of the Hebrew word *ertz* would have been understood as "land" and not the whole world. In that sense, all the land that Noah could have normally seen would have been under water in that part of Mesopotamia and that would have been in effect his whole world. During large floods in Iraq (Mesopotamia), commonly abundant rain falling in the adjacent Zagros Mountains goes underground in solution tunnels in limestone beds and emerges as gushing water in springs in Iraq. These springs are likely the "fountains of the great deep" as described in Genesis 7:11 (Hill 2015).

Lorence G. Collins is a retired professor of geology from California State University Northridge. He has thirty-six articles on the website Opposition to Creationism (http://www.csun.edu/~vcgeo005/creation.html) that describe various views of young-Earth creationists and their scientific errors in interpretations. Among these are three articles that discuss a bogus fossilized Noah's Ark in eastern Turkey.

References

Collins, L. G. 2006. Time to accumulate chloride ions in the world's oceans. Creationism's young Earth not supported. *Reports of the National Center for Science Education* 26(5): 16–24. Available online at http://www.csun.edu/~vcgeo005/collins.pdf.

————. 2009. Yes, Noah's flood may have happened but not over the whole earth. *Reports of the National Center for Science Education* 29(5): 38–41. Available online at http://www.csun.edu/~vcgeo005/Collins2.pdf.

————. 2012. More geologic reasons Noah's flood never happened. *Reports of the National Center for Science Education* 32(6): 1–11. Available online at http://www.csun.edu/~vcgeo005/Collins3.pdf.

————. 2015a. Can flood geology and catastrophic plate tectonics explain sedimentary rocks? Available online at http://www.csun.edu/~vcgeo005/Collins5.pdf.

————. 2015b. When was Grand Canyon carved—millions of years ago or thousands of years ago? How do we know? *Reports of the National Center for Science Education* 35(4): 2.1–2.8. Available online at http://www.csun.edu/~vcgeo005/GrandCanyon.pdf.

————. 2017. When was the Grand Canyon formed? 4,360 years ago during Noah's flood? Or during millions of years by natural geologic processes. Power point presentation available online at http://www.csun.edu/~vcgeo005/PwrPt1.pdf and http://www.csun.edu/~vcgeo005/PwrPt2.pdf.

Hill, C. 2002. The Noachian flood: Universal or local? *Perspectives in Science and Christian Faith* 54(3): 170–183. Available online at http://www.csun.edu/~vcgeo005/Carol%201.pdf.

————. 2015. *A World View Approach to Science and Scripture* (unpublished book).

Hill, C., G. Davidson, T. Helble, et al. (editors). 2016. *The Grand Canyon—Monument to an Ancient Earth—Can Noah's Flood Explain the Grand Canyon?* Kregel Publications.

Hill, C., and S. Moshier. 2009. Flood geology and the Grand Canyon: A critique. *Perspectives in Science and Christian Faith* 61(2): 99–115. Available online at http://www.csun.edu/~vcgeo005/Carol%202.pdf.

Nelson, S. A. 2012. Geology 212, petrology. *Occurrence, Mineralogy, Texture, and Structures of Sedimentary Rocks.* Available online at http://www.tulane.edu/~sanelson/eens212/sedrxintro.htm.

Senter, P. 2011. The defeat of flood geology by flood geology. The ironic demonstration that there is no trace of the Genesis flood in the geologic record. *Reports of the National Center for Science Education* 31(3): 1–14. Available online at http://www.csun.edu/~vcgeo005/Flood%20geology.pdf.

Acknowledgments

The editors offer our heartfelt appreciation to all the authors represented here. Any anthology is only as good as its contributors, and *Skeptical Inquirer* has been fortunate to attract such high-level research, investigation, and writing talent. Thanks also to Julia Lavarnway and Nicole Scott, whose editorial acumen invisibly yet invariably improved these articles when originally published.